CORROSION SCIENCE and TECHNOLOGY

CRC *Series in*
Materials Science and Technology

Series Editor
Brian Ralph

CORROSION SCIENCE and TECHNOLOGY

David Talbot
James Talbot

CRC Press

Boca Raton Boston New York Washington London

Library of Congress Cataloging-in-Publication Data

Talbot, David
    Corrosion science and technology/David Talbot and James Talbot
        p. cm. (CRC series in materials science and technology)
        Includes bibliographical references and index.
        ISBN 0-8493-8224-6
        1. Chemical engineering—materials science. 2. Mechanical engineering—materials science. I. Talbot, James. II. Title. III. Series

    H749.H34B78 1997
    616'.0149—dc20   97-57109
                                                                          CIP

This book contains information obtained from authentic and highly regarded sources. Reprinted material is quoted with permission, and sources are indicated. A wide variety of references are listed. Reasonable efforts have been made to publish reliable data and information, but the author and the publisher cannot assume responsibility for the validity of all materials or for the consequences of their use.

Visit the CRC Press Web site at www.crcpress.com

No claim to original U.S. Government works
International Standard Book Number 0-8493-8224-6
Library of Congress Card Number 97-57109
Printed in the United States of America   3 4 5 6 7 8 9 0
Printed on acid-free paper

Contents

Preface

Engineering metals are unstable in natural and industrial environments. In the long term, they inevitably revert to stable chemical species akin to the chemically combined forms from which they are extracted. In that sense, metals are only borrowed from nature for a limited time. Nevertheless, if we understand their interactions with the environments to which they are subjected and take appropriate precautions, degradation can be arrested or suppressed long enough for them to serve the purposes required. The measures that are taken to prolong the lives of metallic structures and artifacts must be compatible with other requirements, such as strength, density, thermal transfer, and wear resistance. They must also suit production arrangements and be proportionate to the expected return on investment. Thus, problems related to corrosion and its control arise within technologies, but solutions often depend on the application of aspects of chemistry, electrochemistry, physics, and metallurgy that are not always within the purview of those who initially confront the problems.

Corrosion is the transformation of metallic structures into other chemical structures, most often through the intermediary of a third structure, i.e., water and a first task is to characterize these structures and examine how they determine the sequences of events that result in metal wastage. These matters are the subjects of Chapters 2, 3, 4, and 5. The information is applied in Chapter 6 to examine the options available for the most usual strategy to control corrosion, the application of protective coatings. Chapters 7 through 9 examine the attributes and corrosion behavior of three groups of metallic materials, plain carbon steels and irons, stainless steels, and aluminum alloys.

The final chapters deal with some practical implications. Corrosion control is only one aspect of the technologies within which it is exercised and the approaches adopted must accommodate other requirements in the most economic way. For this reason, some total technologies are selected to illustrate how the approach to corrosion control is conditioned by their particular circumstances. Aviation is a capital intensive industry in which the imperatives are flight safety, the protection of investment and uninterrupted operation of aircraft over a long design life. In automobile manufacture, the design life is less but retail sales potential through positive customer perception is vitally important. Food handling introduces aspects of public health, biological contributions to corrosion problems, and the mass production of food cans that are low-value corrosion-resistant artifacts. Building construction has a menu of different approaches to corrosion control from which solutions are selected to suit client requirements, local government ordinances and changing patterns of business under the pressures of competitive tendering.

The form of the present text has evolved from long experience of lectures and seminars arranged for students and graduates drawn into corrosion-related work from a wide variety of different backgrounds.

The Authors

David Talbot graduated with B.Sc. and M.Sc. from the University of Wales and Ph.D. from Brunel University for research on gas-metal equilibria. From 1949 to 1966 he was employed at the Research Laboratories of the British Aluminium Company Ltd., contributing to research promoting the development of manufacturing processes and to customer service. From 1966 to 1994 he taught courses on corrosion and other aspects of chemical metallurgy at Brunel University, maintaining an active interest in research and development, mainly in collaboration with manufacturing industries in the U.K. and U.S.A. He is a member of the Institute of Materials with Chartered Engineer status and has served as a member of Council of the London Metallurgical Society. He has written many papers on chemical aspects of metallurgy, a review on metal–hydrogen systems in *International Metallurgical Reviews* and a section on gas–metal systems in *Smithells Metals Reference Book*.

James Talbot graduated with B.Sc., ARCS from Imperial College, London, M.Sc. from Brunel University and Ph.D. from the University of Reading for research on the physical chemistry of aqueous solutions and its application to natural waters. He is currently employed at the River Laboratory of the Institute of Freshwater Ecology, East Stoke, Wareham, Dorset, U.K. to assess and predict physical chemical changes that occur in river management. He has written papers on the speciation of solutes in natural waters.

Acknowledgments

The authors wish to acknowledge their gratitude to Professor Brian Ralph and Professor Colin Bodsworth for their interest, encouragement, and valuable suggestions.

They also wish to thank the following people for the courtesy of their expert advice:

Mr. Mick Morris, Manager, Aircraft Structures, British Airways — Corrosion control in airframes.

Mr. David Bettridge, Rolls-Royce Limited — Corrosion prevention in gas turbine engines.

Mr. Alan Turrell and Mr. John Creese, The Rover Group — Corrosion protection for automobiles.

Mr. Ray Cox, U. K. Building Research Establishment — Corrosion control in building.

Mr. Derek Bradshaw, Alpha Anodizing Ltd. — Surface cleaning and chromate treatment of aluminum alloys.

Mr. Alan Mudie, Guinness Brewery — Corrosion control in brewing.

1

Overview of Corrosion and Protection Strategies

Metals in service often give a superficial impression of permanence, but all except gold are chemically unstable in air and air-saturated water at ambient temperatures and most are also unstable in air-free water. Hence almost all of the environments in which metals serve are potentially hostile and their successful use in engineering and commercial applications depends on protective mechanisms. In some metal/environment systems the metal is protected by passivity, a naturally formed surface condition inhibiting reaction. In other systems the metal surface remains active and some form of protection must be provided by design; this applies particularly to plain carbon and low-alloy irons and steels, which are the most prolific, least expensive, and most versatile metallic materials. Corrosion occurs when protective mechanisms have been overlooked, break down, or have been exhausted, leaving the metal vulnerable to attack.

Practical corrosion-related problems are often discovered in the context of engineering and allied disciplines, where the approach may be hindered by unfamiliarity with the particular blend of electrochemistry, metallurgy, and physics which must be brought to bear if satisfactory solutions are to be found. This brief overview is given to indicate the relevance of these various disciplines and some relationships between them. They are described in detail in subsequent chapters.

1.1 Corrosion in Aqueous Media

1.1.1 Corrosion as a System Characteristic

Some features of the performance expected from metals and metal artifacts in service can be predicted from their intrinsic characteristics assessed from their compositions, structures as viewed in the microscope, and past history of thermal and mechanical treatments they may have

1

received. These characteristics control density, thermal and electrical conductivity, ductility, strength under static loads in benign environments, and other physical and mechanical properties. These aspects of service-ability are reasonably straightforward and controllable, but there are other aspects of performance which are less obvious and more difficult to control because they depend not only on intrinsic characteristics of the metals but also on the particular conditions in which they serve. They embrace susceptibility to corrosion, metal fatigue, and wear, which can be responsible for complete premature failure with costly and sometimes dangerous consequences.

Degradation by corrosion, fatigue and wear can only be approached by considering a metal not in isolation but within a wider system with the components, metal, chemical environment, stress, and time. Thus a metal selected to serve well in one chemical environment or stress system may be totally inadequate for another. Corrosion, fatigue, and wear can interact synergistically, as illustrated in Chapter 5 but, for the most part, it is usually sufficient to consider corrosion processes as a chemical system comprising the metal itself and its environment.

1.1.2 The Electrochemical Origin of Corrosion

From initial encounters with the effects of corrosion processes it may seem difficult to accept that they can be explained on a rational basis. One example, among many, concerns the role of dissolved oxygen in corrosion. It is well known that unprotected iron rusts in pure neutral waters, but only if it contains dissolved oxygen. Based on this observation, standard methods of controlling corrosion of steel in steam-raising boilers include the removal of dissolved oxygen from the water. This appears to be inconsistent with observations that pure copper has good resistance to neutral water whether it contains oxygen or not. Moreover, copper can dissolve in acids containing dissolved oxygen but is virtually unattacked if the oxygen is removed whereas the complete reverse is true for stainless steels.

These and many other apparently conflicting observations can be reconciled on the basis of the electrochemical origin of the principles underlying corrosion processes and protection strategies. The concepts are not difficult to follow and it is often the unfamiliar notation and conventions in which the ideas are expressed which deter engineers.

At its simplest, a corroding system is driven by two spontaneous coupled reactions which take place at the interface between the metal and an aqueous environment. One is a reaction in which chemical species from the aqueous environment remove electrons from the metal; the other is a reaction in which metal surface atoms participate to replenish the electron deficiency. The exchange of electrons between the two reactions constitutes an electronic current at the metal surface and an important effect is to impose an electric potential on the metal surface of such a value that the supply and demand for electrons in the two coupled reactions are balanced.

The potential imposed on the metal is of much greater significance than simply to balance the complementary reactions which produce it because it is one of the principal factors determining what the reactions shall be. At the potential it acquires in neutral aerated water, the favored reaction for iron is dissolution of the metal as a soluble species which diffuses away into the solution, allowing the reaction to continue, i.e., the iron corrodes. If the potential is depressed by removal of dissolved oxygen the reaction is decelerated or suppressed. Alternatively, if the potential is raised by appropriate additions to the water, the favored reaction can be changed to produce a solid product on the iron surface, which confers effective corrosion protection. Raising the alkalinity of the water has a similar effect.

1.1.3   Stimulated Local Corrosion

A feature of the process in which oxygen is absorbed has two important effects, one beneficial and the other deleterious. In still water, oxygen used in the process must be re-supplied from a distant source, usually the water surface in contact with air; the rate-controlling factor is diffusion through the low solubility of oxygen in water. The beneficial effect is that the absorption of oxygen controls the overall corrosion rate, which is consequently much slower than might otherwise be expected. The deleterious effect is that difficulty in the re-supply of oxygen can lead to differences in oxygen concentration at the metal surface, producing effects which can stimulate intense metal dissolution in oxygen-starved regions, especially crevices. This is an example of a local action corrosion cell. There is much more to this phenomenon than this brief description suggests and it is discussed more fully in Chapter 3.

Another example of stimulated corrosion is produced by the bi-metallic effect. It comes about because of a hierarchy of metals distinguished by their different tendencies to react with the environment, measured by the free energy changes, formally quantified in electrochemical terms in Chapter 3. Metals such as iron or aluminum with strong tendencies to react are regarded as less noble and those with weaker tendencies, such as copper, are considered more noble. For reasons given later, certain strongly passive metals, such as stainless steels, and some non-metallic conductors, such as graphite can simulate noble metals. The effect is to intensify attack on the less noble of a pair of metals in electrical contact exposed to the same aqueous environment. Conversely the more noble metal is partially or completely protected. These matters are very involved and are given the attention they merit in Chapter 4.

1.2   Thermal Oxidation

The components of clean air which are active towards metals are oxygen and water vapor. Atmospheric nitrogen acts primarily as a diluent because although metals such as magnesium and aluminum form nitrides in pure nitrogen gas, the nitrides are unstable with respect to the corresponding oxides in the presence of oxygen.

At ordinary temperatures, most engineering metals are protected by very thin oxide films, of the order of 3 to 10 nm (3 to 10 m^{-9}) thick. These films form very rapidly on contact with atmospheric oxygen but subsequent growth in uncontaminated air with low humidity is usually imperceptible. It is for this reason that aluminum, chromium, zinc, nickel, and some other common metals remain bright in unpolluted indoor atmospheres.

1.2.1   Protective Oxides

At higher temperatures, the oxides formed on most common engineering metals, including iron, copper, nickel, zinc, and many of their alloys, remain coherent and adherent to the metal substrate but reaction continues because reacting species can penetrate the oxide structure and the oxides grow thicker. These oxides are classed as protective oxides because the rate of oxidation diminishes as they thicken, although the protection is incomplete. The oxide grows by an overall reaction driven by two electrochemical processes, an anodic process converting the metal to cations and generating electrons at the metal/oxide interface, coupled with a cathodic process converting oxygen to anions and consuming electrons at the oxygen oxide/atmosphere interface. The natures of these ions and the associated electronic conduction mechanisms are quite different from their counterparts in aqueous corrosion. A new unit of oxide is produced when an anion and cation are brought together. To accomplish this, one or the other of the ions must diffuse through the oxide. The ions diffuse through defects on an atomic scale, which are characteristic features of oxide structures. Associated defects in the electronic structure provide the electronic conductivity needed for the transport of electrons from the metal/oxide to the oxide/air interfaces. These structures, reviewed in Chapter 2, differ from oxide to oxide and are crucially important in selecting metals and formulating alloys for oxidation resistance. For example, the oxides of chromium and aluminum, have such small defect populations that they are protective at very high temperatures. The oxidation resistance afforded by these oxides can be conferred on other metals by alloying or surface treatment. This is the basis on which oxidation-resistance is imparted to stainless steels and to nickel-base superalloys for gas turbine blades.

1.2.2 Non-Protective Oxides

For some metals, differences in the relative volumes of an oxide and of the metal consumed in its formation impose shear stresses high enough to impair the formation of cohesive and adhesive protective oxide layers. If such metals are used for high temperature service in atmospheres with a real or virtual oxygen potential, they must be protected. An example is the need to can uranium fuel rods in nuclear reactors because of the unprotective nature of the oxide.

1.3 Environmentally-Sensitive Cracking

Corrosion processes can interact with a stressed metal to produce fracture at critical stresses of only fractions of its normal fracture stress. These effects can be catastrophic and even life-threatening if they occur, for example, in aircraft. There are two different principal failure modes, corrosion fatigue and stress-corrosion cracking, featured in Chapter 5.

Corrosion fatigue failure can affect any metal. Fatigue failure is fracture at a low stress as the result of cracking propagated by cyclic loading. The failure is delayed, and the effect is accommodated in design by assigning for a given applied cyclic stress, a safe fatigue life, characteristically the elapse of between 10^7 and 10^8 loading cycles. Cracking progresses by a sequence of events through incubation, crack nucleation, and propagation. If unqualified, the term, fatigue, relates to metal exposed to normal air. The distinguishing feature of corrosion fatigue is that failure occurs in some other medium, usually an aqueous medium, in which the events producing fracture are accelerated by local electrochemical effects at the nucleation site and at the crack tip, shortening the fatigue life.

Stress corrosion cracking is restricted to particular metals and alloys exposed to highly specific environmental species. An example is the failure of age-hardened aluminum aircraft alloys in the presence of chlorides. A disturbing feature of the effect is that the onset of cracking is delayed for months or years but when cracks finally appear, fracture is almost imminent. Neither effect is fully understood because they exhibit different critical features for different metals and alloys but, using accumulated experience, both can be controlled by vigilant attention.

1.4 Strategies for Corrosion Control

1.4.1 Passivity

Aluminum is a typical example of a metal endowed with the ability to establish a naturally passive surface in appropriate environments.

Paradoxically, aluminum theoretically tends to react with air and water by some of the most energetic chemical reactions known but provided that these media are neither excessively acidic nor alkaline and are free from certain aggressive contaminants, the initial reaction products form a vanishingly thin impervious barrier separating the metal from its environment. The protection afforded by this condition is so effective that aluminum and some of its alloys are standard materials for cooking utensils, food and beverage containers, architectural use, and other applications in which a nominally bare metal surface is continuously exposed to air and water. Similar effects are responsible for the utility of some other metals exploited for their corrosion resistance, including zinc, titanium, cobalt, and nickel. In some systems, easy passivating characteristics can also be conferred on an alloy in which the dominant component is an active metal in normal circumstances. This approach is used in the formulation of stainless steels, that are alloys based on iron with chromium as the component inducing passivity.

1.4.2 Conditions in the Environment

Unprotected active metals exposed to water or rain are vulnerable but corrosion can be delayed or even prevented by natural or artificially contrived conditions in the environment. Steels corrode actively in moist air and water containing dissolved air but the rate of dissolution can be restrained by the slow re-supply of oxygen, as described in Section 1.1.3 and by deposition of chalky or other deposits on the metal surface from natural waters. For thick steel sections, such as railroad track, no further protection may be needed.

In critical applications using thinner sections, such as steam-raising boilers, nearly complete protection can be provided by chemical scavenging to remove dissolved oxygen from the water completely and by rendering it mildly alkaline to induce passivity at the normally active iron surface. This is an example of protection by deliberately conditioning the environment.

1.4.3 Cathodic Protection

Cathodic protection provides a method of protecting active metals in continuous contact with water, as in ships and pipelines. It depends on opposing the metal dissolution reaction with an electrical potential applied by impressing a cathodic current from a DC generator across the metal/environment interface. An alternative method of producing a similar effect is to couple a less noble metal to the metal needing protection. The protection is obtained at the expense of the second metal, which is sacrificed as explained in Section 1.1.3. The application of these techniques is considered in Chapter 4.

1.4.4 Protective Coatings

When other protective strategies are inappropriate or uneconomic, active metals must be protected by applied coatings. The most familiar coatings are paints, a term covering various organic media, usually based on alkyd and epoxy resins, applied as liquids which subsequently polymerize to hard coatings. They range from the oil-based, air-drying paints applied by brush used for civil engineering structures, to thermosetting media dispersed in water for application by electrodeposition to manufactured products, including motor vehicle bodies. Alternatively, a vulnerable but inexpensive metal can be protected by a thin coating of an expensive resistant metal, usually applied by electrodeposition. One example is the tin coating on steel food cans; another is the nickel/chromium system applied to steel where corrosion resistance combined with aesthetic appeal is required, as in bright trim on motor vehicles and domestic equipment. An important special use of a protective metal coating is the layer of pure aluminum mechanically bonded to aluminum aircraft alloys, which are strong but vulnerable to corrosion.

1.4.5 Corrosion Costs

Estimates of the costs of corrosion are useful in drawing attention to wasteful depletion of resources but they should be interpreted with care because they may include avoidable items more correctly attributed to the price of poor design, lack of information or neglect. The true costs of corrosion are the unavoidable costs of dealing with it in the most economic way. Such costs include the prices of resistant metals and the costs of protection, maintenance and planned amortization.

An essential objective in design is to produce structures or manufactured products which fulfil their purposes with the maximum economy in the overall use of resources interpreted in monetary terms. This is not easy to assess and requires an input of the principles applied by accountants. One such principle is the "present worth" concept of future expenditure, derived by discounting cash flow, which favors deferred costs, such as maintenance, over initial costs; another is a preference for tax-deductible expenditure. The results of such assessments influence technical judgments and may determine, for example, whether it is better to use resources initially for expensive materials with high integrity or to use them later for protecting or replacing less expensive more vulnerable materials.

1.4.6 Criteria for Corrosion Failure

The economic use of resources is based on planned life expectancies for significant metal structures or products. The limiting factor may be corrosion but more often it is something else, such as wear of moving parts, fatigue failure of cyclically loaded components, failure of associated accessories, obsolescent technology, or stock replenishment cycles. The criterion for corrosion failure is therefore premature termination of the useful function of the metal by interaction with its environment, before the planned life has elapsed. Residual life beyond the planned life is waste of resources.

Failure criteria vary according to circumstances and include:

1. Loss of strength inducing failure of stressed metal parts.
2. Corrosion product contamination of sensitive material, e.g., food or paint.
3. Perforation by pitting corrosion, opening leaks in tanks or pipes.
4. Fracture by environmentally sensitive cracking.
5. Corrosion product interference with thermal transfer.
6. Loss of aesthetic appeal.

Strategies for corrosion control must be considered not in isolation but within constraints imposed by cost-effective use of materials and by other properties and characteristics of metallic materials needed for particular applications. Two very different examples illustrate different priorities.

1. The life expectancy for metal food and beverage cans is only a few months and during that time, corrosion control must ensure that the contents of the cans are not contaminated; any surface protection must be non-toxic and amenable to consistent application at high speed for a vast market in which there is intense cost-conscious competition between can manufacturers and material suppliers. The metal selected and any protective surface coating applied to it must withstand the very severe deformation experienced in fabricating the can bodies.
2. Aircraft are designed for many years of continuous capital-intensive airline operations. Metals used in their construction must be light, strong, stiff, damage tolerant, and corrosion resistant. They must be serviceable in environments contaminated with chlorides from marine atmospheres and de-icing salts which can promote environmentally sensitive cracking. Reliable long-term corrosion control and monitoring schedules are essential to meet the imperative of passenger safety and to avoid disruption of schedules through unplanned grounding of aircraft.

1.4.7 Material Selection

In the initial concept for a metallic product or structure, it is natural to consider using an inexpensive, easily fabricated metal, such as a plain carbon steel. On reflection, it may be clear that unprotected inexpensive materials will not resist the prevailing environment and a decision is required on whether to apply protection, control the environment or to choose more expensive metal. The choice is influenced by prevailing metal prices.

Metal prices vary substantially from metal to metal and are subject to fluctuations in response to supply and demand as expressed in prices fixed in the metal exchanges through which they are traded. The prices also vary according to purity and form, because they include refining and fabricating costs. Table 1.1 gives some recent representative prices.

Table 1.1 illustrates the considerable expense of specifying other metals and alloys in place of steels. This applies especially to a valuable metal such as nickel or tin even if it is used as a protective coating or as an alloy component. For example, the influence of nickel content on the prices of stainless steels is evident from the information in the table.

TABLE 1.1
Representative Selection of Metal Prices

Metal	Form	Price $/tonne
Pure metals*		
Aluminum	Primary metal ingot	1486
Copper	Primary metal ingot	2323
Lead	Primary metal ingot	663
Nickel	Primary metal ingot	6755
Tin	Primary metal ingot	5845
Zinc	Primary metal ingot	1021
Steels†		
Mild steel	Continuously cast slab	215
	6 mm thick hot-rolled plate, 1m wide coil	628
	2 mm thick cold-rolled sheet, 1m wide coil	752
	0.20 mm electrolytic tinplate, 1 m wide coil	1520
Stainless steels†		
AISI 409	2 mm sheet	2383
AISI 304	6 mm thick hot-rolled plate, 1m wide coil	2937
	2 mm thick cold-rolled sheet, 1m wide coil	3333
AISI 316	6 mm thick hot-rolled plate, 1m wide coil	3663
	2 mm thick cold-rolled sheet, 1m wide coil	4059

Sources: * Representative Metal Exchange Prices, December, 1996.
　　　　† Typical price lists, December, 1996.
Note: Pure metal prices vary with market conditions and prices of fabricated products are adjustable by premiums and discounts by negotiation.

The use of different metals in contact can be a corrosion hazard because in some metal couples, one of the pair is protected and the other is sacrificed, as described earlier in relation to cathodic protection. Examples of adverse metal pairs encountered in unsatisfactory designs are aluminum/brass and carbon steel/stainless steel, threatening intensified attack on the aluminum and carbon steel respectively. The uncritical mixing of metals is one of the more common corrosion-related design faults and so it is featured prominently in Chapter 4, where the overt and latent hazards of the practice are explained.

1.4.8 Geometric Factors

When the philosophy of a design is settled and suitable materials are selected, the proposed physical form of the artifact must be scrutinized for corrosion traps. Provided that one or two well-known effects are taken into account, this is a straightforward task. Whether protected or not, the less time the metal spends in contact with water, the less is the chance of corrosion and all that this requires is some obvious precautions, such as angle sections disposed apex upwards, box sections closed off or fitted with drainage holes, tank bottoms raised clear of the floor, and drainage taps fitted at the lowest points of systems containing fluids. Crevices must be eliminated to avoid local oxygen depletion for the reason given in Section 1.1.3. and explained in Chapter 3. This entails full penetration of butt welds, double sided welding for lap welds, well-fitting gaskets etc. If they are unavoidable, adverse mixed metal pairs should be insulated and the direction of any water flow should be from less noble to more noble metals to prevent indirect effects described in Chapter 4.

1.5 Some Symbols, Conventions, and Equations

From the discussion so far, it is apparent that specialized notation is required to express the characteristics of corrosion processes and it is often this notation which inhibits access to the underlying principles. The symbols used in chemical and electrochemical equations are not normal currency in engineering practice and some terms, such as electrode, potential, current, and polarization are used to have particular meanings which may differ from their meanings in other branches of science and engineering. The reward in acquiring familiarity with the conventions is access to information accumulated in the technical literature with a direct bearing on practical problems.

1.5.1 Ions and Ionic Equations

Certain substances which dissolve in water form electrically conducting solutions, known as electrolytes. The effect is due to their dissociation into electrically charged entities centered on atoms or groups of atoms, known as ions. The charges are due to the unequal distribution of the available electrons between the ions, so that some have net positive charge and are called cations and some have net negative charge and are called anions. Faraday demonstrated the existence of ions by the phenomenon of electrolysis in which they are discharged at positive and negative poles of a potential applied to a solution. Symbols for ions have superscripts showing the polarity of the charge and its value as charge numbers, i.e., multiples of the charge on one electron. A subscript, (aq), may be added where needed to distinguish ions in aqueous solution from ions encountered in other contexts, such as in ionic solids. The symbols are used to describe the solution of any substances yielding electrolytes on dissolution, e. g:

$$HCl \quad \rightarrow \quad H^+_{(aq)} \quad + \quad Cl^-_{(aq)} \qquad (1.1)$$

hydrochloric acid gas   hydrogen cation   chloride anion

$$FeCl_2 \quad \rightarrow \quad Fe^{2+}_{(aq)} \quad + \quad 2Cl^-_{(aq)} \qquad (1.2)$$

solid iron(II) chloride   iron cation   chloride anions

Symbols like H^+, Cl^- and Fe^{2+} used to represent ions in equations do not indicate their characteristic structures and properties that have very significant effects on corrosion and related phenomena. These structures are described in Chapter 2.

1.5.2 Partial Reactions

Equations 1.1 and 1.2 represent complete reactions but sometimes the anions and cations in solution originate from neutral species by complementary partial reactions, exchanging charge at an electronically conducting surface, usually a metal. To illustrate this process, consider the dissolution of iron in a dilute air-free solution of hydrochloric acid, yielding hydrogen gas and a dilute solution of iron chloride as the products. The overall reaction is:

$$Fe(metal) + 2HCl(solution) = FeCl_2(solution) + H_2(gas) \qquad (1.3)$$

Strong acids and their soluble salts are ionized in dilute aqueous solution as illustrated in Equations 1.1 and 1.2 so that the dominant species present in dilute aqueous solutions of hydrochloric acid and iron(II) chloride are not HCl and $FeCl_2$ but their ions $H^+ + Cl^-$ and $Fe^{2+} + 2Cl^-$. Equation 1.1 is therefore equivalent to:

$$Fe + 2H^+ + 2Cl^- = Fe^{2+} + 2Cl^- + H_2 \qquad (1.4)$$

The Cl⁻ ions persist unchanged through the reaction, maintaining electric charge neutrality, i.e., they serve as counter-ions. The effective reaction is the transfer of electrons, e⁻, from atoms of iron in the metal to hydrogen ions, yielding soluble Fe^{2+} ions and neutral hydrogen atoms which combine to be evolved as hydrogen gas. The electron transfer occurs at the conducting iron surface where the excess of electrons left in the metal by the solution of iron from the metal are available to discharge hydrogen ions supplied by the solution:

$$\text{Fe(metal)} \rightarrow \text{Fe}^{2+}\text{(solution)} + 2e^-\text{(in metal)} \tag{1.5}$$

$$2e^-\text{(in metal)} + 2H^+\text{(solution)} \rightarrow 2H\text{(metal surface)} \rightarrow H_2\text{(gas)} \tag{1.6}$$

Processes like those represented by Equations 1.5 and 1.6 are described as *electrodes*. Electrodes proceeding in a direction generating electrons, as in Equation 1.5. are *anodes* and electrodes accepting electrons, as in Equation 1.6 are *cathodes*. Any particular electrode can be an anode or cathode depending on its context. Thus the nickel electrode:

$$\text{Ni} \rightarrow \text{Ni}^{2+} + 2e^- \tag{1.7}$$

is an anode when coupled with Equation 1.6 to represent the spontaneous dissolution of nickel metal in an acid but it is an cathode when driven in the opposite direction by an applied potential to deposit nickel from solution in electroplating:

$$\text{Ni}^{2+} + 2e^- \rightarrow \text{Ni(metal)} \tag{1.8}$$

1.5.3 Representation of Corrosion Processes

The facility with which the use of electrochemical equations can reveal characteristics of corrosion processes can be illustrated by comparing the behavior of iron in neutral and alkaline waters.

Active Dissolution of Iron with Oxygen Absorption

Iron rusts in neutral water containing oxygen dissolved from the atmosphere. The following greatly simplified description illustrates some general features of the process. The concentration (strictly the activity defined later in Section 2.1.3) of hydrogen ions in neutral water is low so that the evolution of hydrogen is replaced by the absorption of dissolved oxygen as the dominant cathodic reaction and the coupled reactions are:

Anodic reaction: $\text{Fe} \rightarrow \text{Fe}^{2+} + 2e^-$ (1.9)

Cathodic reaction: $\frac{1}{2}O_2 + H_2O + 2e^- \rightarrow 2OH^-$ (1.10)

The use of the half quantity of oxygen in Equation 1.10 is a formal convention to match the mass balance in the two equations. The two reactions simultaneously introduce the ions Fe^{2+} and OH^- into the solution, which co-precipitate, again with simplifying assumptions, as the sparingly soluble compound $Fe(OH)_2$:

$$Fe^{2+}(solution) + 2OH^-(solution) \rightarrow Fe(OH)_2(precipitate) \quad (1.11)$$

In this system, the transport of Fe^{2+} and OH^- ions in the electrolyte between the anodic and cathodic reactions constitutes an ion current. The example illustrates how a corrosion process is a completed electric circuit with the following component parts:

1. An anodic reaction.
2. A cathodic reaction.
3. Electron transfer between the anodic and cathodic reactions.
4. An ion current in the electrolyte.

Methods for controlling corrosion are based on inhibiting one or another of the links in the circuit.

The $Fe(OH)_2$ is precipitated from the solution but it is usually deposited back on the metal surface as a loose defective material which fails to stifle further reaction, allowing rusting to continue. In the presence of the dissolved oxygen it subsequently transforms to a more stable composition in the final rust product. The rusting of iron is less straightforward than this simplified approach suggests and is described more realistically in Chapter 7.

Passivity of Iron in Alkaline Water

Iron responds quite differently in mildly alkaline water. The anodic reaction yielding the unprotective soluble ion, Fe^{2+} as the primary anodic product, is not favored and is replaced by an alternative anodic reaction which converts the iron surface directly into a thin, dense, protective layer of magnetite, Fe_3O_4, so that the partial reactions are:

Anodic reaction: $3Fe + 8OH^- = Fe_3O_4 + 4H_2O + 8e^-$ (1.12)

Cathodic reaction: $\frac{1}{2}O_2 + H_2O + 2e^- = 2OH^-$ (1.10)

Information on conditions favoring protective anodic reactions of this kind is important in corrosion control. Pourbaix diagrams, explained in

Chapter 3, give such information graphically and within their limitations they can be useful in interpreting observed effects.

Further Reading

Economic Effects of Metallic Corrosion in the United States, National Bureau of Standards Special Publication, 1978.

2

Structures Participating
in Corrosion Processes

2.1 Origins and Characteristics of Structure

Conventional symbols are convenient for use in chemical equations, as illustrated in the last chapter, but they do not indicate the physical forms of the atoms, ions, and electrons they represent. This chapter describes these physical forms and the structures in which they exist because they control the course and speed of reactions.

There is an immediate problem in describing and explaining these structures because they are expressed in the conventional language and symbols of chemistry. Atoms can be arranged in close-packed arrays, open networks or as molecules, forming crystalline solids, non-crystalline solids, liquids, or gases, all with their own specialized descriptions. The configurations of the electrons within atoms and assemblies of atoms are described in terms and symbols derived from wave mechanics, that are foreign to many applied disciplines that need the information.

A preliminary task is to review some of this background as briefly and simply as possible, for use later on. At this point it is natural for an applied scientist to enquire whether an apparently academic digression is really essential to address the practical concerns of corrosion. The answer is that, without this background, explanations of even basic underlying principles can only be given on the basis of postulates that seem arbitrary and unconvincing. With this background, it is possible to give plausible explanations to such questions as, why water has a special significance in corrosion processes, why some dissolved substances inhibit corrosion whereas others stimulate it, what features of metal oxides control the protection they afford and how metallurgical structures have a key influence on the development of corrosion damage. Confidence in the validity of fundamentals is an essential first step in exercising positive corrosion control.

2.1.1 Phases

The term, *phase*, describes any region of material without internal boundaries, solid liquid or gaseous, composed of atoms, ions or molecules organized in a particular way. The following is a brief survey of various kinds of phase that may be present in a corroding system and applies to metals, environments, corrosion products, and protective systems.

2.1.1.1   Crystalline Solids

Many solids of interest in corrosion, such as metals, oxides and salts are crystalline; the crystalline nature of bulk solids is not always apparent because they are usually agglomerates of microscopic crystals but under laboratory conditions, single crystals can be produced that reveal many of the features associated with crystals, regular outward geometrical shapes, cleavage along well-defined planes and anisotropic physical and mechanical properties. The characteristics are due to the arrangement of the atoms or ions in regular arrays generating indefinitely repeated patterns throughout the material. This long-range order permits the relative positions of the atoms or ions in a particular phase to be located accurately by standard physical techniques, most conveniently by analysis of the diffraction patterns produced by monochromatic X-rays transmitted through the material. The centers of the atoms form a three-dimensional array known as the *space lattice* of the material.

Space lattices are classified according to the symmetry elements they exhibit. A space lattice is described by its *unit cell*, that is the smallest part of the infinite array of atoms or ions that completely displays its characteristics and symmetry. The lattice dimensions are specified by quoting the *lattice parameters*, that are the lengths of the edges of the unit cell. The complete structure is generated by repetition of the unit cell in three dimensions. Crystallographic descriptions embody the assumption that atoms and ions are hard spheres with definite radii. Strictly, atomic and ion sizes are influenced by local interactions with other atoms, but the assumption holds for experimental determination of atomic arrangements and lattice parameters in particular solids.

A wide range of space lattices is needed to represent all of the structures of crystalline solids of technical interest. Geometric considerations reveal fourteen possible types of lattice, fully described in standard texts.* All that is needed here is a brief review of structures directly concerned with metals and solid ionic corrosion products. These are the close-packed cubic structures and the related hexagonal close-packed structure.

The closest possible packing for atoms (or ions) of the same radius is produced by stacking layers of atoms so that the whole system occupies the minimum volume, as follows. Spheres arranged in closest packing in a single layer have their centers at the corners of equilateral triangles, as shown in Figure 2.1. For a stack of such layers to occupy minimum volume, the spheres in every successive layer are laid in natural pockets between contiguous atoms in the underlying layer. Simple geometry

---

* e.g., Taylor, cited in "Further Reading."

shows that the atoms of a layer in a stack are sited in alternate pockets of the underlying layer, e.g., at the centers of *either* the upright triangles *or* the inverted triangles in Figure 2.1. This option leads to two diffent simple stacking sequences; in one, the positions of the atoms are in register at every second layer in the sequence *ABAB . . .* , generating the hexagonal close-packed lattice and in the other they are in register at every third layer in the sequence *ABCABC . . .* , generating the face-centered cubic lattice.

FIGURE 2.1
Assembly of spheres representing the closest packed arrangement of atoms in two dimensions. The pockets at the centers of the equilateral triangles are sites for atoms in a similar layer superimposed in the closest-packed three-dimensional arrangement.

The Hexagonal Close-Packed (HCP) Lattice

The hexagonal symmetry is derived from the fact that an atom in a close-packed plane is coordinated with six other atoms, whose centers are the corners of a regular hexagon. In three dimensions, every atom in the HCP lattice is in contact with twelve equidistant neighbors. Geometric considerations show that the axial ratio of the unit cell, i.e., the ratio of the lattice parameters normal to the hexagonal basal plane and parallel to it is 1.633.

The Face-Centered Cubic (FCC) Lattice

The *ABCABC . . .* stacking sequence confers cubic symmetry that is apparent in the unit cell, illustrated in Figure 2.2(a), taken at an appropriate angle to the layers. Although the atoms are actually in contact, unit cells are conventionally drawn with small spheres indicating the lattice points to reveal the geometry. As the name suggests, the unit cell has one atom at every one of the eight corners of a cube and another atom at the center of every one of the six cube faces. Every atom is in contact with twelve equidistant neighbors, i.e., its *coordination number* is 12. Every one of the eight corner atoms in the FCC unit cell is shared with seven adjacent unit cells and every face atom is shared with one other cell, so that the cell contains the equivalent of four atoms, $(8_{corner\ atoms} \times \frac{1}{8}) + (6_{face\ atoms} \times \frac{1}{2})$.

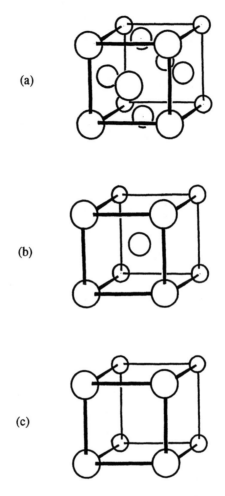

FIGURE 2.2
Unit cells: (a) face-centered cubic; (b) body-centered cubic; and (c) simple cubic.

The FCC lattice completely represents the structures of many metals and alloys but its use is extended to provide convenient crystallographic descriptions of some complex structures, using its characteristic that the spaces between the atoms, the *interstitial sites*, have the geometry of regular polyhedra. The concept is to envision atoms or ions of one species arranged on FCC lattice sites with other species occupying the interstices. Considerable use is made of this device later, especially in the context of oxidation, where a class of metal oxides collectively known as spinels have significant roles in the oxidation-resistance of alloys. This application depends on the geometry and number of interstices.

Inspection of the FCC unit cell illustrated in Figure 2.2(a) reveals that there are two kinds of interstitial sites, *tetrahedral* and *octahedral*. A tetrahedral site exists between a corner atom of the cell and the three adjacent face atoms and there are eight of them wholly contained within every cell. Octahedral sites exist both at the center of the cell between the six face atoms and at the middles of the twelve edges, every one of which is shared with three adjacent unit cells, so that the number of octahedral interstices per cell is $1 + (12 \times \frac{1}{4}) = 4$. Since the cell contains the equivalent of four atoms, the FCC structure contains two tetrahedral and one octahedral spaces per atom. The ratios of the radii of spheres that can be inscribed the interstitial sites to the radius of atoms on the lattice is 0.414 for tetrahedral sites and 0.732 for the octahedral sites. These radius ratios indicate the sizes of interstitial atoms or ions that can be accommodated.

The Body-Centered Cubic (BCC) Lattice

The BCC structure is less closely packed than the HCP and FCC structures. The unit cell, illustrated in Figure 2.2(b), has atoms at the corners of a cube and another at the center. It has the equivalent of 2 atoms and there are 12 tetrahedral and 3 octahedral interstitial sites with the geometries of irregular polyhedra. Corner atoms at opposite ends of the cell diagonals are contiguous with the atom at the center so that the coordination number is 8.

The Simple Cubic Lattice

The simple cubic lattice, illustrated in Figure 2.2(c), can be formed from equal numbers of two different atoms or ions if the ratio of their radii is between 0.414 and 0.732, as explained later in describing oxides. The structure is geometrically equivalent to an FCC lattice of the larger atoms or ions with the smaller ones in octahedral interstitial sites.

Some characteristic metal structures are summarized in Table 2.1

TABLE 2.1
Crystal Structures of Some Commercially
Important Pure Metals

Crystal Structure	Metal
Face-centered cubic	Aluminum, Nickel, α-Cobalt, Copper, Silver, Gold, Platinum, Lead, Iron ($T > 910°C$ and $< 1400°C$).
Body-centered cubic	Lithium, Chromium, Tungsten, Titanium ($> 900°C$), Iron ($T < 910°C$ and $> 1400°C$).
Hexagonal close-pack	Magnesium, Zinc, Cadmium, Titanium ($< 900°C$).
Complex structures	Tin (tetragonal), Manganese (complex), Uranium (complex).

Example 1: *Description of Perovskite Structure*

The structure of perovskite with the empirical formula, $CaTiO_3$, can be described as either:

1. A body-centered cubic (BCC) lattice with calcium ions at the corners of the unit cell, a titanium at the center and all octahedral vacancies occupied by oxygen ions,

2. A face-centred cubic (FCC) lattice with calcium ions at the corners of the unit cell, oxygen ions at the centers of the faces and every fourth octahedral vacancy (the one entirely between oxygen ions) occupied by titanium.

Show that these descriptions are compatible with the numbers of atoms in the formula.

SOLUTION:

There are two ions in the BCC unit cell. The eight corner sites, all shared eight-fold, together contribute one calcium ion and the unshared center site contributes one titanium ion. The oxygen ions occupy the octahedral vacancies at the centers of the six faces, all shared two-fold, contributing three oxygen ions. Hence the description yields the same relative numbers of calcium, titanium and oxygen ions in the structure, i.e., 1:1:3, as atoms in the formula. Incidentally, the basic BCC structure can also be envisioned as two interpenetrating simple cubic lattices, one of calcium ions and the other of titanium ions.

There are four ions in the FCC unit cell. The eight shared corner sites together contribute one calcium ion and the six shared face sites together contribute three oxygen ions. One quarter of the octahedral sites are occupied by titanium ions and since there are four such sites per unit cell, they contribute one titanium ion. This description also yields the same relative numbers of the ions in the structure as atoms in the formula.

2.1.1.2   Liquids

Liquid phases are arrays of atoms with short-range structural order and the atoms or groups of atoms can move relatively without losing cohesion, conferring fluidity. Liquid structures are less amenable to direct empirical study than solid structures. X-ray diffraction studies reveal the average distribution of nearest neighbor atoms around any particular atom but other evidence of structure can be acquired, particularly for solutions. Liquid metal solutions can show discontinuities in properties at compositions that correspond to changes in the underlying solid phases. The most familiar liquid, water, is highly structured because hydrogen-oxygen bonds have a directional character. Its bulk physical properties, excellent solvent powers and behavior at surfaces are striking manifestations of its structure, as explained later in this chapter.

2.1.1.3   Non-Crystalline Solids

Certain solid materials, including glasses and polymeric materials have short-range order but the atoms or groups of atoms lack the easy relative mobility characteristic of liquids.

2.1.1.4   Gases

In gaseous phases, attractions between atoms or small groups of atoms (molecules) are minimal, so that they behave independently as random entities in constant rapid translation. The useful approximation of the hypothetical ideal gas assumes that the atoms or molecules are dimensionless points with no attraction between them. It is often convenient to use this approximation for the "permanent" real gases, oxygen, nitrogen and hydrogen and for mixtures of them. Certain other gases of interest in corrosion, e.g., carbon dioxide, sulfur dioxide and chlorine deviate from the approximation and require different treatment.

2.1.2   The Role of Electrons in Bonding

A phase adopts the structure that minimizes its internal energy within constraints imposed by the characteristics of the atoms that are present. In general, this is achieved by redistributing electrons contributed by the individual atoms. The resulting attractive forces set up between the individual atoms are said to constitute *bonds* if they are sufficient to stabilize a structure. A description of the distributions of electrons among groups or aggregates of atoms that constitute bonds between them must be preceded by a description of the electron configurations in isolated atoms.

An isolated atom comprises a positively charged nucleus surrounded by a sufficient number of electrons to balance the nuclear charge. The order of the elements in the Periodic Table, reproduced in Table 2.2, is also the order of increasing positive charge on the atomic nucleus in increments, e^+, equal but opposite to the charge, e^- on a single electron. The positive charges on the nuclei are balanced by the equivalent numbers of electrons, that adopt configurations according to the energies they possess.

Classical mechanics break down when applied to determine the energies of electrons moving within the very small dimensions of potential fields around atomic nuclei. The source of the problem was identified by the recognition that electrons have a wave character with wavelengths comparable with the small dimensions associated with atomic phenomena and an alternative approach, *wave mechanics*, pioneered by de Broglie and Schrodinger, was developed to deal with it. This approach abandons any attempt to follow the path of an electron with a given total energy moving in the potential field of an atomic nucleus and addresses the conservation of energy using a time-independent wave function as a replacement for classical momentum. It turns out that the probability that the

TABLE 2.2
Simplified Periodic Table of the Elements

1	2		3	4	5	6	7	0
H								He
Li	Be		B	C	N	O	F	Ne
Na	Mg		Al	Si	P	S	Cl	Ar
K	Ca	Sc Ti V Cr Mn Fe Co Ni Cu Zn First transition series (development of 3*d* shell)	Ga	Ge	As	Se	Br	Kr
Rb	Sr	Y Zr Nb Mo Tc Ru Rh Pd Ag Cd Second transition series (development of 4*d* shell)	In	Sn	Sb	Te	I	Xe
Cs	Ba	La* Hf Ta W Re Os Ir Pt Au Hg Third transition series (development of 5*d* shell)	Tl	Pb	Bi	Po	At	Rn
Fr	Ra	Ac†						

* Lanthanide elements (La, Ce, Pr Nd ... etc.) developing the 4*f* shell
† Actinide elements (Ac, Th, Pa, U ... etc.) developing the 5*f* shell

electron is present at any particular point is proportional to the square of the wave amplitude and this introduces the energy and symmetry considerations with far-reaching consequences described below. The validation of the wave mechanical approach is that it delivers results that account with outstanding success for observations that cannot be explained otherwise. The formulation, solution and interpretation of wave equations is a severe, specialized discipline, beyond the scope of this book, but the conclusions that emerge have far-reaching consequences. Classic monographs by Coulson, Pauling, and Hume-Rothery* give reader-friendly explanations. For now, a qualitative description of the approach and conclusions together with enough terminology to explain structures is all that is needed.

2.1.2.1 *Atomic Orbitals*

When continuity and other constraints are applied, it is found that only certain solutions to a wave equation have any valid physical significance, yielding the following interpretations for isolated atoms:

1. The energies allowed for electrons do not vary continuously but can have only discrete values, referred to as *energy levels*.

2. The allowed energy levels are associated with characteristic symmetries for the electron probability distributions. This geometric feature is important in the formation of structures because it can determine whether bonds between atoms have directional character.

---

* Cited in "Further Reading."

It is important to appreciate that this is not a collection of arbitrary assumptions designed to *explain* observations but the inevitable result of using the wave equations, that *predict* the observations.

The standard notation used to classify the energy levels and for any electrons that might occupy them is derived from (1) the sequence of allowed energies and (2) the symmetries associated with them.

The allowed energy levels are arranged in *shells*, numbered outward from the nucleus by the *principal quantum number*, $n = 1, 2, 3 \ldots$ etc. The types of symmetry correspond with a letter code, *s, p, d, f*, originally devised to describe visible light spectra exited from atoms:

s Spherically symmetrical distribution around an atomic nucleus.

p Distribution in two diametrically opposite lobes, about an atomic nucleus. By symmetry, there are three mutually perpendicular distributions with the same energy values, designated p_x, p_y and p_z.

d Distribution in four lobes centered on the nucleus. By symmetry, there are five independent distributions with the same energy values, designated d_{xy}, d_{yz}, d_{xz}, $d_{x^2-y^2}$, d_{z^2}.

f f orbitals are occupied only in the heavier elements and need not be considered here.

Solutions to the wave equations show that the first shell can contain only *s* electrons, the second shell can contain *s* and *p* electrons and the third shell can contain *s, p*, and *d* electrons. It is usual to refer to the allowed energy levels and their associated electron density probability distributions in an isolated atom as *atomic orbitals* although there is no question of any kind of orbital motion associated with them. The term persists from early attempts to apply classical mechanics to electron energies. The existence of an orbital does not imply that it is necessarily occupied by an electron but describes an allowed discrete energy that an electron *may* occupy if it is present. Any particular orbital that an electron occupies corresponds to its *quantum state*. A further constraint applied is the *Pauli exclusion principle*, explained e.g., by Coulson and by Pauling, that limits the occupation of any orbital to two electrons, that must differ in a further quality, called *spin*.

The nominal sequence of allowed energies is (1*s*) (2*s*, 2*p*) (3*s*, 3*p*, 3*d*) (4*s*, 4*p*, 4*d*, 4*f*) (5*s* 5*p* . . .) etc. However, there is a difference in the actual sequence because solutions to the wave equation yield energy values that require occupation of the 3*d* orbital to be deferred until the 4*s* orbital is occupied and there is a similar reversal in sequence for the 4*d*, 4*f* and 5*s* orbitals. Therefore the actual sequence is (1*s*) (2*s*, 2*p*) (3*s*, 3*p*) (4*s*, 3*d*, 4*p*) (5*s*, 4*d*, 5*p* . . .) etc. These changes in the sequence have far-reaching consequences because two series of elements in the Periodic Table, the first and second *transition series*, including most of the commercially important strong metals, are created as the 3*d* and 4*d* orbitals are filled progressively underneath the $4s$ and $5s$ orbitals, respectively. The underlying partly filled d orbitals in these metals confer special characteristics on them, that govern their interactions with water, the structures of their oxides, their mechanical and physical properties and their alloying behavior, all of which are of crucial importance in determining their resistance to corrosion.

Applying the Pauli exclusion principle and taking account of the three- and five-fold multiplicities of p and d orbitals, the total number of electrons that can be accommodated in the first four shells are:

First shell $(n = 1)$: $2_{(1s)} = 2$
Second shell $(n = 2)$: $2_{(2s)} + (3 \times 2)_{(2p)} = 8$
Third shell $(n = 3)$: $2_{(3s)} + (3 \times 2)_{(3p)} = 8$
Fourth shell $(n = 4)$: $2_{(4s)} + (5 \times 2)_{(3d)} + (3 \times 2)_{(4p)} = 18$

A shell with its full complement of electrons is said to be closed, even when it is provisional, as for the third and fourth shells, where the next shell is started before the d orbitals are occupied. The significance of a closed shell is apparent when electron configurations for the elements, given in Table 2.3, is compared with their order in the Periodic Table.

The elements with all shells closed are the noble gases, helium, neon, argon etc., that are remarkably stable, as illustrated by their existence as monatomic gases and by their chemical inertness. In contrast, the other elements, that have partly filled outer shells, react readily. It is therefore apparent that stability is associated with closed shells.

2.1.2.2 *Molecular Orbitals and Bonding of Atoms*

Stable assemblies of atoms form when the energy of the system is reduced by the combination of atomic orbitals to yield *molecular orbitals*. This can be explained by one or other of two main approaches, the *molecular orbital* (MO) and *valence bond* (VB) theories, that are regarded as equivalent, except in their treatment of the small electron-electron interactions, as explained e.g., by Coulson. The *molecular orbital* (MO) theory is easier to apply to simple molecules and inorganic complexes, including those in which metal atoms bind to other discrete entities such as water, hydroxide or chloride ions. Bonding in metallic phases is easier to appreciate using the *valence bond* (VB) theory. In this section we shall deal mainly with molecular orbital theory.

The fundamental principle of molecular orbital theory is that a bonding orbital between two atoms is derived by constructive linear combination of atomic wave functions for the component atoms, yielding a *molecular orbital*, in which electrons contributed by the atoms are accommodated with reduced energy. The geometric aspect of atomic orbitals, referred to earlier, is carried over into molecular orbitals derived from them, that can impart directionality in the interaction of an atom with its neighbors. The

TABLE 2.3
Electron Configurations of the Lighter Elements

	Electron shells						
	1	2		3			4
Element	s	s	p	s	p	d	s
H	1						
He	2						
Li	2	1					
Be	2	2					
B	2	2	1				
C	2	2	2				
N	2	2	3				
O	2	2	4				
F	2	2	5				
Ne	2	2	6				
Na	2	2	6	1			
Mg	2	2	6	2			
Al	2	2	6	2	1		
Si	2	2	6	2	2		
P	2	2	6	2	3		
S	2	2	6	2	4		
Cl	2	2	6	2	5		
Ar	2	2	6	2	6		
K	2	2	6	2	6		1
Ca	2	2	6	2	6		2
Sc	2	2	6	2	6	1	2
Ti	2	2	6	2	6	2	2
V	2	2	6	2	6	3	2
Cr	2	2	6	2	6	4	2
Mn	2	2	6	2	6	5	2
Fe	2	2	6	2	6	6	2
Co	2	2	6	2	6	7	2
Ni	2	2	6	2	6	8	2
Cu	2	2	6	2	6	10	1
Zn	2	2	6	2	6	10	2

tendency for the molecule to minimize its energy favors interactions yielding orbitals that electrons can populate at the lowest energy.

Criteria contributing to a minimum energy associated with a molecular orbital are:

1. Maximum constructive overlap of the atomic orbital wave functions.

2. Minimum electrostatic repulsions between adjacent filled orbitals.

The second of these criteria arises because electron-filled orbitals experience mutual electrostatic repulsion and in the absence of other influences they assume orientations with maximum separation.

Sometimes these criteria are best met by transforming some of the dissimilar 2*s*, 2*p*, and 3*d* atomic orbitals, into a corresponding set of mutually equivalent atomic orbitals called *hybrid* orbitals. For example, the 2*s*, 2*p*$_x$, 2*p*$_y$ and 2*p*$_z$ orbitals in the free oxygen atom hybridize to form four *sp*^{3} orbitals, with their axes directed towards the corners of a tetrahedron, when it combines with hydrogen to form a water molecule. This phenomenon makes a major contribution to the unique structure and character of liquid water described in the Section 2.2.3.

2.1.3   The Concept of Activity

In describing the structures of liquid water, oxides and metals and the interactions within and between them, a physical quantity, *activity*, is applied to the chemical species that are present. Activity is a general concept in chemistry and is rigorously defined.* It is used extensively to determine the energies of chemical reactions and the balance that is struck (the *equilibrium*) between the reactants and the products of chemical reactions, as explained later in Chapter 3. At this point, the concept is provisionally described in a non-rigorous way to assist in describing interactions that atoms and ions experience in solution hence to explain the structural features that they introduce.

When a substance is dissolved in a suitable solvent, its ability to react chemically is attenuated. The property of the substance that is thereby diminished is its activity. The diminished activity is not necessarily, or even generally, proportional to the degree of dilution, because the substance almost always interacts with the solvent or with other solutes that may be present.

The symbol for activity is *a* with a subscript to denote the component to which it refers, e.g., $a_{C_{12}H_{22}O_{11}}$ and a_{H_2O}, respectively represent the activities of sucrose and of water in an aqueous sugar solution. The activity of a species in a solution is a dimensionless quantity referred to its activity in an appropriate *standard state*, in which it is conventionally assigned the value, unity. The standard selected is arbitrary but two approaches are in common use.

In one approach, every component of a solution is treated in the same way and the standard states selected are the pure unmixed substances. There is a hypothetical concept, the *ideal solution* (or *Raoultian solution*, named for its originator) in which, because there is no interaction between the components, the activity of any particular component, i, is proportional to its *mole fraction*, N_i, that expresses the fraction present as the ratio of the number of individual entities of the component to the total number of molecular entities in the solution. Few real solutions approach this ideal. More generally, interactions between the components raise or lower

---

* e.g., by Lewis, Randall, Pitzer, and Brewer cited in Further Reading list.

their activities from the ideal values. The ratio of the real to the ideal activity is called the *activity coefficient, f,* and hence:

$$a_i = f_i . N_i \qquad (2.1)$$

The alternative approach, that is often more convenient for dilute solutions treats the solutes in a different way; whereas the standard state of the solvent is the pure material, the standard state for every solute is some definite composition, usually unit *molality*, a quantity equal to its relative molar mass (formerly the atomic or molecular weight in grams) dissolved in 1 kg of solution. For species such as ions in aqueous solution, that cannot exist as "pure" substances, such a standard state is not only convenient, but unavoidable.

Many real solutions of interest depart from ideality in reasonably regular ways and various models and devices are available to address them, as summarized, e. g. by Bodsworth.* For example, although the activity of a species is not generally equal to the mole fraction, it is always a near-linear function of it when the solution is very dilute, and in these circumstances the solute is said to exhibit *Henrian activity.* This linear function is often exploited to produce a scale suitable for application to dilute solutions, by defining a standard based on an *infinitely dilute solution,* in which the numerical values of activities and molalities are interchangeable. In this scale, the activity coefficient, the *Henry's law coefficient,* is constant and is denoted by the symbol, γ.

2.2 The Structure of Water and Aqueous Solutions

2.2.1 The Nature of Water

Liquid water is by far the most efficient medium to sustain metallic corrosion at ordinary temperatures because it is endowed with the following properties and characteristics:

1. Physical stability and low viscosity over a wide temperature range.
2. An ability to dissolve and disperse foreign ionic species.
3. An ability to dissolve small quantities of oxygen and some other gases.
4. Versatility in providing neutral, acidic or alkaline environments.
5. Some intrinsic electrical conductivity.

---

* Cited in Further Reading list.

The value of this information in explaining corrosion processes and their control is clarified by relating it to the structure of liquid water. Contrary to common perceptions, liquid water is a highly organized substance as described in detail by Franks, Davies and, Bockris and Reddy.* It is essentially a mobile three-dimensional network of coherent hydrogen and oxygen atoms based on the geometric arrangement and distribution of electric charge within the basic structural unit, the water molecule.

2.2.2 The Water Molecule

The molecule, formally written, H_2O, contains three atomic nuclei, one of oxygen with a charge, $8e^+$, two of hydrogen each with a charge, e^+, all balanced by the collective charge of the associated electrons, $10e^-$. Its essential characteristics can be formulated using wave mechanics and confirmed by experimental observations. They include molecular geometry, the strengths of the oxygen-hydrogen bonds, the vibrational modes and the distribution of internal charge that is particularly useful in explaining the solvent function of liquid water.

Figure 2.3 is a schematic diagram of the geometrical arrangement that emerges from calculations for the isolated molecule. There are four hybrid orbitals derived from the oxygen $2s$, $2p_x$, $2p_y$, and $2p_z$ atomic orbitals, directed outwards from the oxygen nucleus towards the corners of a tetrahedron. Every one of the four orbitals has its full complement of two electrons; the oxygen $1s$ inner orbital, containing the remaining two electrons, persists virtually unchanged in the water molecule. Two of the hybrid orbitals interact with the $1s$ atomic orbitals of the hydrogen atoms to form molecular orbitals and the other two orbitals are uncommitted to bonding and the electrons within them are described as *lone pairs*. This configuration produces an asymmetric distribution of electric charge across the molecule and this characterizes it as a *polar* molecule.

The hydrogen $1s$ orbital and the single charge on the hydrogen nucleus produces only a small effect on the shape of the molecular orbital, so that the bonding and non-bonding orbitals are roughly similar but the differential electrostatic repulsion between them yields 104° 27′ and 114° 29′ for the angles between the axes of the two bonding and the two non-bonding orbitals respectively, instead of 109° 28′ as expected for a regular tetrahedral configuration.

2.2.3 Liquid Water

The *polar* character of the water molecule is responsible for the strong bonding between water molecules in the liquid phase. The magnitude of the effect is strikingly illustrated in Table 2.4, where the physical stability

---

* Cited in "Further Reading."

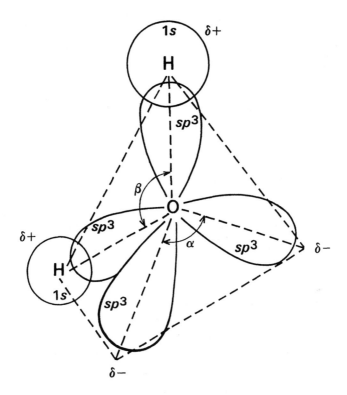

FIGURE 2.3
The tetrahedral configuration of the water molecule. $\alpha = 114° 29'$ $\beta = 104° 27'$

of water is compared with that of liquids with comparable molar masses formed from non-polar molecules, that cohere only by weak *Van der Waals forces* derived from attraction between induced dipoles. Such liquids have low boiling points and exist only within narrow temperature ranges.

2.2.3.1 Hydrogen Bonding

The strong coherence of water molecules is due to *hydrogen bonding*, which is a special mechanism exploiting the polar character of the water molecules. The small mass of the hydrogen nucleus allows it to oscillate and interact with the negative charge on the lone pair of an adjacent water

TABLE 2.4
Physical Stability of Water

Liquid	Liquid range °C	Boiling point °C
Water	100	100
Methane	18	−164
Neon	2.8	−245.9

molecule, producing a bond with about a tenth the energy of the molecular O–H bond, without losing cohesion in its own molecule. It is true that some other small polar molecules, e.g., hydrofluoric acid and ammonia can also link by hydrogen bonding but this is usually restricted. The unique feature of water molecules is that they can form hydrogen bonds at each of *two* tetrahedral corners, producing a coherent *three dimensional* network. The detailed arrangement of molecules in liquid water is uncertain but it is probably based on the hexagonal ice structure in which every water molecule is linked to four neighbors in a very open network.

The liquid is often considered to have short-range order derived from the ice structure but with some of the spaces in the network occupied by other extra molecules constantly interchanging with those in the hydrogen-bonded network, illustrated in Figure 2.4. This increases the average number of nearest neighbors surrounding a water molecule from 4 in ice to 4.4 in the liquid at its freezing point. This underlies the anomalous *increase* in density when ice melts to form water.

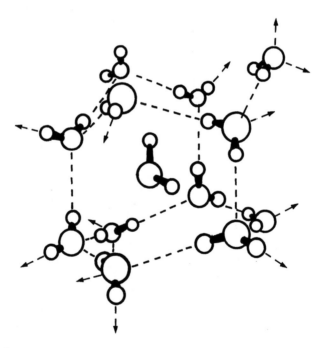

FIGURE 2.4
Schematic diagram of a possible structure of water, envisaged as a dynamic interchange between some molecules in an ice-like hydrogen bonded network, shown as dotted lines, and others temporarily occupying the spaces.

2.2.3.2 Dielectric constant

By virtue of its polarity and hydrogen-bonding interactions, water is a very effective medium for reducing electrostatic forces contained within it.

This is a bulk property described by the *dielectric constant*, ε_r, a dimensionless parameter appearing in the Coulombic force equation:

$$F = \frac{q_1 q_2}{4\pi\varepsilon_0 \varepsilon_r r^2} \tag{2.2}$$

where F is the force between two charges q_1 and q_2, in the medium at a distance, r. ε_0 is the permittivity of a vacuum.

Water has one of the highest known dielectric constants, i.e., 78 at 25°C. The higher is the dielectric constant in a medium, the lower is the energy needed to separate opposite electric charges and this is one of the factors that make water such an effective solvent for ionic species.

2.2.3.3   Viscosity

The viscosity of a solvent controls *inter alia* the transport of solutes within it. In liquid water it is related to the *self-diffusion*, i.e., the mobility, of the molecules through the hydrogen-bonded matrix.

Other things being equal, the viscosity of a liquid is an inverse function of the effective molecular size of the solvent, taking account of any coherence between adjacent molecules, yet despite the exceptionally high degree of molecular association in water, its viscosity has the low value, 1.002 Pa s at 20°C. This is attributable to the small size of the unassociated water molecule and the relatively open structure that provides easy pathways for diffusing molecules.

2.2.4   Autodissociation and pH of Aqueous Solutions

The autodissociation of water and the concept of pH as an indicator of acidity or alkalinity are not always clearly explained. It is convenient to describe the dissociation by the symbolic equation:

$$H_2O(l) = H^+(aq) + OH^-(aq) \tag{2.3}$$

although some texts use the symbols H_3O^+ or $H_9O_4^+$ for the hydrogen ion, indicating its association with one or more water molecules.

It is much better to describe auto-dissociation as disorder in the structure of the bulk liquid. In this description, a hydrogen ion is centered on a water molecule with an excess proton, thus carrying a net positive charge e^+. Charge neutrality within the bulk liquid is conserved by the creation of a complementary hydroxyl ion, centered on another water molecule with a proton deficiency. In both cases, the local excess charge is dissipated over a volume of the surrounding hydrogen-bonded water matrix containing several molecules.

The auto-dissociation is driven by an energy advantage in creating disorder (due to a positive entropy change*). This is opposed by the energy disadvantage (due to a positive enthalpy change*) in breaking O-H bonds and separating the charged fragments. The balance struck between these opposing energy terms leads to definite populations of hydrogen and hydroxyl ions related by an *equilibrium constant*, K_w given by

$$K_w = a_{H^+} \cdot a_{OH^-} \tag{2.4}$$

where a_{H^+}, a_{OH^-} are the activities of the H^+ and OH^- ions, respectively.

Following normal chemical convention, the activity of a pure solvent, in this case hypothetical undissociated water, is equal to unity and is virtually unchanged by the activities of solutes in dilute solution so that its effect is insignificant and is ignored in Equation 2.4. K_w is temperature-dependent and examples of its values are given in Table 2.5.

TABLE 2.5
Temperature-Dependence of K_w and pH Neutrality

Temperature °C	K_w/mol²dm⁻⁶	pH Neutrality
0	10^{-15}	7.5
25	10^{-14}	7.0
60	10^{-13}	6.5

2.2.5 The pH Scale

As Equation 2.3 shows, auto-dissociation produces equal populations of H^+ and OH^- ions so their activities are also equal and for pure water at ambient temperatures close to 25°C:

$$a_{H^+} = a_{OH^-} = 10^{-7} \text{ at } 25°C. \tag{2.5}$$

The ratio of hydrogen and hydroxyl ions is altered if acid or alkali is added to water but the activity product in Equation 2.4 remains equal to K_w, so if for example a_{H^+} is raised from 10^{-7} to 10^{-4} by adding a small quantity of acid to water at 25°C, a_{OH^-} is reduced to 10^{-10}.

This effect suggests a practical scale for the degree of acidity or alkalinity related to the measured activity of the hydrogen ion but it is difficult to measure directly the activity of a single ion species in isolation because a counter ion species must be present to preserve charge neutrality. The difficulty has been circumvented by defining a scale, the pH scale, referred to internationally recommended arbitrary standards based on measurements in solutions that inevitably contain counter ions. The scale is chosen so that for dilute solutions, in which the hydrogen ion activities lie

---

* The significance of these terms is explained in Section 2.3.5.

between 10^{-2} and 10^{-12}, the following expression applies without significant error:

$$pH = - \log_{10} a_{H^+} \qquad (2.6)$$

Thus for pure water at 25°C and for aqueous solutions in which there is no net effect on auto-dissociation, $a_{H^+} = 10^{-7}$, yielding a value for pH of 7.0. This value corresponds to equal hydrogen and hydroxyl ion activities and is often referred to as *pH neutrality*. If the pH value is <7.0, the solution has an excess of hydrogen ions and is acidic; if it is >7.0 the solution has an excess of hydroxyl ions and is alkaline. A word of caution is needed because the value of pH varies with temperature as shown in Table 2.5, corresponding to the temperature-dependence of K_w. Furthermore, the description a_{H^+} strictly denotes the activity of the entities carrying excess protons in a water matrix. Hence any factor that disturbs the water matrix, such as close spacing of ions in concentrated solutions may influence the pH.

2.2.6 Foreign Ions in Solution

Unlike the products of auto-dissociation a foreign ion in aqueous solution introduces distinct structural changes in the water. Foreign ions have finite sizes and perturb the structure. A description of the effect is facilitated by the simplifying assumption that ions are spherical or occupy spherical holes in the water matrix.

The charge on a foreign ion imposes a radial electric field in the surrounding water matrix and it re-orientates the adjacent water molecules around the ion to a greater or lesser extent. The molecules around an *anion* are arranged with the hydrogen atoms facing the ion. Those around a *cation* have diametrically opposite configurations, with the non-bonding orbitals containing the lone pair electrons facing the ion. This spherically symmetric reorientation of the water molecules is incompatible with the tetrahedral hydrogen-bonded matrix of undisturbed water. The consequent local modification of the water structure is illustrated in Figure 2.5 and is known as *solvation*.

Some metal cations with electron-deficient orbitals can use the uncommitted lone pair electrons facing them to increase the cohesion by a tendency to form chemical bonds, so that cations are often more strongly solvated than anions.

Water molecules directly attached to an ion comprise the *primary solvation shell*. Around this shell, there is a region in which the surrounding water molecules are more associated with the primary solvation shell than with the surrounding solvent. This is usually termed secondary solvation. Modern theories also include a third concentric region in which properties of the solvent such as the dielectric constant are influenced by the electric field centered on the ionic charge.

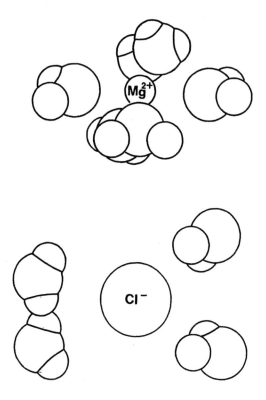

FIGURE 2.5
Typical structures of inner solvation spheres in aqueous solution: (top) cations, e.g., magnesium; (b) anions, e.g., chloride.

It is convenient to classify foreign ions in water as structure-forming and structure-breaking ions according to whether the induced structure around the ion is more or less ordered than the matrix it displaces.

Typical structure-forming ions have high surface charge densities. They include small singly charged ions such as lithium, Li^+, and fluoride, F^- and larger multiply-charged ions such as calcium, Ca^{2+}, magnesium, Mg^{2+}, aluminum, Al^{3+}, zinc, Zn^{2+}, copper(II), Cu^{2+}, and iron(III), Fe^{3+}.

Typical structure-breaking ions have low surface charge densities. They include larger ions such as potassium, K^+, rubidium, Rb^+, chloride, Cl^-, bromide, Br^-, iodide, I^-, and nitrate, NO_3^-. Many salts with structure breaking ions, e.g., those with K^+ or NO_3^-, are very soluble because the cohesive forces in solids composed of ions with low surface charge densities are often not as large as those for other ions and so there is little tendency for these solids to precipitate from solution.

In the dilute solutions associated with corrosion, the ions are remote from interaction with other charged species and exhibit Henrian activity, so that their activities are conveniently referred to the infinitely dilute solution standard. This does not apply to more concentrated solutions of ionic materials, because the electrical potentials of neighboring ions in close proximity mutually interfere.

2.2.7 Ion Mobility

Corrosion rates are influenced by the electrical conductivity of aqueous solutions and the diffusion of ionic species within them. Both are related to the limiting velocity of ions in motion, called the ion mobility.

The physical significance of the ion mobility is that an ion in an electrical or chemical potential gradient accelerates to a terminal velocity at which the viscous drag of the solvent balances the accelerating potential. Thus, for most ions, molar conductivity, diffusivity, and solvent viscosity are all related. Theoretical equations relating these parameters are derived as explained e.g., by Atkins, cited at the end of the chapter but the significant ones are:

The Einstein equation: $\qquad D_i = u_i kT / ez_i$ $\qquad$ (2.7)

The Nernst-Einstein equation: $\quad \lambda_i = (z_i^2 F^2 / RT) D_i$ $\qquad$ (2.8)

The Stokes-Einstein equation: $\quad D_i = kT / 6\pi\eta r_i$ $\qquad$ (2.9)

where: D_i is the diffusion coefficient for an ion, i.
$\qquad$ u_i is the ion mobility.
$\qquad$ λ_i is the molar conductance.
$\qquad$ z_i is the charge number for the ion.
$\qquad$ r_i is the effective (or hydrodynamic) radius.
$\qquad$ η is the viscosity of the solvent.
$\qquad$ e is the charge on an electron.
$\qquad$ R, k, and F are the gas, Boltzmann and Faraday constants,
$\qquad\qquad$ respectively.

Equations 2.7 and 2.8 give ion mobilities in terms of molar conductances, λ_i, since $R/F = k/e$:

$$u_i = \lambda_i / z_i F \qquad (2.10)$$

D_i and λ_i strictly depend on the ionic strength of the medium and the nature of the counter ion. The implied assumption that ions exhibit Stokes behavior is also a simplification. Nevertheless these equations are useful in comparing ion transport rates and limiting ion mobilities at 25°C calculated in this way for some common ions in aqueous solution amenable to measurement are given in Table 2.6.

The most striking feature of Table 2.6 is that the values for limiting conductance of ions produced by autodissociation of water, H+ and OH-, are several times greater than the values for other ions. The exceptionally high mobility of these ions is because they move in a *Grotthus-chain mechanism*, illustrated in Figure 2.6. In this mechanism, the individual ion assemblies do not move bodily through the liquid but the positions of excess protons

TABLE 2.6

Limiting Ionic Mobilities, u_i
for Some Ions in water at 25°C

Ion	Limiting ion mobility $u_i/10^{-9}m^{-2}s^{-1}V^{-1}$
H^+	362.5
OH^-	204.8
Na^+	51.9
Ca^{2+}	61.7
Mg^{2+}	55.0
Zn^{2+}	54.7
Mn^{2+}	55.5
Cu^{2+}	56.0
Cl^-	79.1
NO_3^-	74.0
SO_4^{2-}	82.7

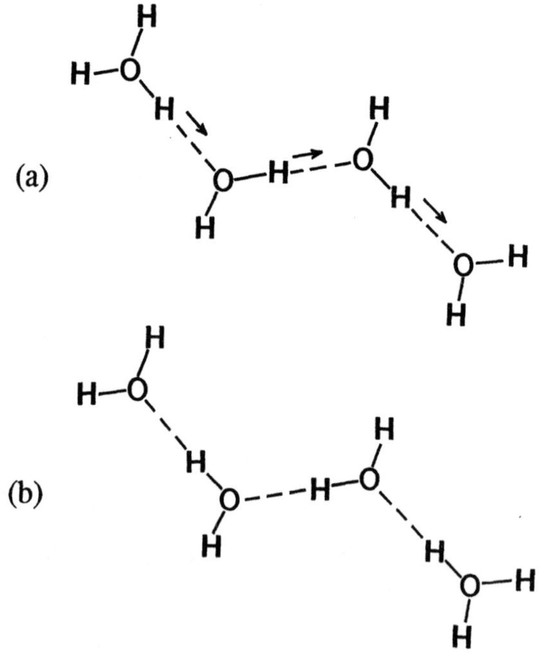

FIGURE 2.6

The Grotthus chain mechanism for proton migration in water. Solid lines indicate covalent bonds and broken lines indicate hydrogen bonds. A hydrogen ion moves from the left in (a) to the right in (b) by successive reorientation of hydrogen bonds between water molecules along the path.

(H+) or proton holes, that form the charge centers of hydrogen (H+) and hydroxyl (OH−) ions, are propagated by successive re-orientation of hydrogen bonds along the path through which the charge centers move, transporting the ions without needing to move atomic species with attached water molecules. As a consequence, mechanisms involving the transport and interaction of protons or hydroxyl ions with other species or surfaces are usually very fast. These include interactions that can contribute to the establishment of the protective passive condition on metal surfaces.

2.2.8 Structures of Water and Ionic Solutions at Metal Surfaces

The structures of water and of aqueous solutions are disturbed in contact with a metal surface. The disturbed structure extends for only a few water molecules into the bulk liquid but it has a special role in corrosion processes, because it controls the rates of interactions between the metal and solutions of ions and it can be a precursor of a surface passive condition for the metals concerned. Bokris and Reddy, cited in Further Reading, give a comprehensive treatment.

The interfacial structure is due to the polar character of water molecules, and the charge centers in ions that induce electric charges on the metal surface establishing an *electrical double layer*, comprising electrically charged parallel planes separated by distances of the order of magnitude of a few water molecules. The evidence for this state of affairs over such small distances is necessarily indirect and is based mainly on the results of experiments to determine the capacity of the surfaces to hold electric charge. It is difficult to generalize from the accumulated evidence[*] but Figure 2.7 is a simplified representation of the structure. The figure shows how access of solvated ions to the metal surface is restricted by their inner solvation spheres and by a row of water molecules in contact with the metal, the *hydration sheath*, re-orientated from the bulk water structure under the influence of the surface charge. The locus of the ion centers at their closest approach is known as the *outer Helmholtz plane*. Certain unsolvated ions that are small enough to displace water molecules in the hydrogen sheath, mainly anions such as Cl− and Br−, can be adsorbed directly on the metal surface by chemical rather than electrostatic interactions and the locus of their centers is known as the *inner Helmholtz plane*. All of these structures are of practical as well as theoretical interest because they have important consequences in surface processes. The activation energy needed to transport an atom from the metal surface to become an ion in the outer Helmholtz plane is one of the rate-controlling factors in the dissolution of a corroding metal, as explained in Section 3.2. Structural differences can also account for differences between the abilities of chloride and fluoride ions to depassivate stainless steels, as explained in Chapter 8.

---

[*] Reviewed by Bokris and Reddy cited in "Further Reading."

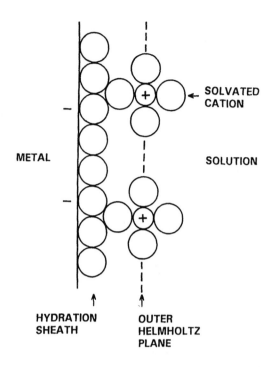

FIGURE 2.7
Schematic diagram representing the arrangement of cations in aqueous solution at a negatively charged metal surface.

2.2.9 Constitutions of Hard and Soft Natural Waters

Natural waters are common environments for corrosion. They exist in such variety that generalizations are difficult but one frequently encountered aspect of composition is the quality of being "soft" or "hard". The terms are used loosely but it is generally understood that on evaporation, a hard water leaves a calcareous deposit whereas a soft water does not. It is well known that corrosion problems are generally less severe in hard than in soft waters implying that calcareous deposits afford some protection and that rainwater can also be acidic and therefore mildly aggressive. These aspects of water quality are influenced by equilibria between the water, atmospheric carbon dioxide and calcium-bearing material with which the water comes into contact. They are considered in detail by Butler, cited in Further Reading.

2.2.9.1   Rainwater

In the absence of pollutants, rainwater is almost pure water that absorbs quantities of oxygen and carbon dioxide in equilibrium with the air as it falls in small droplets to earth. It has an oxygen potential conferred by the dissolved oxygen and a particular pH value derived from the carbon dioxide in solution. It can be calculated as in the following example.

Example 2: *pH of rainwater*

The pH of the water can be determined from the carbon dioxide content, by a *speciation calculation*. Because the calculation depends on balancing the total electric charges, it deals both with equilibrium *constants* that relate to activities denoted in the usual way, i.e., a_{X^-} for the activity of an ion, X^-, and also with equilibrium *quotients*, that relate to corresponding concentrations, denoted by square brackets, i.e., $[X^-]$ for the ion X^-. These quantities are related by:

$$a_{X^-} = (\gamma_{X^-})[X^-]$$

where γ is the activity coefficient. The symbols for the mean activity coefficients for singly and doubly charged ions are, $\gamma_\pm$, and $\gamma_{2\pm}$ respectively. The activity coefficient for uncharged species in dilute solution, e.g., CO_2 is approximately 1. An ideal gas is referred to unit activity for the gas at atmospheric pressure.

The pH of the water is determined by four mutually interacting equilibria:

1. The dissolution of CO_2 in water from the gas phase:

$$CO_2(g) = CO_2(aq) \tag{2.11}$$

Equilibrium quotient, $\qquad K_H = [CO_2]/p_{CO_2}$ (2.12)

2. The reaction of CO_2 with water, producing bicarbonate and hydrogen ions:

$$CO_2(aq) + H_2O(l) = HCO_3^- + H^+ \tag{2.13}$$

Equilibrium constant, $\qquad K_1 = (a_{H^+})(a_{HCO_3^-})/(a_{CO_2})$

$$= [H^+][HCO_3^-]\gamma_\pm^2/[CO_2] \tag{2.14}$$

3. The dissociation of the bicarbonate ion into carbonate and hydrogen ions:

$$HCO_3^- = CO_3^{2-} + H^+ \tag{2.15}$$

Equilibrium constant, $\quad K_2 = (a_{H^+})(a_{CO_3^{2-}})/(a_{HCO_3^-})$

$$= [H^+][CO_3^{2-}]\gamma_{2\pm}/[HCO_3^-] \tag{2.16}$$

4. The autodissociation of water:

$$H_2O = H^+ + OH^- \tag{2.17}$$

Equilibrium constant, $K_w = (a_{H^+})(a_{OH^-})$

$$= [H^+][OH^-]\gamma_\pm^2 \tag{2.18}$$

The concentration of the solvent remains virtually constant at unity because it is in a large excess, so it is omitted from Equations 2.14 and 2.18.

To preserve electrical neutrality, the sums of the charges on the positive and negative ions must be equal. The charge balance equation is:

$$[H^+] = [HCO_3^-] + [OH^-] + 2[CO_3^{2-}] \tag{2.19}$$

In practice, since K_2 is much smaller than K_1, the concentration of CO_3^{2-} is very small and can be neglected. The $[CO_3^{2-}]$ term can be dropped and $[HCO_3^-]$ and $[OH^-]$ substituted by the following expressions:

From 2.14 and 2.12, $[HCO_3^-] = K_1[CO_2]/[H^+]\gamma_\pm^2 = K_H K_1 p_{CO_2}/[H^+]\gamma_\pm^2$ (2.20)

From 2.18, $[OH^-] = K_w/[H^+]\gamma_\pm^2 \tag{2.21}$

Combining Equations 2.19, 2.20 and 2.21 and omitting $[CO_3^{2-}]$ gives:

$$[H^+] = (K_H K_1 p_{CO_2} + K_w)/[H^+]\gamma_\pm^2 \tag{2.22}$$

The hydrogen activity $a_{H^+} = [H^+]\gamma_\pm$. Multiplying both sides of equation 2.22 by $[H^+]\gamma_\pm^2$ and rearranging:

$$a_{H^+} = [H^+]\gamma_\pm = (K_H K_1 p_{CO_2} + K_w)^{1/2} \tag{2.23}$$

whence: $pH = -\log a_{H^+} = -0.5 \log(K_H K_1 p_{CO_2} + K_w) \tag{2.24}$

The normal atmosphere contains 0.03% CO_2 (3×10^{-4} atm). At a typical ambient temperature of 15°C the values of the equilibrium quotients and constants are:

$$K_H = 4.67 \times 10^{-2} \text{ mol dm}^{-3} \text{ atm}^{-1}.$$
$$K_1 = 3.81 \times 10^{-7} \text{ mol dm}^{-3}.$$
$$K_2 = 3.71 \times 10^{-11} \text{ mol dm}^{-3}.$$
$$K_w = 4.60 \times 10^{-15} \text{ mol}^2 \text{dm}^{-6}.$$

Inserting these values into Equation 2.24 yields pH = 5.6 for water in equilibrium with atmospheric carbon dioxide, which is consistent with the results of practical measurements. Use is made of this result in considering corrosion tendencies in Chapter 3. Equation 2.24 also shows that in the absence of CO_2, $p_{CO_2} = 0$ and $pH = -0.5 \log K_w$, in agreement with values given in Table 2.5.

Oxygen Potential of Rainwater
Oxygen dissolves as molecules in the open network of the water structure. Air contains 21% oxygen and since water falling as rain approaches equilibrium with air and oxygen is nearly ideal at ambient temperatures, the oxygen activity in solution is 0.21, refered to the pure gas as the standard state. This value is assumed in later calculations.

2.2.9.2 Hard Waters

Hard water can often be regarded as a solution in simultaneous equilibrium with both solid limescale (calcite) and atmospheric carbon dioxide, even though other ions such as SO_4^{2-} and Mg^{2+} may be present. Its pH can be calculated as in the following example.

Example 3: *pH of a representative hard water*
The principal solution reaction is:

$$CaCO_3(s) + CO_2(g) + H_2O(l) = Ca^{2+}(aq) + 2HCO_3^-(aq) \qquad (2.25)$$

The charge balance equation for all of the ionic species, including ions from autodissociation of water is:

$$2[Ca^{2+}] + [H^+] = [HCO_3^-] + 2[CO_3^{2-}] + [OH^-] \qquad (2.26)$$

The relative values of K_H, K_1, K_2, and K_W, given above and K_s, given below, justify the approximation:

$$2[Ca^{2+}] = [HCO_3^-] \qquad (2.27)$$

The solubility of $CaCO_3$ is determined by the solubility product:

$$K_s = (a_{Ca^{2+}})(a_{CO_3^{2-}})$$
$$= [Ca^{2+}][CO_3^{2-}]\gamma_{2\pm}^2 \qquad (2.28)$$

and the value of K_s is 3.90×10^{-9} mol^2dm^{-6}
From Equation 2.16

$$[CO_3^{2-}] = K_2[HCO_3^-]/[H^+]\gamma_{2\pm} \qquad (2.29)$$

Substituting $[Ca^{2+}] = \frac{1}{2}[HCO_3^-]$ from Equation 2.27 and for $[CO_3^{2-}]$ from Equation 2.29 into Equation 2.28 gives:

$$K_s = \frac{1}{2}K_2[HCO_3^-]^2\gamma_{2\pm}/[H^+] \qquad (2.30)$$

Substituting for $[HCO_3^-]$ from Equation 2.20 and rearranging yields:

$$[H^+]^3\gamma_\pm^3 = (K_H^2\,K_1^2\,K_2\gamma_{2\pm}/2K_s\gamma_\pm)\cdot(p_{CO_2})^2 \qquad (2.31)$$

Since $pH = -\log a_{H^+} = -\log[H^+]\gamma_\pm$, Equation 2.31 gives:

$$pH = -\frac{1}{3}\log(K_H^2 K_1^2 K_2/2K_s) - \frac{1}{3}\log\gamma_{2\pm}/\gamma_\pm - \frac{2}{3}\log p_{CO_2} \qquad (2.32)$$

For dilute solutions such as fresh waters, $\frac{1}{3}\log\gamma_{2\pm}/\gamma_\pm$ is small and can be ignored. Inserting the values for K_H, K_1, K_2, and K_s at 15°C given above:

$$pH = 5.94 - \frac{2}{3}\log p_{CO_2} \qquad (2.33)$$

Equation 2.33 yields pH = 8.3 for water in equilibrium with calcite and atmospheric carbon dioxide at its normal pressure in the atmosphere, i.e., $p_{CO_2} = 3.0 \times 10^{-4}$ atm. For environments such as artesian pumps, where water in equilibrium with a chalk phase is exposed to high pressures, p_{CO_2} can be higher and the pH of the system is lower.

Example 4: *Preventing limescale deposition*
Can limescale be prevented from forming from water containing 120 mg dm^{-3} of calcium (i.e., 3.3×10^{-3} mol dm^{-3} of Ca^{2+} ions) by treating it with a waste gas containing 10.5% carbon dioxide (i.e., $p_{CO_2} = 0.105$ atm) at 15°C?

SOLUTION:
Applying Equation 2.33 gives the pH at equilibrium:

$$pH = 5.94 - \frac{2}{3}\log p_{CO_2} = 5.94 - \frac{2}{3}\log(0.105) = 6.59$$

$$a_{H^+} = 10^{-(pH)} = 10^{-6.59}$$

Applying Equations 2.16 and 2.20:

$$[CO_3^{2-}] = (K_H K_1 K_2 p_{CO_2})/[H^+]^2$$

$$= (4.67 \times 10^{-2})(3.81 \times 10^{-7})(3.71 \times 10^{-11})(0.105)/(10^{-6.59})^2$$

$$= 1.05 \times 10^{-6} \text{ mol } dm^{-3}$$

Inserting this value for $[CO_3^{2-}]$ in Equation 2.28:

$$[Ca^{2+}] = K_S/[CO_3^{2-}] = (3.55 \times 10^{-9})/(1.05 \times 10^{-6}) = 3.4 \times 10^{-3} \text{ mol dm}^{-3}$$

Hence the proposition is feasible because the existing concentration of calcium ions, $[Ca^{2+}]$, 3.3×10^{-3} mol dm^{-3}, is below this value and cannot deposit limescale.

2.3 The Structures of Metal Oxides

Oxides are not only the corrosion products, to which metals are converted during oxidation in gaseous environments, but they are also the media through which the reaction usually proceeds. The oxide film initially formed on a metal physically separates it from the gaseous environment but even when it remains coherent and adherent to the metal surface, it does not necessarily prevent continuing attack because the primary reactants, the metal and/or oxygen, may be able to diffuse through the oxide, allowing reaction to proceed. The rate at which reaction continues and the growth pattern of the oxide depend on the oxide structure, especially on its characteristic *lattice defects* that provide the pathways for the diffusing reactants. Lattice defects of certain kinds are accompanied by complementary *electronic defects*, to compensate for charge imbalances that they would otherwise introduce, conferring some electronic conduction on the oxides in which they occur, recognized in the description, *semiconductors*.

The natures and populations of lattice and corresponding electronic defects vary from oxide to oxide and the differences are an essential part of explanations for differences in the ability of engineering metals and alloys to resist oxidation at elevated temperatures.

Descriptions of oxide structures relevant to the oxidation of metals are based on assessments of the bonds between ions, illustrations of lattice structures and explanations for various characteristic lattice and electronic defects they contain. Much of this is explained by the ideas of electronegativity, the partial ionic character of oxides and the influence of electron energy bands.

Solid metal oxides are assemblies of positively charged metal cations and negatively charged oxygen anions formed by transfer of electrons from the metal atoms to the oxygen atoms from which they are composed. The electron transfer is substantial but incomplete so that the bonds between anions and cations have only partial ionic character. Nevertheless electrostatic forces between the oppositely charged ions make a major contribution to the cohesion of an oxide and is one of the dominant factors determining its structural geometry.

The origin of partial ionic character can be explained as follows. In a single bond between two atoms of the same element, the atomic wave functions obviously contribute equally to the molecular wave function, but in a bond between two atoms of different elements, the most energetically favorable molecular orbital may have a greater contribution from the atomic orbital of one atom than from the other. This implies that the electrons in the molecular orbital are associated more with one atom than with the other. In a hypothetical limiting case, the contribution to the molecular orbital is wholly from the atomic wave function of one atom with virtually no contribution from the other. Because the two atoms each formally donate one electron to the molecular orbital, this would be equivalent to the donation of an electron by one atom, A, to the other, B, forming ions:

$$A + B = A^+ + B^- \tag{2.34}$$

In practice, the limiting case is never completely realized and real bonds between atoms of different elements have aspects of both ionic and covalent character. To facilitate further consideration of this dual character a property of atoms called electronegativity is now introduced.

2.3.1 Electronegativity

Electronegativity is generally accepted as a qualitative idea but a quantitative approach is elusive. It is based on comparisons of the strengths of single bonds between pairs of atoms of the same element, *like atoms*, with the strengths of bonds between pairs of atoms of different elements, *unlike atoms*. The strength of a bond is identified with the energy, E, released when it is formed and if a pair of unlike atoms, A and B, formed the same kind of bond as pairs of like atoms, the bond energy, $E_{(A-B)}$, would have some mean value, such as the geometric mean, between bond energies, $E_{(A-A)}$ and $E_{(B-B)}$, for like pairs of the constituent atoms:

$$E_{(A-B)} = [E_{(A-A)} \times E_{(B-B)}]^{1/2} \tag{2.35}$$

The actual energies of bonds between unlike atoms, determined experimentally, are found to exceed the predicted bond energies by a quantity, ΔE, which is small for bonds between some unlike atom pairs, e.g., nitrogen and chlorine, N – Cl or bromine and chlorine, Br – Cl but large for others, e.g., hydrogen and chlorine, H – Cl or hydrogen and bromine, H – Br. The inference is that the extra energy indicates some characteristic of the bond that varies from one pair of unlike atoms to another. By comparing such results with other chemical evidence, such as the ability of substances to ionize in aqueous solution, it is possible to associate the extra energy of bonds with the extent of a partial transfer of electrons from one of the atoms in the unlike pair to the other and to attribute it to a difference between them in their attraction for electrons, a concept called *electronegativity*. This property of electronegativity represents the attraction of a neutral atom in a stable molecule for electrons. There are alternative ways of putting the concept on a semi-quantitative basis. Pauling* based a dimensionless scale on the values of excess bond energy found for many pairs of unlike atoms, in which the numbers are obtained by manipulation of the excess energy values to be self-consistent and convenient to use. Mulliken,* devised an alternative scale based on the averages of measured ionization potentials and electron affinities of the elements. The two scales produce different numbers but agree on the sequence of the elements in order of electronegativity. For illustration, Table 2.7 gives values on Pauling's scale for elements of frequent interest.

TABLE 2.7
Electronegativity Values on Pauling's Scale for Representative Elements

Element	Fluorine	Oxygen	Chlorine	Carbon	Sulfur
Electronegativity	4.0	3.5	3.0	2.5	2.5
Element	Hydrogen	Copper	Tin	Iron	Nickel
Electronegativity	2.1	1.9	1.8	1.8	1.8
Element	Zinc	Titanium	Aluminum	Magnesium	Cesium
Electronegativity	1.6	1.6	1.5	1.2	0.7

Pauling's values range from 0.7 for the least electronegative (or most electropositive) natural element, caesium, to 4.0 for the most electronegative element, fluorine. When two atoms of different elements interact, the more electronegative atom acquires an anionic character and the less electronegative atom acquires a cationic character. In the present context, it is of particular interest to note that oxygen is the second most electronegative element, with a value of 3.5, that is much greater than the values for metals, which are all <2.4, with most of them <1.9.

2.3.2 Partial Ionic Character of Metal Oxides

The electronegativity concept was developed for single bonds between pairs of unlike atoms in isolation. Its benefits can be fully realized only if it can be used to assess the characters of bonds in real assemblies of atoms. In the particular case of metal oxides, it can be applied, for example, to explain why one kind of lattice defect predominates in some oxides and different kinds predominate in others, as will be seen later. However, this presents two problems that have concerned the originators of the concept. The first problem is to what extent an electronegativity scale derived for single bonds can be transferred to assemblies of ions in a solid with multiple bond contacts. Pauling's inference is that the electronega-

---

* Chapter 3 in Pauling's text cited in "Further Reading."

tivity values can be transferred without serious error. The second problem is how to devise a quantitative scale of partial ionic character from the differences in electronegativity between the constituent atom precursors. Pauling, cited at the end of the chapter suggested the empirical equation:

$$\%(\text{ionic character}) = 100\left(1 - \exp\left\{\frac{-(x_A - x_B)^2}{4}\right\}\right) \qquad (2.36)$$

where x_A and x_B are the electronegativities of atoms A and B.

There is an alternative empirical equation, more suitable for large differences in electronegativity:

$$\%(\text{ionic character}) = 16\ |x_A - x_B| + 3.5\ |x_A - x_B|^2 \qquad (2.37)$$

Equation 2.36 yields the general relation between ionic character and electronegativity difference given in Table 2.8 and the ionic characters of some particular metal-oxygen single bonds given in Table 2.9. Despite uncertainties in the absolute values of the numbers given in Tables 2.8 and 2.9, they are derived logically and taking them as relative values, they can be used to support other evidence of differences between various oxides to be described.

2.3.3   Oxide Crystal Structures

Atoms in solids are arranged to minimize the energy of the system. The factors that determine the minimum energy can be very complex and may include not only coulombic forces but also the options available for mixing the wave functions and the strength of possible covalent contributions to the bonding.

If the coulombic forces predominate, as they do for the oxides of most metals of commercial interest, the lowest energy state is realized by closely packing the ions in a space lattice relaxing the coulombic forces. The particular close-packed arrangement in any given oxide depends on the relative numbers of anions and cations, *the stoichiometric ratio* and on their relative sizes, the cation/anion *radius ratio*. Table 2.10 gives sizes of some selected ions determined from their closest distance of approach to other ions in appropriate series of compounds.

The close-packed structures with minimum coulombic energies are those in which the smaller metal ions are most efficiently accommodated between the larger oxide ions and so oxides with the same stoichiometric and radius ratios often exhibit the same structure. These include the simple cubic, rhombohedral-hexagonal, rutile and spinel structures described below. Brucite is included because several metal hydroxides that can form as corrosion products or their intermediaries adopt this structure.

TABLE 2.8
Ionic Character of Bonds as a Function of Difference in Electronegativity
Between Unlike Atoms

Electronegativity difference	0	0.4	0.8	1.2	1.6	2.0	2.4	3.0
% Ionic character	0	4	15	30	47	63	76	86

TABLE 2.9
The Ionic Character of Some Metal-Oxygen Single Bonds

Bond	Electronegativity		Difference	Ionic Character %
	Cation	Oxygen		
Mg–O	1.2	3.5	2.3	74
Cu–O	1.9	3.5	1.6	47
Zn–O	1.6	3.5	1.9	60
Fe–O	1.8	3.5	1.7	50
Ni–O	1.8	3.5	1.7	50
Sn–O	1.9	3.5	1.6	47
Cr–O	1.6	3.5	1.9	60
Al–O	1.5	3.5	2.0	63
Ti–O	1.5	3.5	2.0	63
Si–O	1.8	3.5	1.7	50

TABLE 2.10
Molecular Formulae, Cation/Anion Radius Ratios and Structures
for Metal Oxides

Metal	Oxide Formula	Ion	Ion Radius/nm	Cation/Anion Radius Ratio*	Oxide Structure
Aluminum	Al_2O_3	Al^{3+}	0.050	0.36	Corundum
Magnesium	MgO	Mg^{2+}	0.065	0.46	Simple cubic
Titanium	TiO_2	Ti^{4+}	0.068	0.48	Rutile
Chromium	Cr_2O_3	Cr^{3+}	0.069	0.49	Corundum
Iron II	FeO	Fe^{2+}	0.076	0.54	Simple cubic
Iron III	Fe_2O_3	Fe^{3+}	0.064	0.45	Corundum
Nickel	NiO	Ni^{2+}	0.072	0.51	Simple cubic
Copper I	Cu_2O	Cu^+	0.096	0.69	Cubic
Copper II	CuO	Cu^{2+}	0.072	0.51	Monoclinic
Zinc	ZnO	Zn^{2+}	0.069	0.49	Würtzite
Tin	$\alpha\text{-}SnO_2$	Sn^{4+}	0.071	0.50	Rutile
Magnesium and aluminum	$MgAl_2O_4$	Mg^{2+}	0.065	0.46	Spinel
		Al^{3+}	0.050	0.36	
Iron(II) and chromium	$FeCr_2O_4$	Fe^{2+}	0.076	0.54	Spinel
		Cr^{3+}	0.069	0.49	
Iron(II) and iron(III)	Fe_3O_4	Fe^{2+}	0.076	0.54	Inverse spinel
		Fe^{3+}	0.064	0.45	

* Radius of oxygen ion = 0.140 nm

2.3.3.1   Simple Cubic Structures (or rock salt, NaCl structure)

The simple cubic structure is favorable for ionic compounds with equal numbers of anions and cations and with cation/anion radius ratios between 0.414 and 0.732, which is the range for octahedral coordination. As Table 2.10 illustrates, the radius ratios for several metal oxides lie within this range. In the simple cubic structure, every ion is surrounded by six contacting ions of opposite charge. Oxides with simple cubic structures include magnesium oxide, wüstite FeO and nickel(II) oxide NiO.

2.3.3.2   Rhombohedral-Hexagonal Structures (Corundum Structure)

Trivalent metals form oxides with the general formula M_2O_3 and different structures are needed to accommodate the ratio of two metal ions to three oxide ions. One of these is the rhombohedral structure, which is best described as a hexagonal close-packed lattice of oxygen ions with metal ions occupying two-thirds of the octahedral spaces. Since the corundum arrangement is based on a hexagonal close-packed lattice the structure also has hexagonal geometry. Oxides with the corundum structure include corundum itself, α-Al_2O_3, normal hematite Fe_2O_3 and chromium oxide, Cr_2O_3.

2.3.3.3   Rutile Structures

The name rutile is derived from its generic member, a natural mineral of TiO_2. The structure is commonly adopted by metal oxides of the form MO_2 where the (metal ion)/(oxygen ion) radius ratio is in the range 0.414 to 0.732. A ratio within this range favors the cation in octahedral coordination with every metal ion coordinated with six oxygen ions. There are twice as many oxygen ions as metal ions in MO_2 and so every oxygen ion is coordinated with three metal ions in a triangular configuration. The crystal structure is essentially series of octahedra centered on the metal ions, interconnected by the triangular configurations centered on the oxygen ions and aligned in two sets of mutually perpendicular parallel planes. The overall geometry is tetragonal. Oxides with the rutile structure include titanium oxide, TiO_2, tin(IV) oxide SnO_2, lead(IV) oxide PbO_2 and manganese dioxide, MnO_2.

2.3.3.4   Spinel Structures

A further arrangement is the 2:3 spinel structure in which two metals participate, M(II) and M(III), yielding the divalent and trivalent cations, M^{2+} and M^{3+}, respectively. The name is derived from the generic member, Spinel, $MgAl_2O_4$. The structure can be described as a face-centered cubic lattice of oxygen ions in which divalent and trivalent metal ions, M^{2+} and M^{3+}, respectively occupy tetrahedral and octahedral interstices. Only sufficient interstices are occupied to give the correct stoichiometric correspondence and hence charge balance between the oxygen and metal ions. This is accomplished when half of the octahedral interstices are occupied by ions of the trivalent metal and one eighth of the tetrahedral sites are occupied by the divalent metal yielding the general formula, $M^{II}M^{III}_2O_4$. Members of this group that are important in thermal oxidation are *Chromite*, $Fe^{II}Cr^{III}_2O_4$ and *Spinel*, $Mg^{II}Al^{III}_2O_4$, that can form on iron-chromium alloys and aluminum-magnesium alloys, respectively. *Magnetite*, $Fe^{II}Fe^{III}_2O_4$, one of the oxides that form on pure iron is an inverse 2:3 spinel, so-called because the *divalent* ions, Fe^{2+}, are predisposed to occupy octahedral sites, displacing half of the *trivalent* ions, Fe^{3+}, to the tetrahedral sites. There are other variants of the structure, for example, 1:3 spinel-type oxides in which monovalent and trivalent metals participate, such as β-alumina, $Na_2Al_2O_4$.

2.3.3.5 Brucite Structures

The Brucite structure is adopted by several hydroxides, including magnesium hydroxide, nickel hydroxide and iron(II) hydroxide, all important species in aqueous corrosion processes. These hydroxides have the general formula $M(OH)_2$ where M is a divalent metal with a moderate ion size. In this structure, every metal atom is in octahedral coordination with six hydroxyl, OH^-, ions. These octahedra are linked in two-dimensions by sharing edges in which every hydroxyl ion is coordinated with three metal ions. The layers so formed are weakly bound together in the third dimension to give a complex layered structure with hexagonal symmetry.

2.3.3.6 Other Structures

The oxides of some engineering metals are exceptional because of peculiarities of their ions. Despite its favorable radius for octahedral, i.e., six-fold, coordination with oxygen ions, the zinc ion is predisposed to tetrahedral, i.e., four-fold coordination and so zinc oxide adopts the hexagonal structure called würtzite.

This brief survey of oxide structures is far from comprehensive but it is sufficient to proceed with a discussion of defect structure.

2.3.4 Conduction and Valence Electron Energy Bands

In approaching the subject of oxide lattice defects, some attention must be paid to the distribution of electrons within oxides because perturbations in this distribution constitute the electronic defects inevitably accompanying certain lattice defects, that were referred to earlier. The configurations of electrons in assemblies of ions in solid compounds is a matter of great complexity because they are accommodated in orbitals formed by all of the atoms that constitute the material so that the solid contains an immense number of energy states for possible occupation by the available electrons.

The relevant result for the present discussion is that in contrast with the sharp energy levels in free atoms, the energy states are grouped into energy ranges or bands and in oxides with distinct ionic character, the bands are separated by an energy gap, the *band gap*.* The band below the band gap, the *valence band*, is associated with wave functions contributed by the oxygen atomic orbitals and the band above the gap, the *conduction band*, is associated with wave functions contributed by the metal atomic orbitals. Comparison of the number of energy states in the valence band with the number of available electrons shows that if they were all in the valence band, it would be full. In this event, the oxide would be an electrical insulator because under a small applied potential, there would be no unoccupied energy states to allow net transfer of electrons in the valence band and no electrons in the conduction band to act as charge carriers.

This condition is closely approached in oxides with strong ionic character, such as MgO, where the band gap is so wide that the energy required to promote electrons from the valence band to the conduction band is prohibitively high. In oxides with less ionic character, such as ZnO, the band gap is narrower and the conduction band contains a few electrons, conferring a small but significant degree of electronic conductivity, i.e., the oxide is a semiconductor. The ability to accept electrons into the conduction band stabilizes the particular kinds of lattice defect that prevail in such oxides, as described in the next section.

2.3.5 The Origins of Lattice Defects in Metal Oxides

The term *lattice defects* in metal oxides or other solids does not mean accidental or sporadic faults that ought not to be present but has a very specific reference to imperfections in the arrangement of ions on a lattice that are entirely natural features of the material in its normal condition. In metal oxides, the defects can take the form of ions missing from lattice positions, called *lattice vacancies*, additional ions inserted between the ions on their normal lattice sites, called *interstitials* or various combinations of them. The existence of the defects is due to the tendency of the material to attain its lowest energy by allowing a degree of disorder in the structure. The prevailing number of defects of any particular kind can be quantified by assessing the energy released in a notional process by which the defects are progressively introduced into a hypothetical perfect lattice.

The energy that can be released in a process is the sum of two terms; one describes the heat taken in or given out during the process and the other describes internal changes in the state of order of the system. The release of energy is expressed in one or another of a set of equations, selected to suit the conditions under which the process occurs. When there are no

---

* e.g., Hume-Rothery or Glasstone cited in "Further Reading."

pressure changes, such as in processes occurring in the ambient atmosphere, the particular equation is:

$$\Delta G = \Delta H - T \Delta S \tag{2.38}$$

where: ΔG is the energy released (strictly the change in *Gibbs function, G,* which is the decrease in the energy of a system when the only work done by the system is expansion against a constant external pressure).

ΔH is the change in *enthalpy, H,* which is numerically equal to the heat taken into the system when the pressure on the system is constant.

ΔS is the change in *entropy, S,* of the system.

T is temperature on the Kelvin (absolute) scale.

The quantities G, H, and S are *extensive state properties*, that is, properties of a prescribed mass of a given material or system of materials with definite values, irrespective of the previous history of the material.

Entropy, S, is a cornerstone of science and is rigorously defined. The full implications are explained in standard classic texts.* For now, the relevant aspect of entropy is a quantitative correlation between the entropy of a system and the statistical distribution of energy and mass within it at the atomic level that includes:

1. The distribution of the total thermal energy within the system among its atoms, molecules or ions and among energy modes available within them, i.e., translation, rotation and vibration, described as the *degrees of freedom*.

2. The distribution of the atoms, molecules or ions in the three dimensions of the space allocated to the system. Entropy arising from the spacial distribution is described as *configurational entropy*.

2.3.5.1 General Approach

A dimensionless quantity, the *thermodynamic probability, W,* introduced below, expresses the number of ways in which a system in a given state can be implemented. It is related to the entropy of the system by the Boltzmann-Planck equation:

$$S = k \ln W \tag{2.39}$$

where k is a constant, *Boltzmann's constant*, equal to the universal gas constant, R, divided by Avogadro's number N_A.

---

* e.g., Lewis, Randall, Pitzer, and Brewer or Glasstone cited in "Further Reading."

Equation 2.39 allows S to be calculated statistically from first principles. In the particular case of interest here, it provides a means for calculating the defect populations corresponding to the minimum energy and hence to the most stable state of an oxide.

A statistical method is employed in which the system is represented by a multi-dimensional *phase space* with coordinates representing the three space dimensions and sufficient additional dimensions to represent the momenta of the fundamental particles in the available degrees of freedom. The phase space is divided into convenient identical small *phase cells* and the number of possible ways of distributing the atoms within the cells is counted. Every possible distribution is a *microscopic state* of the system but since atoms of any one kind are identical, many microscopic states correspond to the same state the system as a whole and the number of these microscopic states is the thermodynamic probability of the state. When the statistics are applied to the immense number of atoms, ions or molecules in a real system, the spectrum of probabilities peaks so sharply at one particular state of the system that this state is a virtual certainty. An often quoted illustrative example is a near-ideal gas, such as hydrogen at ordinary temperatures and pressures, for which the statistics predict correctly that the gas will fill the volume available to it at uniform concentration and temperature.

2.3.5.2   *Configurational Entropy of Atoms or Ions on a Lattice*

The object is to calculate the increase in entropy, ΔS, produced by creating defects of a particular kind in a hypothetical perfect lattice of N lattice sites, ignoring entropy due to the distribution of the thermal energy within the system, that does not enter into the calculation. In this case, the phase cell chosen includes one lattice site. There is only one microscopic state corresponding to a perfect lattice and so the probability, W_1 is one and by Equation 2.39:

$$S_1 = k \ln 1 = 0 \tag{2.40}$$

If now n defects are created and distributed among the N lattice sites, application of standard statistics gives the new probability W_2:

$$W_2 = \frac{N!}{(N - n)! n!} \tag{2.41}$$

Taking one mole of material, N equals Avogadro's number (6.022×10^{23}). Using Stirling's approximation ($\ln N! = N \ln N - N$) and assuming n/N is small, substituting for W_2 in Equation 2.39, yields:

$$S_2 = -R(n/N) \ln (n/N)$$

$$= - RX \ln X \tag{2.42}$$

where X is the mole fraction of defects, i.e., the ratio of the number of defect sites to the total number of lattice sites and R is the gas constant, 8.314 J K mol^{-1}.

The entropy change due to the creation of the defects is therefore:

$$\Delta S = S_2 - S_1 = - RX \ln X \qquad (2.43)$$

2.3.5.3 Equilibrium Number of Defects

The formation of lattice defects is favored by the release of energy in introducing disorder but constrained by the energy that must be supplied to create the defects. These two energy quantities are represented by the ΔH and $T\Delta S$ terms respectively in Equation 2.38. Provided that n/N is small so that the creation of new defects is not influenced by those already present, the energy needed to create defects is a linear function of their mole fraction:

$$\Delta H = \chi NX \qquad (2.44)$$

where χ is the energy required to create one defect.

Inserting expressions for ΔS and ΔH from Equations 2.43 and 2.44 in Equation 2.38 gives the decrease in energy of the material by the creation of a mole fraction X of defects:

$$\Delta G = \chi NX + RTX \ln X \qquad (2.45)$$

The value of X at the minimum for G, where $d\Delta G/dX = 0$ is:

$$X = \exp(- \chi N/RT - 1) \qquad (2.46)$$

Equation 2.46 shows that at a given temperature there is a definite value of X for the minimum energy of the system. This is the most stable state of the material.

Two features of Equation 2.46 are important:

1. The number of defects increases with rising temperature.
2. At a given temperature, the number of defects depends on the energy needed to create them and recalling earlier remarks, this includes energy both to break bonds with adjacent ions and to create complementary electronic defects, as explained shortly.

An alternative schematic illustration of the minimum value for G given in Figure 2.8, illustrates graphically how the number of defects depends on temperature and on the enthalpy for creating defects. Raising T displaces the $T \Delta S$ curve to the right, moving the minimum for G to a larger value of N. Increasing ΔH steepens the plot of ΔH against N, shifting the minimum to a smaller value of N.

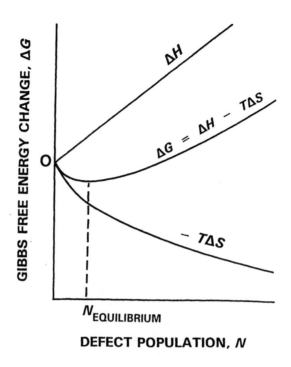

FIGURE 2.8
Gibbs free energy change, ΔG, produced by introducing a defect population, N, into a hypothetical perfect metal oxide. The minimum in the plot of ΔG against N that corresponds to the equilibrium value for N, is derived from the sum of the enthalpy and entropy terms ΔH and $-T\Delta S$.

2.3.5.4 Natures of Defects in Oxide Lattices

Four kinds of lattice defect are significant in metal oxides:

1. Schottky defects (vacant cation and anion sites in equal numbers).
2. Vacant metal cation sites.
3. Metal cation interstitials.
4. Vacant oxygen anion sites.

Another type of defect, paired anion interstitials and vacancies (Frenkel defects) that prevail in certain other solids, are generally excluded from metal oxides by the large size and high negative charge density of the oxygen ion.

Except when paired as Schottky or Frenkel defects, the defects disturb the balance of charge on the lattice and electrical neutrality is preserved by compensating *electronic defects*. Cation interstitials or anion vacancies introduce a net positive charge, which is compensated by excess electrons, described as *interstitial electrons*; the term is perhaps ill-chosen because these electrons are not associated with the interstitial spaces in the lattice. Conversely, cation vacancies introduce a net negative charge, which is compensated by a deficit of electrons, described individually as *electron holes*.

2.3.5.5 Representation of Defects

The usual symbols for defects in an oxide with a metal cation, M^{z+} are:

Cation vacancies: $M^{z+}\square$

Cation interstitials: $M^{z+}\bullet$

Anion vacancies: $O^{2-}\square$

Interstitial electrons: $e\bullet$

Electron holes: $e\square$

Their use can be illustrated by equations describing the formation the defects:

1. Creation of a cation interstitial by inserting a metal atom into the lattice.

$$M \quad = \quad M^{z+}\bullet \quad + \quad z\,e\bullet \qquad (2.47)$$
metal atom $\quad$ cation interstitial $\quad$ interstitial electrons

2. Creation of a cation vacancy by removing a metal atom from the lattice.

$$M^{z+} \quad = \quad M \quad + \quad M^{z+}\square \quad + \quad ze\square \qquad (2.48)$$
cation $\quad$ metal atom $\quad$ cation vacancy $\quad$ electron holes

3. Creation of an anion vacancy by removing an oxygen atom from the lattice.

$$O^{2-} = \quad \tfrac{1}{2}O_2 \quad + \quad O^{2-}\square \quad + \quad 2e\bullet \qquad (2.49)$$
anion $\quad$ ½ oxygen molecule $\quad$ anion vacancy $\quad$ interstitial electrons

2.3.6 Classification of Oxides by Defect Type

An equation in the form of Equation 2.45 applies to every kind of defect in all oxides but usually one kind predominates in any particular oxide because the energy needed to create it is less than that for the others and its influence determines much of the character of the oxide. For this reason, it is convenient to classify oxides by their predominant defects, as in Table 2.11.

TABLE 2.11
Classification of Some Metal Oxides by Defect Structure

Defect	Conduction	Oxides
Schottky	Ionic	MgO, Al_2O_3 (T<825 °C)
Cation vacancies	p-type	FeO, NiO, MnO, CoO, Cu_2O, $FeCr_2O_4$, UO_2
Cation excess	n-type	ZnO, CdO, BeO, Al_2O_3 (T>825°C), $MgAl_2O_4$, UO_3, U_3O_8
Anion vacancies	n-type	TiO_2, ZrO_2, Fe_2O_3*
Mixed	Mixed	Fe_3O_4

* With some cation interstitials.

2.3.6.1 *Stoichiometric Oxides*

Oxides of strongly electronegative metals, such as magnesium oxide, are characterized by a large band gap, so that the energy needed to insert electrons into the conduction band to compensate for cation interstitials or anion vacancies as required in Equations 2.47 and 2.49, is very high. The energy for creating electron holes in the valence band to compensate for cation vacancies as required in Equation 2.48, is also very high, for reasons given earlier. As a consequence, the only significant defects are Schottky defects, which are paired cation and anion vacancies with no requirement for compensating electronic defects. Therefore the compositions of these oxides compositions correspond closely to their molecular formula, recognized by the term *stoichiometric*; for example, the ratio of metal ions to oxygen ions is 1:1 and 2:3, respectively for MgO and Al_2O_3. Figure 2.9 is a schematic illustration of a Schottky defect pair in magnesium oxide.

$$O^{2-} \quad Mg^{2+} \quad O^{2-} \quad Mg^{2+} \quad O^{2-}$$

$$Mg^{2+} \quad \square \quad Mg^{2+} \quad O^{2-} \quad Mg^{2+}$$

$$O^{2-} \quad Mg^{2+} \quad O^{2-} \quad \square \quad O^{2-}$$

$$Mg^{2+} \quad O^{2-} \quad Mg^{2+} \quad O^{2-} \quad Mg^{2+}$$

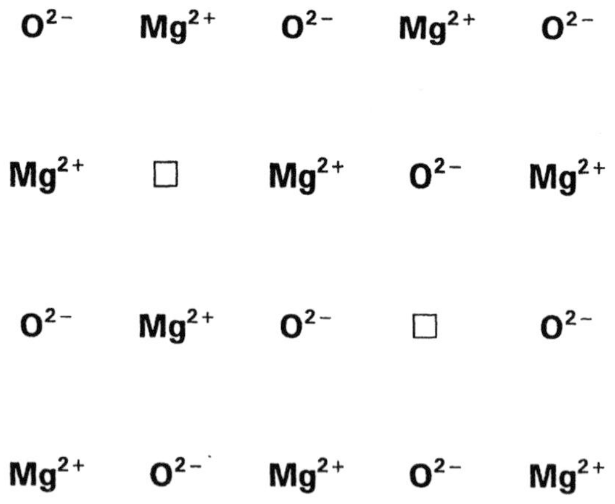

FIGURE 2.9
Example of an ionic oxide. Schematic diagram of a Schottky defect pair in magnesium oxide.

2.3.6.2 *Cation Excess Oxides*

Oxides of less electronegative metals, such as zinc oxide, are characterized by a smaller band gap, so that less energy is needed to insert electrons into the conduction band to compensate for cation interstitial defects. It is again difficult to remove electrons from the valence band so that compensation for cation vacancies is difficult. As a consequence, cation interstitials are the predominant defects. Figure 2.10 is a schematic illustration of a cation interstitial in zinc oxide.

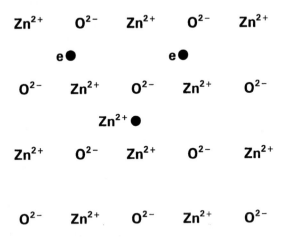

FIGURE 2.10
Example of a cation excess oxide. Schematic diagram of an interstitial cation and electrons in zinc oxide.

The small population of electrons in the otherwise empty conduction band confers some electron mobility and hence cation excess oxides are semi-conductors of electricity. The term *n-type semiconductor* is applied to recognize that the conductivity is due to *negative* charge carriers.

2.3.6.3 Cation Deficit Oxides

Recalling the brief discussion in Section 2.1.2.1, the partly-filled $3d$ orbitals underlying the nominal $4s$ valency electrons confer special characteristics on metals in the first transition series. For some of these metals, e.g., manganese, iron, cobalt, nickel, and copper, the energies of the d electrons are close enough to the energies of the valency electrons to provide a source from which electrons can be withdrawn to compensate for charge imbalance introduced by cation vacancies. For example, it is relatively easy to convert the normal valency of nickel from 2 as in Ni^{2+} to 3 as in Ni^{3+}. The increase in valency corresponds to the creation of electron holes, $e\square$, in the d orbitals. Hence the predominant defects in the lower oxides of these metals are cation vacancies. The effect is usually represented schematically by assigning the increased valency to arbitrary cations as illustrated for nickel oxide in Figure 2.11, but it is important to note that the electron deficit applies to the oxide as a whole and does not reside on particular cations.

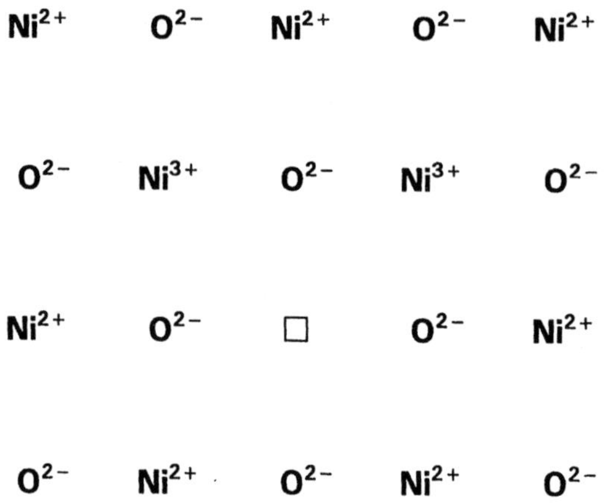

Ni^{2+} O^{2-} Ni^{2+} O^{2-} Ni^{2+}

O^{2-} Ni^{3+} O^{2-} Ni^{3+} O^{2-}

Ni^{2+} O^{2-} □ O^{2-} Ni^{2+}

O^{2-} Ni^{2+} O^{2-} Ni^{2+} O^{2-}

FIGURE 2.11
Example of a cation vacant oxide. Schematic diagram of a cation vacancy and electron holes (Ni^{3+}) in nickel oxide.

The electron holes confer mobility on the electrons, which in the schematic diagram in Figure 2.11 would be represented by electron transfer between the majority Ni^{2+} ions and the minority Ni^{3+} ions, exchanging their oxidation states but in reality all of the nickel ions are equivalent with all of them being mainly of Ni^{2+} character with some Ni^{3+} character. Cation deficient oxides are therefore semiconductors. The term *p-type semiconductor* is applied to recognize that the conductivity is due to *positive* charge carriers.

2.3.6.4   *Anion Deficit Oxides*

As an alternative to cation interstitials, some metal oxides with the potential to compensate for excess positive charge exhibit anion vacancies. They are mainly higher oxides of metals in the transition series, including TiO_2, ZrO_2 and Fe_2O_3. The excess positive charge is compensated by absorbing extra electrons into the cation *d*-orbitals. In schematic representations, the effect can be indicated either by assigning *decreased* valency to arbitrary cations or by introducing symbolic interstitial electrons, as illustrated for titanium dioxide in Figure 2.12. These oxides are, of course, n-type semiconductors.

The formation of anion vacancies in an oxide is thus related to the existence of two relatively stable oxidation states for the metal concerned, e.g., Ti^{4+} and Ti^{3+}, Fe^{3+}, and Fe^{2+} and this is also reflected in the ability of these metals to exercise more than one valency in their aqueous chemical behavior, as considered later with reference to aqueous corrosion.

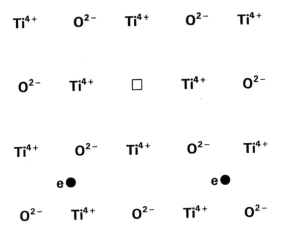

FIGURE 2.12
Example of an anion vacant oxide. Schematic diagram of an anion vacancy and interstitial electrons in titanium oxide.

2.3.6.5 Degrees of Non-Stoichiometry

The creation of Schottky defects needs energy to remove both cations and anions from the lattice, so that the enthalpy of formation is high and the defect population in stoichiometric oxides is small but significant.

Cation deficient oxides can exhibit a high degree of non-stoichiometry, that is often measurable by chemical analysis. This is because, where the appropriate charge compensation mechanism exists, as for the lower oxides of transition metals, cation vacancies have the lowest enthalpies of formation and are thus the easiest to form. One of the most non-stoichiometric oxides known is nominally FeO, that typically exists over the composition range $Fe_{0.95}O$ to $Fe_{0.99}O$, which does not even include stoichiometric composition. Incidentally, the degree of non stoichiometry of sulfides can be even greater, e. g. nominal FeS can exist at a chemical composition corresponding to $Fe_{0.92}S$, indicating that about 8% of the cation sites are vacant.

The compositions of cation excess oxides are much closer to the true stoichiometric ratio and although the non-stoichiometry can be detected indirectly, e.g., by diffusion and electrical conductivity measurements, it is not easy to detect directly by chemical analysis.

2.4 The Structures of Metals

Considerable detail of the electron configurations in bonding was needed to describe the structures of liquid water and of oxides in sufficient detail to explain corrosion phenomena. There is less need to describe the electronic counterpart for metals but for completeness it is referred to very briefly in Section 2.4.1. Section 2.4.2 describes crytallographic features. Section 2.4.3 summarizes the main features of phase equilibria in alloys to provide a basis for descriptions of particular alloy systems in later chapters. Section 2.4.4 deals with an aspect of structure in its widest sense, which is often overlooked, concerned with characteristics imparted to metals by manufacturing procedures.

2.4.1 The Metallic Bond

Bonding in metals is clearly quite different from that in covalent molecules or in ionic solids. The electronic configuration responsible for the cohesion of assemblies of metal atoms must be consistent with the following features of the metallic state:

1. Metals have a unique combination of properties, including *inter alia* strength, ductility, and high electrical and thermal conductivities.

2. In general, metallic elements have considerable mutual miscibility even when their chemical valencies differ, as for aluminum and magnesium. They also form intermetallic compounds which appear to disregard valency rules.

The metallic elements are *hypoelectronic* elements, meaning that their outer electron shells are less than half filled so that there is an insufficient pool of electrons to satisfy either atom-to-atom covalent bonding or the formation of ions. The problem of accounting for the cohesion of metals was solved by Pauling, who conceived that a special orbital, the *metallic orbital* is set aside by means of which bonding electrons can move from atom to atom, thereby permitting the bonds between atoms to switch in turn between all of their neighbors in an unsynchronized way. Without going into detail, this represents *resonance energy* that stabilizes the structure. Metals such as lithium, magnesium, and aluminum, which can provide only 1, 2, and 3 valency electrons respectively for this kind of bonding are not intrinsically strong and have relatively low melting points. Metals from the transition series can augment their valency electrons by up to 4 extra electrons promoted from the underlying *d* shell and for this reason, metals such as iron, nickel, and copper are much stronger and have higher melting points.

The resonant bonding accounts for the following metallic characteristics:

1. Metals crystalize in close-packed structures, usually FCC, BCC, or hexagonal close pack, which are characteristic of non-directional bonding for assemblies of atoms of the same radius.

2. Metals are ductile because the assembly of atoms coheres as a whole and not on an atom-by-atom basis. Thus atoms can move relatively to one another, without losing cohesion and planes of

atoms can slide past each other by successive propagation of *dislocations* in the lattice.

3. Metals have high electrical and thermal conductivities because resonant bonding *delocalizes* the bonding electrons, freeing them to move throughout the metal, carrying electric charge and thermal energy.

2.4.2 Crystal Structures and Lattice Defects

The crystal structures of common engineering metals are listed in Table 2.1. Iron is especially interesting and important because the structure is BCC at temperatures <910°C, FCC in the temperature range 910 to 1400°C, reverting to BCC at temperatures >1400°C. These transformations are manipulated in alloys to produce a wide range of steels with different characteristics; this is illustrated in later discussion of the formulation of stainless steels to produce the FCC structure for applications where easy formability is required and the BCC structure where resistance to stress corrosion cracking is needed.

Metals contain vacant lattice sites and interstitials but, since the atoms are neutral and chemically similar, there is no counterpart to the electronic defects and non-stoichiometry in oxides. The defect population density rises with temperature and typically approaches 10^{-3}, i.e., one vacancy per thousand atoms, near the melting point. The principal role of the vacancies is to facilitate diffusion of solutes in solid solution by repetitive interchange between vacancies and adjacent atoms.

Manufactured metal products are agglomerates of small crystals of various orientations, which can be revealed by etching polished sections of the material. Generally speaking, randomly orientated microscopic crystals are desirable to give uniform isotropic mechanical properties. The boundaries between the crystals, *grain boundaries*, mark changes in orientation between adjacent crystals. The regions within a few atomic diameters of the boundaries have different characteristics from the bulk of the crystal because of lattice irregularities accommodating the change in orientation; among the effects is a predisposition for impurities to segregate and for minority phases to precipitate there. These are sometimes influential in directing corrosion or stress-corrosion cracking along grain boundary paths.

2.4.3 Phase Equilibria

Most but not all combinations of metals are miscible in the liquid state. On solidification, a liquid mixture transforms to one or more solid phases and there may be further transformations as the metal cools. The transformations may be complete (equilibrium) or partial (non-equilibrium) and the manipulation of alloy compositions and of the transformations forms the traditional basis for alloying.

Five basic kinds of transformation are particularly important, i.e., the separation of solid solutions, eutectic, eutectoid, and peritectic transformations and the formation of intermetallic compounds. The phase relationships for a few alloy systems are completely represented by one or another of these transformations but more usually they are developed from combinations of them.

Phase relationships are depicted in equilibrium *phase diagrams*, in which the relative stabilities of phases in an alloy system are represented as functions of two parameters, alloy composition, and temperature. Lines representing equilibria between phases are plotted from thermodynamic information, confirmed by experiment. The effect is to divide a diagram into *phase fields*, which enclose the coordinates of composition and temperature over which particular phases are stable. If sufficient time is available, a phase change occurs whenever the temperature or composition of an alloy is changed so that the coordinates representing it on the diagram cross a boundary between adjacent phase fields. The basic phase equilibria are briefly described in Sections 2.4.3.1 to 2.4.3.5 and illustrated in Figures 2.13 to 2.17

2.4.3.1   Solid Solutions

Most pairs of metals are partially miscible in the solid state, forming *binary solid solutions*. Some pairs of closely similar metals form complete series of solid solutions and the phase diagram for such a pair of metals, A and B, forming a *binary system*, is given in Figure 2.13. The upper and lower boundaries of the loop are respectively the *liquidus* and *solidus*. Above the liquidus, alloys of the two metals are liquid solutions, below the solidus they are solid solutions and in the liquidus/solidus field they are mixtures of solid with liquid.

The diagram yields information on the mechanism by which liquid alloys solidify, which is used in Section 2.4.4.1 to explain some structural features of commercially-produced castings and ingots. The principle can be appreciated by tracing the progress of solidification when a liquid alloy at a temperature above the liquidus line, represented by the point P in Figure 2.13, is cooled through the liquidus/solidus domain. On crossing the liquidus line at Q, a solid solution of composition R is deposited. On further cooling, if freezing were infinitely slow, the instantaneous composition of the growing fraction of solid metal would follow the compositions traced out by the arc RS on the solidus line, leaving a diminishing fraction of liquid with the corresponding compositions traced out by the arc QT on the liquidus line. If the compositions of the corresponding, i.e., *conjugate*, solid and liquid solutions are assumed to be continuously adjusted to their equilibrium values by inter-diffusion of the components, the final solid would be homogeneous with the same composition, Q, as the original

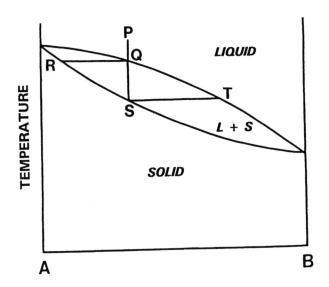

FIGURE 2.13
Solid solutions of two metals A and B completely miscible in the solid and liquid states.

liquid. Real casting processes deviate from this idealized mechanism because freezing rates are finite and transfer of the components between the solid and liquid fractions of the metal is incomplete. The phenomenon is called *selective freezing* and one effect is to introduce differential electrochemical characteristics in the material, that can adversely influence surface-sensitive treatments, such as the anodizing of aluminum.

2.4.3.2   Eutectic Transformations

In a eutectic transformation, a liquid mixture of metals transforms on solidification to form not one but two different phases. In the simple binary system for two metals, A and B, illustrated in Figure 2.14, one is a solution of B in A and the other is a solution of A in B designated by Greek symbols, e.g., α and β. Liquid alloys with less of metal B than the composition S solidify as the single-phase α solid solution. Alloys with compositions between S and E, such as that represented by the point, P, deposit the α solid solution as they cool through the liquidus/solidus field, labelled α+ L. The composition of the deposited solid follows the line RS and that of the remaining liquid follows the line QE. When the composition of the liquid reaches the point E, the remaining liquid solidifies at constant temperature as an intimate mixture of the two phases, α and β, often in a lamellar morphology. The composition and temperature represented at E are the *eutectic composition* and *eutectic temperature*. The α solid solution deposited before the eutectic composition is reached is described as the *primary α phase*. The solidification of alloys with compositions between E and T follows a similar path except that the solid deposited in the

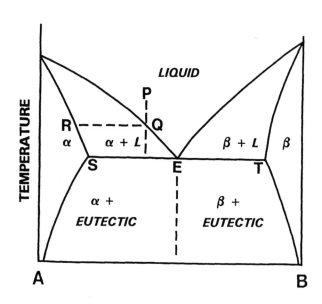

FIGURE 2.14
Eutectic system formed by two metals A and B partially miscible in the solid state.

corresponding liquidus/solidus field, labelled β + L, is the primary β phase. Liquid alloys with more metal B than the composition, T, solidify as a single-phase β solid solution. The resultant alloy structures depend on the relative quantities of primary crystals and eutectic. Alloys with compositions close to the eutectic composition exhibit isolated primary phase dendrites distributed in a eutectic matrix. Alloys with compositions remote from the eutectic composition consist mainly of the primary phase, with isolated pockets of eutectic.

2.4.3.3   Peritectic Transformations

In peritectic transformations the first solid separated from a solidifying liquid alloy reacts with residual liquid at a constant lower temperature to produce a final solid of different structure and composition, illustrated in Figure 2.15. The horizontal line, RQ, marks the peritectic temperature and the point, P, is the peritectic composition. When alloys with compositions between R and Q solidify, they initially deposit δ-solid solution but when the metal has cooled to the peritectic temperature, the solid δ phase reacts with remaining liquid to form the γ–solid solution. When the reaction is complete, alloys with compositions between R and P leave a residue of δ that subsequently transforms to γ as the alloy cools further. Alloys with compositions between P and Q leave a residue of liquid that solidifies to γ. Alloys with metal B in excess of the composition Q do not experience the peritectic reaction and solidify directly to γ. Peritectic transformations

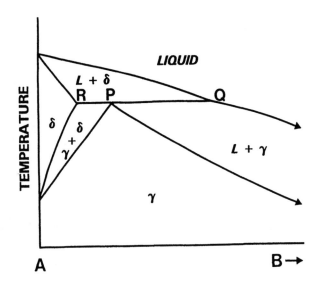

FIGURE 2.15
Part of a phase diagram for metals A and B showing the intervention of a peritectic reaction during solidification.

feature in some iron alloy systems, notably iron-carbon and iron-nickel systems.

2.4.3.4. Eutectoid Transformations

Eutectoid transformations have similar characteristics to eutectic transformations but they take place entirely in the solid state. An important example is the transformation of austenite (a solid solution of carbon in γ–iron) to a eutectoid mixture of ferrite (a solid solution of carbon in α–iron) and cementite (an iron carbide, Fe_3C). This is illustrated in the partial phase diagram given in Figure 2.16. The versatility of carbon steels is due mainly to exploitation of this transformation in various ways, as touched on in Chapter 7.

2.4.3.5 Intermetallic Compound Formation

Figure 2.17 illustrates a diagram for two metals which form an intermetallic compound, AB. The binding energy of the compound is reflected in its high melting point. Often, the intermetallic compound tolerates a degree of non-stoichiometry and characteristically forms a eutectic system with one of the component metals, as illustrated in the figure.

2.4.3.6 Real systems

Some real binary alloy systems exhibit simple phase relationships. Examples are copper–nickel and silver–gold, (solid solutions), lead–tin, and

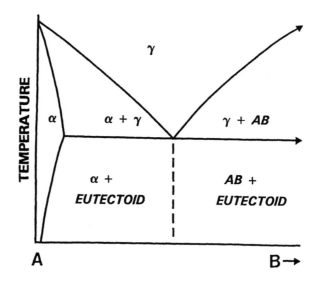

FIGURE 2.16

Part of a phase diagram for metals A and B showing decomposition of the solid phase, γ, into α and a eutectoid of α and an intermetallic compound AB.

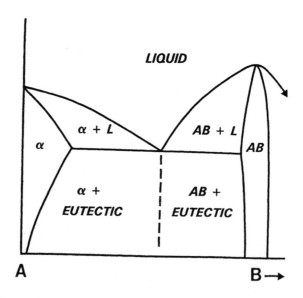

FIGURE 2.17

Part of a phase diagram for metals A and B showing formation of an intermetallic compound, AB.

aluminum–silicon, (single eutectic systems). More generally, alloy systems are complex combinations of solid solutions, eutectics, eutectoids, peritectics, and compounds. Convenient sources of diagrams for binary systems are Hansen and Anderko* and Smithells Metals Reference Book.* Some examples are described in Chapters 7 to 9, dealing with corrosion behavior in specific systems.

2.4.4 Structural Artifacts Introduced During Manufacture

Commercially-produced metals inevitably bear imprints of the manufacturing processes by which they are produced. These are due both to inherent characteristics of the metal and to contamination from external sources. The following brief description summarizes various kinds of artifacts which can occur in commercial metal products, all of which, if uncontrolled, can compromise the expected corrosion resistance. The degree to which such artifacts prevail is usually described by the generic term, *metal quality*. It is as much a feature of real metal structures as the more formal characteristics summarized above and insistence on good metal quality is an important aspect of corrosion control.

2.4.4.1 *Structural Features of Castings and Ingots*

Effects of Selective Freezing

The metal crystals grow from the liquid in a characteristic tree-like morphology with orthogonal axes and are called *dendrites*. In real casting processes, there is insufficient time for equilibrium to be maintained and the solidified metal crystals have composition gradients from their centers to their peripheries, called *coring*. Movement of the liquid during freezing usually also produces differences in composition on a macroscopic scale, known as *segregation*. The concentration effects of selective freezing may alter the proportions of phases present in an alloy and even produce phases in the spaces between the dendrites which would not exist in the alloy at equilibrium.

Contraction During Solidification

In poorly designed castings or ingots, the liquid is unable to flow to compensate for contraction of the metal accompanying solidification, leading to cavities or internal tears. Casting can also impose severe thermal gradients in the metal, causing differential contraction of the solidified metal, which can induce stresses sufficient to crack alloys with insufficient ductility to relax them.

---

* Cited in "Further Reading."

Gases Occluded in the Structure

If, by design or default, liquid metals, or alloys contain dissolved gases or their progenitors, gases evolved in the solidifying structure are trapped in irregular spaces between dendrites, generating *interdendritic porosity;* examples include hydrogen in aluminum alloys, carbon monoxide in some steels and steam in copper.

Oxide and Other Extraneous Inclusions

The surfaces of liquid metals prepared for casting are almost always covered with either a thick oxide dross or a flux or slag cover to prevent oxidation. Unless great care is exercised to prevent the liquid from enveloping its own surface during transfer of the liquid to ingot molds or castings, pieces of oxide, slag or detritus from the surfaces over which the liquid flows become included in the metal structure. If they outcrop through the metal surface, these inclusions can be very potent sites of local corrosion.

Artifacts of the kind described diminish the quality of near net shape castings and if present in ingots for subsequent mechanical working they can persist in modified forms to fabricated products, leading to surface blemishes and internal cracks in sheet, plate, or extrusions.

2.4.4.2 Characteristics Imparted by Mechanical Working

Texture

In products fabricated by extensive unidirectional deformation of cast ingots at high temperatures, such as hot-rolled plates or extruded sections, the metal recrystallizes continuously as hot-working proceeds, replacing the cast dendritic crystals with polygonal crystals which become elongated in the direction of working. At the same time, compression of the structure across the working direction greatly diminishes the distances over which chemical concentration gradients prevail, allowing some dispersion of heterogeneities introduced during solidification. As a consequence, worked materials have a more uniform structures than the ingots from which they are produced.

Thin metal sheet is produced by cold-rolling thicker hot-rolled material. The metal does not recrystallize during cold-reduction and becomes harder, i.e., it *work-hardens.* At the same time, the crystals align with their lattice planes of easiest slip in the working direction, so that the product is anisotropic. It is said to exhibit *preferred orientation* and the degree and direction of the preferred orientation is referred to as the *texture of the material.* Cold-rolled material can be softened for further fabrication by *annealing,* i.e., by reheating it to a moderate elevated temperature. Annealing does not remove texture but may change its orientation with respect to the original working direction.

Further Reading

Taylor, A., *An Introduction to X-Ray Metallography*, Chapman and Hall, London, 1952.

Coulson, C. A., *Valence*, Oxford University Press, London, 1952.

Pauling, L., *The Nature of the Chemical Bond*, Cornell University Press, Ithaca, NY, 1967.

Hume-Rothery, W., *Atomic Theory for Students of Metallurgy*, Institute of Materials, London, 1955.

Lewis, G. N., Randall, M., Pitzer, K. S. and Brewer, L, *Thermodynamics*, McGraw Hill, New York, 1961, chap. 20.

Bodsworth, C., *The Extraction and Refining of Metals*, CRC Press, Boca Raton, FL, 1994, chap. 2.

Franks, F., *Water*, The Royal Society of Chemistry, London, 1984.

Davies, C. W., *Electrochemistry*, George Newnes, London, 1967.

Bokris, J. O'M. and Reddy, A. K. N., *Modern Electrochemistry*, Plenum Press, New York, 1970, chap. 2. and 7.

Butler J. N., *Carbon Dioxide Equilibria and their Applications*, Addison-Wesley, Reading MA, 1982.

Cotton, F. A. and Wilkinson, G., *Advanced Inorganic Chemistry*, John Wiley, New York, 1980, chap. 1.

Swalin, R. A., *Thermodynamics of Solids*, John Wiley, New York, 1962, chap. 13 and 15.

Kubaschewski, O. and Hopkins B. E., *Oxidation of Metals and Alloys*, Butterworths, London, 1962, chap. 1.

Azaroff, L. V., *Introduction to Solids*, McGraw-Hill, New York, 1960.

Smallman, R., *Modern Physical Metallurgy*, Butterworths, London, 1985.

Hansen, M. and Anderko, K. *Constitution of Binary Alloys*, McGraw Hill, New York, 1957.

Brandes, E. A., *Smithells Metals Reference Book*, Butterworth, London, 1983, chap. 11.

3

Thermodynamics and Kinetics
of Corrosion Processes

3.1 Thermodynamics of Aqueous Corrosion

Thermodynamics provides part of the scientific infrastructure needed to evaluate the course and rate of corrosion processes. Its principle value is in yielding information on intermediate products of the complementary anodic and cathodic partial reactions that together constitute a complete process. The structures and characteristics of these intermediate products can control the resistance of a metal surface to attack within such wide limits as to make the difference between premature failure and sufficient life for its practical function.

An illustrative example concerns the behavior of iron in water containing dissolved oxygen. In neutral water, the anodic reaction produces a soluble ion, Fe^{2+} as an intermediate product and although the final product is a solid, it is precipitated from solution and does not protect the metal, whereas in alkaline water, the anodic reaction converts the iron surface directly into a thin protective solid layer. Such effects can be explained by thermodynamic assessments.

Classic works develop the principles of chemical thermodynamics *ab initio* but a concise summary of the background to the basic relations needed is given by Bodsworth in a companion volume of the present CRC series cited in "Further Reading," and is recommended to the reader.

3.1.1 Oxidation and Reduction Processes in Aqueous Solution

Corrosion science is frequently concerned with the exchange of electrons between half-reactions, i.e., anodic reactions that produce them and complementary cathodic reactions that consume them, and some means of accounting for the electrons is essential. Such an accounting system is based on the concept of *oxidation*, that has wider implications in inorganic chemistry than the name suggests. It is approached through the idea of *oxidation states*.

3.1.1.1 Oxidation States

The concept of oxidation states expresses the combining power exercised by elements in their compounds and is thus related to their valencies. It can be illustrated by comparing the ratios of metal atoms to oxygen atoms in some examples of metal oxides. These ratios, *stoichiometric ratios*, for Na_2O, MgO, Al_2O_3, and TiO_2, are 2:1, 1:1, 2:3, and 1:2, respectively. The differences in the relative amounts of oxygen with which these metals are combined are expressed on a scale of *oxidation states* in which the numbers +I, +II, +III, and +IV are assigned to sodium, magnesium, aluminum and titanium when present in their oxides. The number applies to the state of combination and not to the element itself and so the oxidation state of an uncombined element is zero. Many elements, including metals in the transition series, can exercise more than one oxidation state. The oxidation state of iron is +II in FeO but +III in Fe_2O_3. For copper it is +I in Cu_2O but +II in CuO.

Metal oxides are formed by transfer of electrons from the electropositive metal atoms to the electronegative oxygen atoms, converting them to cations and anions respectively. Every increment of one in the oxidation state of a metal represents the loss of an electron and corresponds to *oxidation* of the metal. The electrons lost by the metal are gained by the oxygen. This constitutes complementary *reduction* of the oxygen and by counting electrons it is found that its oxidation state is reduced from zero to −II.

The idea of oxidation and reduction is so useful that the concept of oxidation states is extended to cover other compounds containing elements that differ significantly in electronegativity, whether or not they contain oxygen and whether the transfer of electrons is complete or partial. Thus in the reaction between sodium and chlorine to produce sodium chloride, the sodium is said to be oxidized since it loses its valency electron to chlorine and in the reaction between hydrogen and bromine to form hydrogen bromide, the hydrogen is similarly said to be oxidized although the bond does not have strong ionic character.

The common oxidation states of some commercially important metals and corrosion products in aqueous systems are given in Tables 3.1 and 3.2.

The oxidation state of an element changes when it absorbs or releases electrons by transfer of electrons between chemical species. For example iron (II) hydroxide in an alkaline solution is oxidized by dissolved oxygen to the red-brown hydrated iron (III) oxide, a component of rust, by the reaction:

$$4Fe(OH)_2 + O_2 = 2Fe_2O_3 \cdot H_2O + 2H_2O \qquad (3.1)$$

The overall change in oxidation state in a completed reaction must be zero so that in Equation 3.1 as written, four iron atoms all increase in oxidation state from +II to +III and this is balanced by a reduction in the oxidation state of the two atoms in molecular oxygen from 0 to −II in the form of the oxide. This illustrates the principle that in any complete system an oxidation reaction must be compensated by an equivalent reduction reaction.

TABLE 3.1
Oxidation States of Some Selected Metals

Metal	Oxidation State	Oxides	Hydroxides	Aquo Ions	Others
Aluminium	III	Al_2O_3	$Al(OH)_3$	Al^{3+}, AlO_2^-	$AlO(OH)$
Chromium	III	Cr_2O_3	$Cr(OH)_3$	Cr^{3+}	
	VI	CrO_3	—	CrO_4^{2-}, $Cr_2O_7^{2-}$	
Copper	I	Cu_2O	—	Cu^+ unstable	
	II	CuO	$Cu(OH)_2$	Cu^{2+}	$CuCO_3,Cu(OH)_2$
Iron	II	FeO	$Fe(OH)_2$	Fe^{2+}	
	III	Fe_2O_3	Fe_2O_3,nH_2O	Fe^{3+}	$FeO(OH)$
	II + III	Fe_3O_4	—	—	
Lead	II	PbO	$Pb(OH)_2$	Pb^{2+}	
	IV	PbO_2	—	PbO_3^{2-}	
	II + IV	Pb_3O_4	—	—	
Magnesium	II	MgO	$Mg(OH)_2$	Mg^{2+}	$MgCO_3$
Nickel	II	NiO	$Ni(OH)_2$	Ni^{2+}	
Tin	II	SnO	$Sn(OH)_2$	Sn^{2+}	
	IV	SnO_2	—	SnO_3^{2-}	
Titanium	IV	TiO_2	—	TiO^{2+}	
Zinc	II	ZnO	$Zn(OH)_2$	Zn^{2+}, ZnO_2^{2-}	

TABLE 3.2
Oxidation States of Some Non-Metallic
Elements of Interest in Corrosion

Element	Oxidation State	Ions
Chlorine	− I	Chloride, Cl^-
Nitrogen	− III	Ammonium, NH_4^+
	+ III	Nitrite, NO_2^-
	+ V	Nitrate, NO_3^-
Oxygen	− II	Hydroxide, OH^-
Sulfur	− II	Hydrogen sulfide, HS^-
	+ IV	Sulfite, SO_3^{2-}
	+ VI	Sulfate, SO_4^{2-}

3.1.1.2   *Electrodes*

The oxidation state of species can also be changed by the intervention of an electrically conducting surface at which electrons can be transferred. Such a system is an *electrode*. The process proceeds if the charge transferred is removed by an electron source or sink, such as another electrode transferring an opposite charge to the same surface.

Electrodes coupled in this way sustain the active corrosion of metals. At one electrode, surface atoms of a metal, e.g., iron, dissolve as ions, raising the oxidation state of the iron from 0 to II, leaving electrons as an excess charge in the metal:

$$Fe = Fe^{2+} + 2e^- \tag{3.2}$$

At the other electrode, the electrons are absorbed by a complementary reaction on the same surface, where some other species, e.g., dissolved oxygen, is reduced:

$$\tfrac{1}{2}O_2 + H_2O + 2e^- = 2OH^- \tag{3.3}$$

reducing the oxidation state of oxygen from 0 for the element to –II in the hydroxyl ion. The balance of electrons between Equations 3.2 and 3.3 produces Fe^{2+} and OH^- ions in the correct ratio for precipitation of sparingly soluble $Fe(OH)_2$, that continuously removes the ions from solution, allowing reaction to continue.

$$Fe^{2+} + 2(OH)^- = \downarrow Fe(OH)_2 \tag{3.4}$$

Summation of the Reactions 3.2 through 3.4 yields the complete process:

$$Fe + \tfrac{1}{2}O_2 + H_2O = Fe(OH)_2 \tag{3.5}$$

To proceed further, the characteristics of electrode processes must be examined.

3.1.2 Equilibria at Electrodes and the Nernst Equation

In an isolated electrode process, the charge transferred accumulates until an equilibrium is established. As an example, consider the dissolution of a metal, M, in an aqueous medium to produce ions, M^{z+}:

$$M \rightarrow M^{z+} + ze^- \tag{3.6}$$

The accumulation of electrons in the metal establishes a negative charge on the metal relative to the solution, creating a potential difference opposing further egress of ions and promoting the reverse process, i.e., the discharge of ions and their return to the metal as deposited atoms. A dynamic equilibrium is established when the metal has acquired a characteristic potential relative to the solution. Conditions for equilibrium at a given constant temperature are derived from a form of the Van't Hoff reaction isotherm:

$$\Delta G = \Delta G^{\ominus} + RT \ln J \tag{3.7}$$

where J is the activity quotient corresponding to a free energy change, ΔG:

$$J = \frac{a_{(PRODUCT\ 1)} \times a_{(PRODUCT\ 2)} \times \dots etc.}{a_{(REACTANT\ 1)} \times a_{(REACTANT\ 2)} \times \dots etc.} \qquad (3.8)$$

and $\Delta G^{\ominus}$ is the free energy change for all reactants in their standard states.

In a normal chemical change without charge transfer, Equation 3.7 does not specify an equilibrium condition but in an electrode process, ΔG is balanced by the potential that the electrode acquires. This is expressed in electrical terms by replacing the Gibbs free energy terms, ΔG and $\Delta G^{\ominus}$ in Equation 3.7 with the corresponding equilibrium electrical potential terms, E and $E^{\ominus}$, using the expressions:

$$\Delta G = -zFE \text{ and } \Delta G^{\ominus} = -zFE^{\ominus} \qquad (3.9)$$

where the potential, $E^{\ominus}$, corresponding to the standard Gibbs free energy change, $\Delta G^{\ominus}$, is called the *standard electrode potential*. This yields the *Nernst Equation*:

$$E = E^{\ominus} - \frac{RT}{zF} \ln J \qquad (3.10)$$

The equation is often applied at ambient temperature. Taking this as 298 K (25 °C), inserting the values for R (8.314 J mol^{-1}) and F (96490 coulombs mol^{-1}) and substituting 2.303 log J for ln J, Equation 4 becomes:

$$E = E^{\ominus} - \frac{0.0591}{z} \log J \qquad (3.11)$$

Since any particular electrode process proceeds by the gain or loss of electrons, i.e., by reduction or oxidation, it is often convenient to use the term, *redox potential*, to describe the potential of the system at equilibrium for the prevailing activities of participating species. It is simply an abbreviated form of the expression, **reduction/oxidation** potential.

3.1.3 Standard State for Activities of Ions in Solution

The matter of selecting standard states for the activities of reacting species to be inserted in the activity quotient, J, must now be addressed.

There is no problem in selecting the standard state for the solvent, water, or for pure solid and gaseous phases, e.g., a metal, a solid-reaction product, hydrogen or oxygen, since it is convenient to use the usual definition of the standard state as the pure substance under atmospheric pressure but two problems must be addressed in selecting a standard state for ions in solution:

1. Ions do not exist in isolation as "pure substances" so that an arbitrary solution is needed as the standard state. Consistency with standard states often used for non-ionic solutions might suggest that this should be based on *unit molality*, i.e., one mole of solute in 1 kg of solvent.

2. Since ions are charged particles with repulsion between like ions and attraction between unlike ions, their activities in solutions non-linear functions of the quantities present. Thus selection of the concentrated solutions corresponding to unit molality as the standard state would extend the inconvenience of a composition-dependent activity coefficient to dilute solutions that are frequently of interest.

The problems are resolved by choosing a standard state that is convenient for dilute solutions in which the ions are separated by sufficient distances to limit interaction between them to a negligible extent, so that the activities are linear functions of molality. This facilitates treatment of many solutions encountered in electrochemistry, including corrosion processes.

The formal rigorous definition is that the standard state corresponds to that solution that has the effect that activity equals molality at infinite dilution.

3.1.4 Electrode Potentials

3.1.4.1 *Convention for Representing Electrodes at Equilibrium*

The equation for an electrode process is intuitively visualized as written for the direction in which it proceeds. For example, if interest centers on the active corrosion of a metal such as nickel, it could be written in the direction of oxidation, as an anodic reaction with electrons appearing on the right-hand side:

$$Ni \rightarrow Ni^{2+} + 2e^- \tag{3.12a}$$

Alternatively, if interest centers on the electrodeposition of nickel it might seem more appropriate to write it as a cathodic reaction, in the direction of reduction, with electrons appearing on the left-hand side:

$$Ni^{2+} + 2e^- \rightarrow Ni \tag{3.12b}$$

Similarly the reduction of oxygen complementing the anodic dissolution of metals in aerated water can be written as a cathodic reaction:

$$\tfrac{1}{2}O_2 + H_2O + 2e^- \rightarrow 2OH^- \tag{3.13a}$$

but the reverse process is an anodic reaction generating oxygen in the electrolysis of water:

$$2OH^- \rightarrow \frac{1}{2}O_2 + H_2O + 2e^- \qquad (3.13b)$$

The matter is important because the formal direction of the reaction determines whether the signs for equilibrium electrode potentials, E', and standard electrode potentials, $E^{\ominus}$, are positive or negative. It is resolved by a mandatory international convention that an electrode process at equilibrium is formally written in the direction of reduction, i.e., with the electrons on the left-hand side, e.g:

$$Ni^{2+} + 2e^- = Ni \qquad (3.14)$$

Some caution is needed in reading some excellent textbooks that do not observe the convention because they were written before it was established.

3.1.4.2 Choice of a Potential Scale

It is possible to estimate the absolute values of the potentials established at electrodes indirectly by summation of the free energy changes entailed but such estimates lack the precision needed for both practical and theoretical study. This presents no real difficulty because the important quantities are not absolute potentials but relative potentials between electrodes. These relative potentials are obtained by referring electrodes to a scale in which a standard reference electrode is arbitrarily assigned the value, zero. The requirements of the standard are:

1. It must be internationally agreed.
2. It must be easily reproducible.
3. It must not be susceptible to interference from side reactions.

The standard selected is the *standard hydrogen electrode*. A full explanation of the reasons for selecting this standard and for the arrangements by which it is physically realized must be deferred until after some discussion of electrode kinetics but a provisional description is required now.

3.1.4.3 The Standard Hydrogen Electrode

The standard hydrogen electrode is the reaction:

$$2H^+ + 2e^- = H_2 \qquad (3.15)$$

conducted at 25 °C, with the reacting species in their standard states, i.e., with a_{H_2} and a_{H^+} both equal to unity. It is physically realized by a system in which an inert metal, platinum is partially immersed in an acid solution containing unit activity of hydrogen ions. The acid is hydrochloric acid at 298 K (25°C), corresponding to a molality of about 1.2 to reproduce the standard state for hydrogen ions. The system is enclosed in a glass vessel and a stream of pure hydrogen at atmospheric pressure is passed through the acid near the platinum. The platinum is "platinized", i.e., coated with finely divided platinum powder, giving a very large effective surface on which a layer of hydrogen absorbs. This yields a hydrogen surface exposed to the acid in intimate contact with the platinum that can conduct electrons to or from it.

The standard hydrogen electrode is arbitrarily assigned the potential 0.000 V and other electrode potentials referred to this standard are described as potentials on the *standard hydrogen scale* (SHE). The sign to be given to the potential is determined by the direction of the spontaneous reaction at the electrode when it is coupled to a standard hydrogen electrode. If the reaction is in the direction of reduction, the potential is positive and if it is in the direction of oxidation, it is negative, as the following examples illustrate:

1. $Cu^{2+} + 2e^- = Cu$ proceeds to the right if $a_{Cu^{2+}} = 1$
   $2H^+ + 2e^- = H_2$ proceeds to the left
   so that the sign for $Cu^{2+} \rightarrow Cu$ is positive.
2. $Fe^{2+} + 2e^- = Fe$ proceeds to the left if $a_{Fe^{2+}} = 1$
   $2H^+ + 2e^- = H_2$ proceeds to the right
   so that the sign for $Fe^{2+} \rightarrow Fe$ is negative.

The standard hydrogen electrode cell, though accurate, is tedious in use and sub-standard scales are based on cells that are more convenient for practical work. These include the saturated calomel (mercury I chloride) scale (SCE) and the silver/silver chloride scale (Ag/AgCl). Sub-standard electrodes are selected to suit particular applications on grounds of sensitivity, temperature-dependence, compactness, or non-toxicity and it is important not to omit reference to the scale on which a potential is quoted because the zeros for sub-standard scales are shifted relative to the SHE scale, e.g:

$$0.00 \text{ V (SCE)} \equiv + 0.244 \text{ V (SHE) at } 25°C$$

$$0.00 \text{ V (Ag/AgCl)} \equiv + 0.222 \text{ V (SHE) at } 25°C$$

The potential at an electrode depends not only on its nature but also on the activities of the participating species. Thus a table of potentials for different electrodes is significant only if the potentials are given on the same scale (SHE) and for equal activities. The logical choice is for all reactants in their standard states and potentials for this condition are described as *standard electrode potentials, $E^{\ominus}$*. Table 3.3 gives some standard electrode potentials of interest in corrosion science. Note that the change in sign on passing through the hydrogen electrode in the series is purely a result of

TABLE 3.3
Selected Standard Electrode Potentials

Electrode	Standard Electrode Potential, $E^{\ominus}$ V (SHE)
$Au^{3+} + 3e^- = Au$	+ 1.50
$Cl_2 + 2e^- = 2Cl^-$	+ 1.360
$\frac{1}{2}O_2 + 2H^+ + 2e^- = H_2O$	+ 1.228
$Br_2 + 2e^- = 2Br^-$	+ 1.065
$Ag^+ + e^- = Ag$	+ 0.799
$Hg_2^{2+} + 2e^- = 2Hg$	+ 0.789
$Fe^{2+} + e^- = Fe^{3+}$	+ 0.771
$I_2 + 2e^- = 2I^-$	+ 0.536
$Cu^+ + e^- = Cu$	+ 0.520
$Cu^{2+} + 2e^- = Cu$	+ 0.337
$2H^+ + 2e^- = H_2$	0.000 (by definition)
$Pb^{2+} + 2e^- = Pb$	− 0.126
$Sn^{2+} + 2e^- = Sn$	− 0.136
$Ni^{2+} + 2e^- = Ni$	− 0.250
$Fe^{2+} + 2e^- = Fe$	− 0.440
$Cr^{3+} + 3e^- = Cr$	− 0.740
$Zn^{2+} + 2e^- = Zn$	− 0.763
$Al^{3+} + 3e^- = Al$	− 1.663
$Mg^{2+} + 2e^- = Mg$	− 2.370
$Na^+ + e^- = Na$	− 2.714

choosing it as the arbitrary standard. In fact, although it is not known precisely, the absolute value of the standard hydrogen electrode potential is about −0.9 V. Standard electrode potentials, $E^{\ominus}$ follow the temperature-dependence of the standard Gibbs free energies to which they correspond.

Electrodes such as the saturated calomel and silver/silver chloride cells used as reference standards are often based on the electrochemical reaction of a sparingly soluble salt. Historically, electrodes of this kind have been used to measure their very small solubilities.

Example 1

Calculate the solubility product, K_s, for silver chloride from the measured potential, 0.222 V (SHE) of the silver/silver chloride standard cell, saturated with AgCl.

SOLUTION:
The electrode reaction is:

$$AgCl + e^- = Ag + Cl^- \qquad \text{(Reaction 1)}$$

and can be considered as the sum of the reactions:

$$AgCl(s) = Ag^+ + Cl^- \qquad \text{(Reaction 2)}$$

$$Ag^+ + e^- = Ag \qquad \text{(Reaction 3)}$$

Since standard Gibbs free energies, $\Delta G^{\ominus}$, are additive:

$$\Delta G_1^{\ominus} \text{(Reaction 1)} = \Delta G_2^{\ominus} \text{(Reaction 2)} + \Delta G_3^{\ominus} \text{(Reaction 3)}$$

Applying the Van't Hoff isobar to Reaction 2, for the saturated solution, i.e., $a_{AgCl} = 1$:

$$\Delta G_2^{\ominus} = -RT \ln(a_{Ag^+})(a_{Cl^-})/a_{AgCl} = -2.303RT \log(a_{Ag^+})(a_{Cl^-})/1$$

and replacing $\Delta G_1^{\ominus}$ and $\Delta G_3^{\ominus}$ by the corresponding standard electrode potentials:

$$-zFE_1^{\ominus} = -2.303RT \log\left(a_{Ag^+}\right)\left(a_{Cl^-}\right) - zFE_3^{\ominus}$$

$$\log\left(a_{Ag^+}\right)\left(a_{Cl^-}\right) = \frac{zF}{2.303RT}\left(E_1^{\ominus} - E_3^{\ominus}\right) = \frac{\left(E_1^{\ominus} - E_3^{\ominus}\right)}{0.0591}$$

$E_3^{\ominus}$ is given as 0.222 V (SHE), $E_3^{\ominus}$ is given in Table 3.3 as 0.799 V (SHE) and $z = 1$

Hence, $\log(a_{Ag^+})(a_{Cl^-}) = (0.222 - 0.799)/0.0591 = -9.763$

and $K_s = (a_{Ag^+})(a_{Cl^-}) = 1.726 \times 10^{-10}$

3.1.5   Pourbaix (Potential-pH) Diagrams

3.1.5.1   *Principle and Purpose*

Pourbaix (or potential-pH) diagrams, named for the originator, are graphical representations of thermodynamic information appropriate to electrochemical reactions. The presentation of information in this format facilitates its application to practical problems in a wide variety, including corrosion, electrodeposition, geological processes, and hydrometallurgical extraction processes. A particular diagram is called "the Pourbaix diagram for the iron-water system", "the Pourbaix diagram for the zinc-water system," etc. They are examples of *predominance area diagrams*, discussed in a wider context by Bodsworth, cited in Further Reading list.

The objective is to represent the relative stabilities of solid phases and soluble ions that are produced by reaction between a metal and an aqueous environment as functions of two parameters, the electrode potential, E, and the pH of the environment. The information needed to construct a

Pourbaix diagram are the standard electrode potentials, $E^{\ominus}$, or the equilibrium constants, K, as appropriate, for all of the possible reactions considered. The purpose and construction of these diagrams is best appreciated by considering a particular system in detail.

3.1.5.2 Example of the Construction of a Diagram — The Iron-Water System

The example selected is the iron-water system both for its technical importance and because it exhibits all of the features to be found in the diagrams generally.

Selection of Species and Reactions:

About twenty known reactions can proceed between various species including water, iron metal, Fe^{2+}, Fe^{3+}, Fe_3O_4, Fe_2O_3, $Fe(OH)_2$, FeO_4^{2-}, $HFeO_2^-$ and others. The first task is to select reactions appropriate to the problem in hand. This is not always straightforward. In considering the corrosion of iron, reactions including Fe, Fe^{2+}, Fe^{3+}, Fe_3O_4 and Fe_2O_3 could be selected, on the grounds that they are the stable species in the presence of dissolved oxygen. Alternatively, reactions including Fe, Fe^{2+}, Fe^{3+}, $Fe(OH)_2$ and FeO(OH) or some other combination could be selected, because hydroxides can form as primary corrosion products, even though they are unstable in the presence of dissolved oxygen and ultimately become converted to hydrated oxides. For illustration, the species Fe, Fe^{2+}, Fe^{3+}, Fe_3O_4 and Fe_2O_3, that yield the most usual version are selected. The reactions are written conventionally and in a form that may include H^+ ions but not OH^- ions, e.g., equilibrium between metallic iron and Fe_3O_4 is written:

$$Fe_3O_4 + 8H^+ + 8e^- = 3Fe + 4H_2O \tag{3.16}$$

and not as the mass/charge balance equivalent equation:

$$Fe_3O_4 + 4H^+ + 8e^- = 3Fe + 4OH^- \tag{3.17}$$

Written in this form, the reactions significant for corrosion are:

Reaction 1. $Fe^{2+} + 2e^- = Fe$

Reaction 2. $Fe_3O_4 + 8H^+ + 8e^- = 3Fe + 4H_2O$

Reaction 3. $Fe_3O_4 + 8H^+ + 2e^- = 3Fe^{2+} + 4H_2O$

Reaction 4. $Fe_2O_3 + 6H^+ + 2e^- = 2Fe^{2+} + 3H_2O$

Reaction 5. $3Fe_2O_3 + 2H^+ + 2e^- = 2Fe_3O_4 + H_2O$

Reaction 6. $Fe^{3+} + e^- = Fe^{2+}$

Reaction 7. $2Fe^{3+} + 3H_2O = Fe_2O_3 + 6H^+$

The Approach to Calculations:

In Reactions 1 to 7, the solid phases, Fe_3O_4 and Fe_2O_3 and also the solvent, water, are present at unit activity and do not influence the equilibria. The equilibria are therefore governed by the following factors:

1. The charge transferred, manifest as an electrode potential, E. This applies only to reactions involving electrons, Reactions 1 to 6 but not Reaction 7.
2. The activity of hydrogen ions, i.e., the pH of the solution.
3. The activities of other soluble ions, e.g., Fe^{2+} and Fe^{3+}.

For reactions in which there is charge transfer, these quantities are inter-related by the Nernst equation, given earlier in Equation 3.11:

$$E = E^{\ominus} - \frac{0.0591}{z} \ln J$$

For reactions in which there is no charge transfer, the activities are related simply by the normal equilibrium constant, e.g., for Reaction 7:

$$K = \frac{\left(a_{H^+}\right)^6}{\left(a_{Fe^{3+}}\right)^2} \tag{3.18}$$

Equation 3.11 or 3.18 is applied as appropriate for all of the seven reactions to obtain the equations of lines plotted against E and pH as coordinates, that constitute the Pourbaix diagram.

Calculations:

Reaction 1. $Fe^{2+} + 2e^- = Fe$

Information needed: $E^{\ominus} = -0.440$ V SHE.

Apply the Nernst Eqn: $E = -0.440 - \dfrac{0.0591}{2} \cdot \log \dfrac{1}{a_{Fe^{2+}}}$

$$= -0.440 + 0.0295 \log(a_{Fe^{2+}}) \tag{3.19}$$

Reaction 2. $Fe_3O_4 + 8H^+ + 8e^- = 3Fe + 4H_2O$

Information needed: $E^{\ominus} = -0.085$ V SHE.

Apply the Nernst Eqn: $E = -0.085 - \dfrac{0.0591}{8} \cdot \log \dfrac{1}{\left(a_{H^+}\right)^8}$

$$= -0.085 - 0.0591 \text{ pH} \qquad (3.20)$$

Reaction 3. $Fe_3O_4 + 8H^+ + 2e^- = 3Fe^{2+} + 4H_2O$

Information needed: $E^{\ominus} = +0.980$ V SHE.

Apply the Nernst Eqn: $E = +0.980 - \dfrac{0.0591}{2} \cdot \log \dfrac{\left(a_{Fe^{2+}}\right)^3}{\left(a_{H^+}\right)^8}$

$$= 0.980 - 0.2364 \text{ pH} - 0.0886 \log(a_{Fe^{2+}})^3 \qquad (3.21)$$

Reaction 4. $Fe_2O_3 + 6H^+ + 2e^- = 2Fe^{2+} + 3H_2O$

Information needed: $E^{\ominus} = +0.728$ V SHE.

Apply the Nernst Eqn: $E = +0.728 - \dfrac{0.0591}{2} \cdot \log \dfrac{\left(a_{Fe^{2+}}\right)^2}{\left(a_{H^+}\right)^6}$

$$= 0.728 - 0.1773 \text{ pH} - 0.0591 \log(a_{Fe^{2+}}) \qquad (3.22)$$

Reaction 5. $3Fe_2O_3 + 2H^+ + 2e^- = 2Fe_3O_4 + H_2O$

Information needed: $E^{\ominus} = +0.221$ V SHE.

Apply Nernst Eqn: $E = +0.221 - \dfrac{0.0591}{2} \cdot \log \dfrac{1}{\left(a_{H^+}\right)^2}$

$$= 0.221 - 0.0591 \text{ pH} \qquad (3.23)$$

Reaction 6. $Fe^{3+} + e^- = Fe^{2+}$

Information needed: $E^{\ominus} = +0.771$ V SHE.

Apply the Nernst Eqn: $E = +0.771 - \dfrac{0.0591}{1} \cdot \log \dfrac{\left(a_{Fe^{2+}}\right)}{\left(a_{Fe^{3+}}\right)}$ (3.24)

$$= 0.771 + 0.0591 \log(a_{Fe^{3+}}) - 0.0591 \log(a_{Fe^{2+}})$$

Reaction 7. $2Fe^{3+} + 3H_2O = Fe_2O_3 + 6H^+$

Information needed: $(aH^+)^6/(a_{Fe^{3+}})^2 = 10^2$

Taking logarithms: $6 \log(a_{H^+}) - 2 \log(a_{Fe^{3+}}) = 2$

Hence, rearranging: $pH = \{-\log(a_{Fe^{3+}}) - 1)\}/3$ (3.25)

Plotting Lines on the Diagram:

Equations 3.19 to 3.25 contain, in a suitable form, all of the information needed to plot the diagram. Lines are plotted against coordinates, E and pH, representing the conditions for which the activities of soluble ions, e.g., Fe^{2+} and Fe^{3+}, are at some specified value.

Consider Reaction 1. If the activity of the soluble ion, Fe^{2+}, is arbitrarily chosen to be unity, i.e., 10^0, then inserting $a_{Fe^{2+}} = 1$ into Equation 3.19 gives a value for E of -0.440 V SHE. For other values of $a_{Fe^{2+}}$, i.e., 10^{-2}, 10^{-4} and 10^{-6}, the corresponding values obtained for E are -0.50, -0.56 and -0.62 V (SHE), respectively. Note that these values do not depend on pH, since the hydrogen ion, H^+, does not participate in the reaction and pH therefore does not appear as a variable in Equation 3.19. Thus the plots for Reaction 1 appear as a *family* of *horizontal* lines, corresponding to the various selected values of $a_{Fe^{2+}}$, as illustrated in Figure 3.1. As a convenient shorthand, it is conventional to label the individual lines as 0, -2, -4, and -6 to indicate the values for $a_{Fe^{2+}}$ to which they refer, i.e., 10^0, 10^{-2}, 10^{-4} and 10^{-6}.

The interpretation to be placed on the plots is as follows. Points on a selected line represent combinations of E and pH for which the soluble ion activity *equals* the value to which the line refers, points above the line correspond to conditions for which the activity is *higher* and points below the line correspond to conditions for which the activity is *lower*. Thus the line forms the boundary between two regions, an upper region in which the Fe^{2+} ion is the stable species and a lower on in which metallic iron is the stable species, *with respect to the soluble ion activity for which the line is drawn*. This point is made clearer in Section 3.1.5.4, where application of Pourbaix diagrams to corrosion problems is considered.

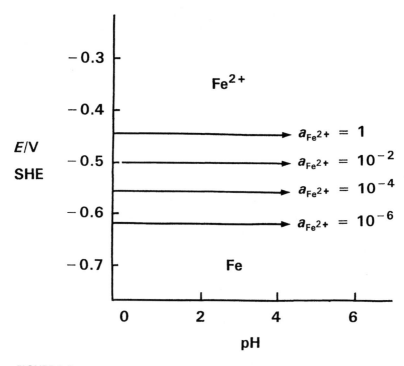

FIGURE 3.1
Potential-pH plots of the equilibrium, $Fe^{2+} + 2e^- = Fe$, for selected. values of $a_{Fe^{2+}}$.

The Complete Diagram:

Corresponding series of lines are drawn for Reactions 1 to 7. Since the reactions are mutually exclusive, they terminate at intersections, as illustrated in the full diagram given in Figure 3.2. The effect of the intersecting lines is to divide the diagram into several *domains*, within which one or another of the species is considered to be stable with respect to the soluble ion activity corresponding to any chosen set of lines.

Figure 3.2 illustrates the following features of Reactions 1 to 7:

1. Equilibria for Reactions 1 and 6 depend on $a_{Fe^{2+}}$ and E but are independent of pH, because H^+ ions do not participate in the reactions. This yields a family of horizontal lines for the prescribed values of $a_{Fe^{2+}}$.

2. Equilibrium for Reaction 7 depends on $a_{Fe^{3+}}$ and pH but is independent of E, because electrons do not participate in the reaction, i.e., it is *not an electrode reaction*. This yields a family of vertical lines for the prescribed values of $a_{Fe^{3+}}$.

3. Equilibria for Reactions 3 and 4 depend on all three of the variables, E, pH and $a_{Fe^{2+}}$. This yields families of sloping lines for the prescribed values of $a_{Fe^{2+}}$. The slopes are negative because H^+ ions are on the left side of the reactions as written, leading to negative pH terms in Equations 3.21 and 3.22.

4. Equilibria for Reactions 2 and 5 depend on E and pH but are independent of $a_{Fe^{2+}}$. This yields *single* sloping lines. The slopes are again negative for the reason given in (3) above.

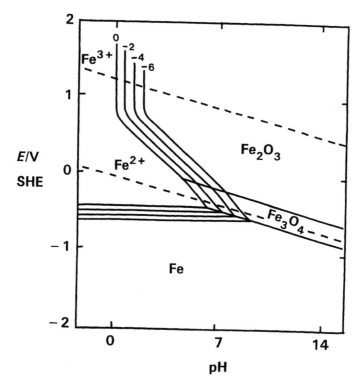

FIGURE 3.2

Pourbaix diagram for the iron-water system at 25°C. Labels: 0 - $a_{Fe^{2+}} = a_{Fe^{3+}} = 1$, -2 - $a_{Fe^{2+}} = a_{Fe^{3+}} = 10^{-2}$, -4 - $a_{Fe^{2+}} = a_{Fe^{3+}} = 10^{-4}$, -6 - $a_{Fe^{2+}} = a_{Fe^{3+}} = 10^{-6}$. Domain for the stability of water shown by dotted lines.

3.1.5.3 The Domain of Stability for Water

The usefulness of a Pourbaix diagram is enhanced by superimposing on it the domain enclosing the combinations of the parameters, E and pH for which water is stable. This is defined by lines on the diagram representing the decomposition of water by evolution of hydrogen by Reaction 1 or of oxygen by Reaction 2:

Reaction 1. $\qquad$ $2H^+ + 2e^- = H_2$ $\qquad$ $E^{\ominus} = 0.000$ V SHE.

Reaction 2. $\quad$ $\frac{1}{2}O_2 + 2H^+ + 2e^- = H_2O$ $\qquad$ $E^{\ominus} = +1.228$ V SHE.

Note that Reaction 2 is an alternative form of the equation introduced earlier:

$$\frac{1}{2}O_2 + H_2O + 2e^- = 2OH^-$$

but re-written in terms of H^+ ions instead of OH^- ions, to permit plotting on the potential-pH diagram. The gases are evolved against atmospheric pressure, so that $a_{H_2} = a_{O_2} = 1$.

Application of the Nernst equation yields:

Reaction 1. $\qquad$ $E = 0.000 - \dfrac{0.0591}{2} \cdot \log \dfrac{1}{\left(a_{H^+}\right)^2}$

$$= -0.0591 \text{ pH} \tag{3.26}$$

Reaction 2. $\qquad$ $E = +1.228 - \dfrac{0.0591}{2} \cdot \log \dfrac{1}{\left(a_{H^+}\right)^2}$

$$= 1.228 - 0.0591 \text{ pH} \tag{3.27}$$

Equations 3.26 and 3.27 are superimposed as dotted lines on Figure 3.2. The two lines enclose a domain within which water is stable. For combinations of potential and pH above the top line, water is unstable and decomposes evolving oxygen. For combinations below the bottom line, it is also unstable and decomposes evolving hydrogen.

3.1.5.4 *Application of Pourbaix Diagrams to Corrosion Problems*

The first requirement is to select the appropriate lines from the Pourbaix diagram. An internationally agreed convention is adopted in which a metal is considered to be actively corroding if the equilibrium activity of a soluble ion derived from it, e.g., $a_{Fe^{2+}}$ or $a_{Fe^{3+}}$ exceeds 10^{-6}. For clarity, the diagram for $a_{Fe^{2+}} = a_{Fe^{3+}} = 10^{-6}$, selected from Figure 3.2, is given in Figure 3.3.

As explained later in discussing kinetics, a metal exposed to an aqueous medium acquires a potential from coupled electrode processes such as

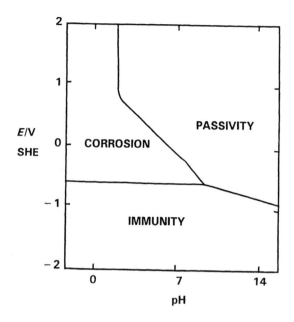

FIGURE 3.3
Pourbaix diagram for the iron-water system at 25°C showing nominal zones of immunity, passivity and corrosion for $a_{Fe^{2+}} = a_{Fe^{3+}} = 10^{-6}$.

those represented in Equations 3.2 and 3.3. This potential and the pH of the medium define a point in the diagram that suggests the likely response of the metal by the domain in which it lies. There are three kinds of domain:

1. *Domains of immunity*, in which the metal is the stable species and is immune to corrosion in the terms for which the diagram is drawn.

2. *Domains of corrosion* in which the stable species is a soluble ion and the metal is expected to corrode *if the kinetics are favorable.*

3. *Domains of passivity* in which the stable species is an insoluble solid that can protect the metal *if it forms an impervious, adherent layer.*

3.1.5.5   *Pourbaix (Potential-pH) Diagrams for Some Common Metals*

Figures 3.4 to 3.8 give simplified Pourbaix diagrams for the aluminum-, zinc-, copper-, tin- and nickel-water systems, with the domain of stability for water superimposed. Examples 2 to 6 in the Appendix to this chapter detail the construction of these diagrams.

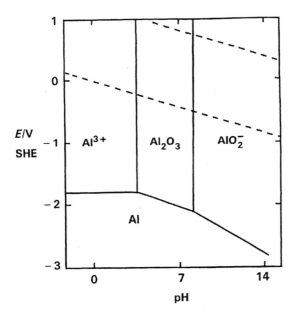

FIGURE 3.4

Pourbaix diagram for the aluminum-water system at 25°C. $a_{Al^{3+}} = a_{AlO_2^-} = 10^{-6}$ Domain for the stability of water shown by dotted lines.

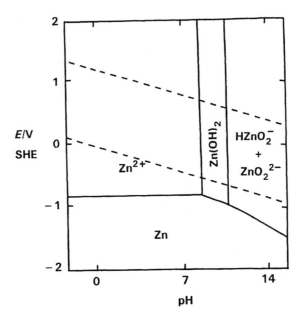

FIGURE 3.5

Pourbaix diagram for the zinc-water system at 25°C.

$a_{Zn^{2+}} = a_{HZnO_2^-} = a_{ZnO_2^{2-}} = 10^{-6}$ Domain for the stability of water shown by dotted lines.

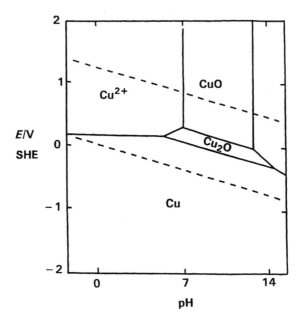

FIGURE 3.6
Pourbaix diagram for the copper-water system at 25°C. $a_{Cu^{2+}} = 10^{-6}$ Domain for the stability of water shown by dotted lines.

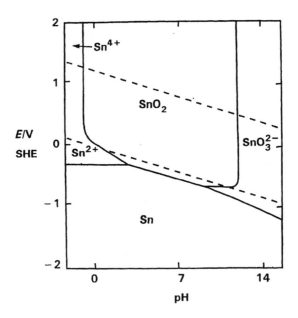

FIGURE 3.7
Pourbaix diagram for the tin-water system at 25°C. $a_{Sn^{2+}} = a_{Sn^{4+}} = a_{SnO_3^{2-}} = 10^{-6}$ Domain for the stability of water shown by dotted lines.

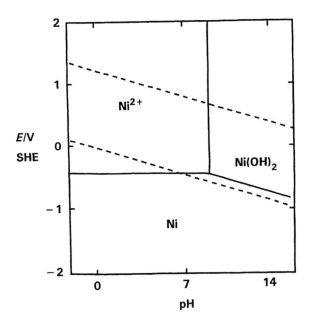

FIGURE 3.8
Pourbaix diagram for the nickel-water system at 25°C. $a_{Ni^{2+}} = 10^{-6}$ Domain for the stability of water shown by dotted lines.

The diagram for the copper-water system, given in Figure 3.6, exhibits an unusual feature. Although there is a domain of stability for Cu(I) in the solid oxide, Cu_2O, there is no domain for the corresponding Cu^+_{aq} ion in solution. This is because the stability of the ion is sensitive to its chemical environment as illustrated in the following Example 7.

Example 7

In the Pourbaix diagram of the copper-water system for $a_{Cu^{2+}} = 10^{-6}$ at 25°C, given in Figure 3.6, the +II oxidation state is represented by both a soluble ion, $Cu^{2+}_{(aq)}$, and an oxide, CuO, but the +I state is represented only by the oxide Cu_2O. Show that the activity of the $Cu^+_{(aq)}$ ion is unsustainable at the value, 10^{-6}, for which the diagram is prepared, whereas the oxide Cu_2O has a domain of stability.

SOLUTION:
Soluble species, $Cu^+_{(aq)}$ and $Cu^{2+}_{(aq)}$:
The objective is to assess the extent to which the $Cu^+_{(aq)}$ ion is destabilized by disproportionation in water to $Cu^{2+}_{(aq)}$ ions and Cu metal by the reaction:

$$2Cu^+_{(aq)} = Cu^{2+}_{(aq)} + Cu_{(s)} \qquad \text{(Reaction 1)}$$

The equilibrium constant is calculated from the standard Gibbs free energy for the reaction, $(\Delta G^{\ominus})_1$, applying the Van't Hoff isobar:

$$\ln K = (-\Delta G^{\ominus})_1 / RT$$

The value of $(\Delta G^{\ominus})_1$ is obtained from the weighted sum of the Gibbs free energies for the reactions:

$$Cu^+ + e^- = Cu \qquad \text{(Reaction 2)}$$

and:

$$Cu^{2+} + 2e^- = Cu \qquad \text{(Reaction 3)}$$

since: (Reaction 1) = 2 × (Reaction 2) – (Reaction 3)

so that: $(\Delta G^{\ominus})_1 = 2(\Delta G^{\ominus})_2 - (\Delta G^{\ominus})_3$

$(\Delta G^{\ominus})_2$ and $(\Delta G^{\ominus})_3$ are obtained from the standard electrode potentials for Reactions 2 and 3, given in Table 3.3, using the relation $\Delta G^{\ominus} = -zFE^{\ominus}$

thus: $(\Delta G^{\ominus})_2 = -(1 \times 96490 \times 0.520) = -50180 \text{ J}$

and: $(\Delta G^{\ominus})_3 = -(2 \times 96490 \times 0.337) = -65040 \text{ J}$

hence: $(\Delta G^{\ominus})_1 = 2 \times (-50180) - (-65040) = -35320 \text{ J}$

The negative value for $(\Delta G^{\ominus})_1$ indicates that the reaction is spontaneous. Inserting this value and $R = 8.314 \text{ J mol}^{-1}$ into the Van't Hoff isobar applied to Reaction 1 at $T = 25°C$ ($\equiv 298$ K), assuming unit activity for the metallic copper:

$$\ln K = \ln \frac{\left(a_{Cu^{2+}}\right)}{\left(a_{Cu^+}\right)^2} = \frac{-(-35320)}{8.314 \times 298} = 14.256$$

whence:

$$\frac{\left(a_{Cu^{2+}}\right)}{\left(a_{Cu^+}\right)^2} = 1.55 \times 10^6$$

Consider a solution in which all of the copper, Cu_T, is assumed to be present initially as Cu^+. On equilibration, every mole of Cu^+ eliminated produces 0.5 mole of Cu^{2+}

i.e., $$[Cu^{2+}] = 0.5(Cu_T - [Cu^+])$$

where the square brackets represent concentrations. In dilute solution, $a_{Cu^+} \rightarrow [Cu^+]$ and $a_{Cu^{2+}} \rightarrow [Cu^{2+}]$ and therefore:

$$0.5(Cu_T - [Cu^+])/[Cu^+]^2 = 1.55 \times 10^6$$

Rearranging: $\quad 3.1 \times 10^6 \cdot [Cu^+]^2 + [Cu^+] - Cu_T = 0$

that is solved to give the concentration of Cu^+ ions in solution for various values of Cu_T. For $Cu_T = 10^{-6}$ mol dm^{-3} the standard quadratic formula gives:

$$[Cu^+] = \{-1 + (1^2 + 4 \times 3.1 \times 10^6 \times 10^{-6})^{\frac{1}{2}}\}/\{2 \times 3.1 \times 10^6\} = 4.3 \times 10^{-7} \, mol \, dm^{-3}$$

Since 57% of the Cu^+ disproportionates at the activity that would be assigned to it in the Pourbaix diagram, it cannot be considered a stable species and no domain can be assigned to it. The disproportionation increases with rising concentrations of copper ions, e.g., it is 80% at $Cu_T = 10^{-3}$ mol dm^{-3}, for which $[Cu^+] = 0.02 \times 10^{-3}$ mol dm^{-3}.

The Solid Oxides Cu_2O and CuO:

In contrast to the environment of solvating water molecules for soluble ions, the environment of O^{2-} ions in the solid oxides stabilizes the Cu(I) state.

A hypothetical disproportionation reaction for Cu^+ in Cu_2O, corresponding to Reaction 1 would be:

$$Cu_2O = CuO + Cu \qquad \qquad \text{(Reaction 4)}$$

that is half the difference between the reactions:

$$Cu_2O + 2H^+ + 2e^- = 2\,Cu + H_2O \qquad \qquad \text{(Reaction 5)}$$

$$2CuO + 2H^+ + 2e^- = Cu_2O + H_2O \qquad \qquad \text{(Reaction 6)}$$

Proceeding as before, using data from Example 4 given in the Appendix to Chapter 3:

$$E'(\text{for Reaction 5}) = 0.471 - 0.0591\text{pH (SHE)}$$

$$E'(\text{for Reaction 6}) = 0.669 - 0.0591\text{pH (SHE)}$$

$$\Delta G(\text{Reaction 4}) = \tfrac{1}{2}(-2 \times 96490 \times \{0.471 - 0.0591\text{pH}\})$$

$$- \tfrac{1}{2}(-2 \times 96490 \times \{0.669 - 0.0591\text{pH}\})$$

$$= +19100 \text{ J mol}^{-1}.$$

Since $\Delta G(\text{Reaction 4})$ is positive and independent of pH, the reaction is not spontaneous so that a domain of stability for Cu_2O is possible.

3.1.5.6 *Limitations of Pourbaix Diagrams*

Pourbaix diagrams are of great utility in guiding consideration of corrosion and other problems but they apply only for the conditions assumed in their construction and they are not infallibly predictive because they have limitations, as follows:

1. The diagrams are derived from thermodynamic considerations and yield no kinetic information. There are situations in which zones of corrosion suggest that a metal dissolves and yet it does not, due, for example, to the formation of a metastable solid phase or to kinetic difficulties associated with a complementary cathodic reaction.

2. Domains in which solid substances are considered to be the stable species relative to arbitrary soluble ion activities $<10^{-6}$ give good indications of conditions in which a metal may be passive. Whether particular metals are actually passivated within these nominal domains and to what extent a useful passive condition can extend beyond their boundaries depends on the nature, adherence and coherence of the solid substance. This reservation is considered further in Sections 9.2.1 and 9.2.2 with respect to the development of passivity on aluminum.

3. The diagrams yield information only on the reactions considered in their construction and take no account of known or unsuspected impurities in the solution or of alloy components in the metal that may modify the reactions. For example, Cl^- or SO_4^{2-} ions present in solution may attack, modify or replace oxides or hydroxides in domains of passivity, diminishing the protective power of these substances and small quantities of alloy components can introduce microsructural features of the metal that resist passivation.

4. The form and interpretation of a Pourbaix diagram are both temperature-dependent, the form because T appears in Equation 3.10 and the interpretation because pH is temperature-dependent, as shown in Table 2.5.

3.2 Kinetics of Aqueous Corrosion

In the long term, the degradation of engineering metals and alloys by corrosion is inevitable and so resistance to it is essentially concerned with the rates of corrosion. Reaction rate theory can be quite complicated and is dealt with in specialized texts* but the following brief summary of the essential principles is sufficient to underpin the derivation of some well-known rate equations for electrochemical processes frequently applied in corrosion problems.

The rate of any transformation is controlled by the magnitude of one or more energy barriers that every particulate entity, e.g., an atom or an ion must surmount to transform. These peaks are the energy maxima of intermediate transition states through which the entity must pass in transforming and the energy that must be acquired is the *activation energy*, ΔG^*. The statistical distribution of energy among the particles ensures that at any instant a small but significant fraction of the particles has sufficient energy to surmount the peaks. This fraction and hence the fraction of particles transforming in a small time span, *the reaction rate, r*, depends on the value of the activation energy. An expression for reaction rate that applies to many reactions over moderate temperature ranges is Arrhenius' equation:

$$r = A \exp \frac{-\Delta G^*}{RT} \tag{3.28}$$

To apply the equation to an electrode process, it must be restated in electrical terms. Ions transported across an electrode carry electric charge, so that the reaction rate, r in Equation 3.28, can be replaced by an electric current, i. The energy of the process is the product of the charge and the potential drop, E, through which it is carried. Thermodynamic quantities are expressed per mole of substance and because the charge on a mole of singly charged ions is the Faraday, $F = 96490$ coulombs, the free energy change, ΔG of an electrode process is:

$$\Delta G = -zFE \tag{3.29}$$

where z is the charge number on the particular ion species transferred in the process, e.g., z is 1 for H^+ or Cl^-, 2 for Fe^{2+}, 3 for Fe^{3+} etc.

---

* e.g., Hinshelwood cited in "Further Reading."

Replacing r in the Arrhenius equation, Equation 3.28 with an appropriate change of constant:

$$i = k \exp \frac{-\Delta G^*}{RT} = k \exp \frac{zFE^*}{RT} \qquad (3.30)$$

3.2.1 Kinetic View of Equilibrium at an Electrode

The equilibrium at an electrode is dynamic and the ionic species are produced and discharged simultaneously at the conducting surface. Taking the dissolution of a metal as a relevant tangible example:

$$M^{z+} + ze^- = M \qquad (3.31)$$

the metal dissolves as ions and ions deposit back on the metal surface at equal rates. Since the electrode is at equilibrium, there is no net change in Gibbs free energy, ΔG, in either the forward or reverse process. The chemical free energy change due to the dissolution or deposition of the metal is balanced by an equivalent quantity of electrical work done by the ions in crossing the electric field imposed by the equilibrium electrode potential. Since they are charged entities, the ion flows constitute two equal and opposite electric currents. The currents leaving and entering the metal, denoted $\vec{i}$ and $\overleftarrow{i}$, respectively, are called the *partial currents*. Their magnitude at equilibrium is called the *exchange current density, i_0*:

$$\vec{i} = \overleftarrow{i} = i_0 \qquad (3.32)$$

The activation energy is the excess energy that must be acquired to transform metal *atoms* at the metal surface into solvated metal *ions*. This is because fully solvated ions cannot approach closer to the metal surface than the outer Helmholtz plane because they are obstructed by their own solvation sheaths and the monolayer of water molecules attached to the metal surface, as illustrated in Figure 2.7. Hence the metal atoms are only partially solvated during the transformation and are therefore in a transient higher energy state. The free energy profile for the reaction is shown schematically in Figure 3.9. Applying Equation 3.30 gives the exchange current density as a function of the activation free energy:

$$i_0 = \vec{i} = \overleftarrow{i} = k \exp \frac{-\Delta G^*}{RT} \qquad (3.33)$$

where: ΔG^* is the activation energy
   T is the temperature
   R is the gas constant
   k is a constant depending on the process and on the ion activity

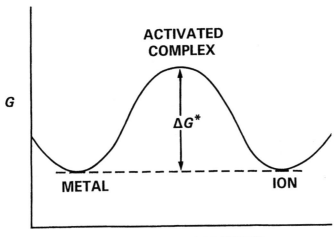

FIGURE 3.9
Schematic energy profile for equilibrium at an electrode. ΔG^* is the activation Gibbs free energy.

3.2.2 Polarization

If equilibrium at an electrode is disturbed, a net current flows across its surface displacing the potential in a direction and to an extent depending on the direction and magnitude of the current. The shift in potential is called *polarization* and its value, η, is the *overpotential*. There are three possible components, *activation*, *concentration*, and *resistance* polarization.

3.2.2.1 Activation Polarization

Activation polarization is a manifestation of the relative changes in the activation energies for dissolution and deposition, when equilibrium is disturbed. It is always a component of the total polarization, whether or not there are also significant contributions from concentration and resistance effects. The polarization is positive, i.e., *anodic*, or negative, i.e., *cathodic*, according to whether the net current is a dissolution or deposition current.

The free energy profile of an electrode subject to activation polarization is shown schematically in Figure 3.10, where the electrode is assumed to be anodically polarized with an overpotential of η. Since $\Delta G = -zFE$, the polarization raises the energy of the metal by $zF\eta$ and that of the activated complex by $\alpha zF\eta$, relative to that of the ions, where α is a symmetry factor defining the position of the maximum in the profile in Figure 3.10 e.g., α = 0.5 if the maximum is equidistant from the two minima.

For the electrode at equilibrium, the activation energies for dissolution and deposition were both equal to ΔG^* but for the polarized electrode, the

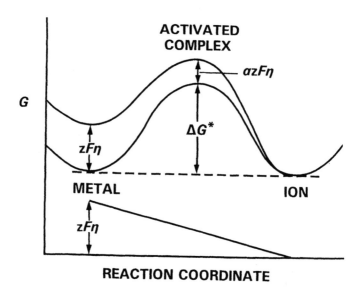

FIGURE 3.10

Schematic energy profile for activation polarization at an electrode: ΔG^* = activation Gibbs free energy; η/V = anodic overpotential.

activation energy for dissolution is decreased to $[\Delta G^* - (1 - \alpha)zF\eta]$ and the activation energy for deposition is increased to $[\Delta G^* + \alpha zF\eta]$ (from the geometry of the profile in Figure 3.10). Hence the two partial currents are no longer equal. The dissolution, $\vec{i}$ current is:

$$\vec{i} = k \exp \frac{-\left\{\Delta G^* - (1 - \alpha)zF\eta\right\}}{RT} \qquad 3.34)$$

$$= k \exp \frac{-\Delta G^*}{RT} \cdot \exp \frac{(1 - \alpha)zF\eta}{RT} \qquad (3.35)$$

$$= i_0 \exp \frac{(1 - \alpha)zF\eta}{RT} \qquad (3.36)$$

and by similar reasoning, the deposition current, $\overleftarrow{i}$, is:

$$\overleftarrow{i} = i_0 \exp \frac{-\alpha zF\eta}{RT} \qquad (3.37)$$

The net current is the difference between the two partial currents:

$$i_{net} = \vec{i} - \overleftarrow{i} = i_0 \left[\exp \frac{(1 - \alpha)zF\eta}{RT} - \exp \frac{-\alpha zF\eta}{RT} \right] \qquad (3.38)$$

This is the *Butler-Volmer* equation expressing the relation between the net anodic current flowing at the electrode, i_{net} and the overpotential, η.

Tafel or High-Field Approximation

For most purposes, a much more user-friendly approximation for Equation 3.38 can be used. If the polarization is anodic, with $\eta > 0.1$ V, the value of $\overleftarrow{i}$ is insignificant and, making the approximation $i_{net} \approx \overrightarrow{i}$, the equation becomes:

$$i_{net} \approx \overrightarrow{i} = i_0 \exp\left[\frac{(1 - \alpha)zF\eta}{RT}\right] \qquad (3.39)$$

Since α and z, are constant for a given electrode process at constant temperature (usually room temperature) this equation can be re-arranged as:

$$\eta_{anodic} = b \log i_0 + b \log i_{anodic} \qquad (3.40)$$

By similar reasoning, if the polarization is cathodic with $\eta > -0.1$ V, $\overrightarrow{i}$ is insignificant and with the approximation $i_{net} \approx \overleftarrow{i}$, it is easily shown that:

$$\eta_{cathodic} = b \log i_0 - b \log i_{cathodic} \qquad (3.41)$$

Equations 3.40 and 3.41 are usually combined in the single expression:

$$\eta = b \log i_0 \pm b \log i \qquad (3.42)$$

This *high-field approximation*, replicates the equation developed empirically in 1905 by Tafel and is hence known as *Tafel's equation*. The constant, b, is the *Tafel slope*, where:

$$b = \frac{2.303\ RT}{(1 - \alpha)zF} \qquad (3.43)$$

The Symmetry Factor, α

If α is 0.5, Equation 3.38 becomes a hyperbolic sine function:

$$i_{net} = i_0\left[\exp\frac{(1 - \alpha)zF\eta}{RT} - \exp\frac{-\alpha zF\eta}{RT}\right] = 2i_0 \sinh\frac{zF\eta}{2RT} \qquad (3.44)$$

Equation 3.44 can be used to show that the symmetry factor, α, is usually close to 0.5 in the following way. The hyperbolic sine function is symmetrical about the origin. This means that if $\alpha \approx 0.5$, the potential-current relationships are the same for forward and reverse currents, so that the electrode cannot act as a rectifier for AC current. This is the usual situation. However, it is found by experiment that certain electrodes do act as rectifiers and the phenomenon is called *Faradaic rectification*. This implies that for those particular electrodes the curve is not symmetrical about the origin, and so α is not equal to 0.5.

Low-Field Approximation

Equation 3.44 can also be used to derive a simple approximation for Equation 3.38 when the polarization is very small, <0.05 V. For small values of the independent variable, the hyperbolic sine function approximates to a linear function, so that:

$$i_{net} = 2i_0 \sinh \frac{zF\eta}{2RT} \approx \frac{i_0 zF\eta}{RT} \tag{3.45}$$

This is the *low-field approximation* for the Butler-Volmer equation.

3.2.2.2   *Concentration Polarization*

As the potential of an electrode is altered further and further away from its equilibrium potential, the net current flowing, whether anodic, i_a, or cathodic, i_c, increases at first according to the Tafel equation, Equation 3.42. However, the current cannot be increased indefinitely because there are limits to the rate at which ions can carry charges through the solution to and from the electrode. This results in an excess potential over that predicted by the Tafel equation. The situation is illustrated in Figure 3.11. The effect arises because ions are produced or consumed at the electrode surface faster than they can diffuse to or from the bulk of the solution. In an anodic reaction, the concentration of ions in the immediate vicinity of the electrode is raised above that in the bulk solution; conversely, in a cathodic reaction, the local concentration is depressed. As a consequence, the polarization for a given current is greater than that predicted by the Tafel equation. The excess potential is called the *concentration polarization*, η_C. The magnitude of the effect can be examined by applying the Nernst Equation, Equation 3.10, assuming that it is valid in the prevailing dynamic situation.

In the dissolution reaction:

$$M \rightarrow M^{z+} + ze^-$$

let $[a_{M^{z+}}]$ and $[a_{M^{z+}}]_P$, respectively represent the ion activity in the bulk of the solution and the enhanced activity at the surface of the polarized electrode.

The potential of the unpolarized electrode is the equilibrium potential, E', for the reaction $M^{z+} + ze^- = M$:

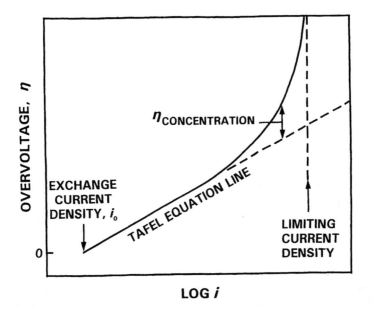

FIGURE 3.11
Logarithmic plot of current density, i, v. overvoltage, η, for a polarized electrode, showing deviation from Tafel line and limiting current density due to concentration polarization. η is positive for an anodic current, $\vec{i}$, and negative for a cathodic current, $\overleftarrow{i}$.

$$E' = E^{\ominus} + \frac{0.0591}{z} \log\left[a_{M^{z+}}\right] \tag{3.46}$$

When the electrode is polarized, its potential must be referred not to the nominal equilibrium potential but to a higher potential, E'_P, corresponding to equilibrium with the enhanced ion activity at the polarized electrode surface, where:

$$E'_P = E^{\ominus} + \frac{0.0591}{z} \log\left[a_{M^{z+}}\right]_P \tag{3.47}$$

For any given current, the concentration polarization, η_C is the difference between the potentials given in Equations 3.47 and 3.46:

$$\eta_C = E'_P - E' = \frac{0.0591}{z} \log\frac{\left[a_{M^{z+}}\right]_P}{\left[a_{M^{z+}}\right]} \tag{3.48}$$

As the current rises, $[a_{M^{z+}}]_P$ and hence η_C also rises. Eventually, the ion activity at the electrode surface reaches saturation and a limiting current density is reached.

For the reverse (cathodic) reaction:

$$M^{z+} + ze^- \rightarrow M$$

in which metal is deposited, the solution at the electrode surface is depleted of M^{2+} ions and the concentration polarization is in the opposite sense:

$$\eta_C = E' - E_P = \frac{0.0591}{z} \log \frac{\left[a_{M^{z+}}\right]}{\left[a_{M^{z+}}\right]_P} \tag{3.49}$$

Again there is a limiting current density, this time because as the current increases, $[a_{M^{z+}}]_P \rightarrow 0$ and therefore $\eta_C \rightarrow \infty$. Limiting current densities are marked on Figure 3.11.

3.2.2.3   *Resistance Polarization*

In discussing activation and concentration polarization, no account was taken of ohmic resistances. For some electrode reactions the effects of ohmic resistance can be very considerable. This is especially significant when the reaction itself or a complementary reaction produces films on the electrode surface. The total potential drop across such resistance is called *resistance polarization*, η_R.

The total polarization at an electrode is therefore the sum of three components, activation, concentration and resistance polarization:

$$\eta_{TOTAL} = \eta_A + \eta_C + \eta_R \tag{3.50}$$

The effects of these forms of polarization are illustrated by the characteristics of hydrogen evolution and oxygen reduction reactions that feature prominently in corrosion processes.

3.2.2.4   *The Hydrogen Evolution Reaction and Hydrogen Overpotential*

The hydrogen evolution reaction:

$$H^+ + e^- = \tfrac{1}{2}H_2 \tag{3.51}$$

provides a good subject for exploring characteristics of activation polarization and it is a common cathodic reaction supporting the corrosion of metals in acidic aqueous solutions.

In Tafel's equation, Equation 3.42, the constant, b is given by:

$$b = \frac{2.303\ RT}{\left(1 - \alpha\right)zF} \tag{3.52}$$

The value of z for the reaction is 1 and provisionally taking the most probable value for α as 0.5, as suggested in Section 3.2.2.1, the theoretical value of b at 298 K is:

$$b = (2.303 \times 8.315 \times 298)/0.5 \times 1 \times 96490 = 0.118 \text{ V/decade} \quad (3.53)$$

The unit, V/decade, appears because η is a logarithmic function of i in the Tafel equation.

Example 8

Table 3.4 gives representative results from measurements of potential, E, versus current density, i, for the evolution of hydrogen on platinum. They apply to 0.1 M hydrochloric acid at 25°C. Do they yield a value for b comparable with the value given in Equation 3.53?

TABLE 3.4
Experimental Potential v Current Relation for Hydrogen Evolution on Platinum

E / V SHE	−0.07	−0.08	−0.12	−0.17	−0.22	−0.27	−0.32	−0.37
η/ V	−0.01	−0.02	−0.06	−0.11	−0.16	−0.21	−0.26	−0.31
i/Acm^{-2}	5.0×10^{-4}	1.0×10^{-3}	3.5×10^{-3}	1.0×10^{-2}	3.0×10^{-2}	9.0×10^{-2}	2.5×10^{-1}	6.0×10^{-1}

SOLUTION:
The equilibrium potential, E', is calculated by applying the Nernst equation. The pH value for 0.1 M hydrochloric acid is close to 1 and at 25°C (T = 298 K), the value of the term, 2.303 RT/F is 0.0591. Hence:

$$E' = E^{\ominus} - (0.0591/z) \log(1/a_{H^+})$$
$$= 0.00 - 0.0591 \text{ pH} = -0.0591 \times 1 = -0.059 \text{ V (SHE)} \quad (3.54)$$

The overvoltage, η, is $E - E'$, yielding the values given in the middle row of Table 3.4. These values are plotted as a Tafel plot, η versus log i in Figure 3.12. The plot is linear for $\eta > 0.06$ V but deviates progressively from linearity as $\eta \to 0$. due to the approximation $i_{net} \approx i$ made in deriving the Tafel equation. This can be exploited to determine the exchange current density, i_0, because extrapolation of the linear part of the plot to $\eta = 0$ disregards the anodic current, i and gives the value of the cathodic current at the equilibrium potential, which by Equation 3.32 is equal to the exchange current density, i_0. From Figure 3.12, this is found to be 1.0×10^{-3} A cm^{-2}. The value of b, found from the slope of the plot is 0.11 V/decade, which is close to the theoretical value given by Equation 3.53, consistent with the assumptions that hydrogen evolution on platinum is a one electron process, i.e., z = 1, and that the symmetry factor is 0.5. Inserting the values found for b and i_0 in Equation 3.41 yields a Tafel equation:

$$\eta_c = -0.33 - 0.11 \log i_c \quad (3.55)$$

This equation applies only to this particular set of results because the value obtained for the exchange current density, i_0 is very sensitive to the state of the metal surface.

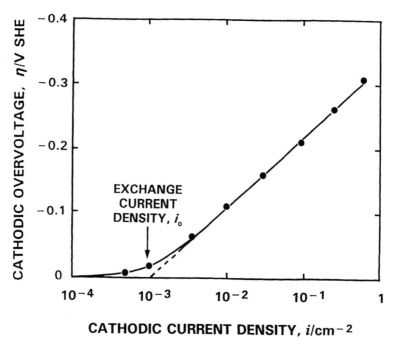

FIGURE 3.12

Tafel plot for the evolution of hydrogen on platinum from 0.1 M hydrochloric acid at 25°C.

For metals other than platinum, the hydrogen overpotential is more difficult to assess and its value varies widely from metal to metal. For practical application, the orders of magnitude for different metals are often compared by quoting the overpotential required to evolve hydrogen at an arbitrary current, e.g., 1 mA cm^{-2}, as in Table 3.5. Hydrogen overpotential has very important effects in technology. In Chapter 6, it is shown how high hydrogen overpotentials prevent hydrogen evolution during the electrodeposition of metals such as tin and zinc, preventing practical difficulties that would otherwise arise and in Chapter 12 it is shown how the high hydrogen overpotential on tin allows its use as a coating on steel sheet used for cans to conserve acidic foods. The low hydrogen overpotential and high exchange current density on platinum are important factors in the choice and construction of the hydrogen electrode for the standard potential scale, because they contribute to its reproducibility and insensitivity to unwanted side reactions.

TABLE 3.5
Typical Hydrogen Overvoltages for Current Density
of 1 mA cm^{-2} on Some Metals

Metal	Platinum	Gold	Nickel	Iron	Copper	Aluminum	Tin	Lead
Overpotential η/ V	0.09	0.15	0.30	0.40	0.45	0.70	0.75	1.0

3.2.2.5   The Oxygen Reduction Reaction

The reduction of oxygen dissolved in water at metal surfaces exemplifies
the influences of concentration and resistance polarization. It is a common
cathodic process supporting corrosion of metals because natural waters
are constantly replenished with dissolved oxygen by recycling through air
as rain. In acidic water, the reduction reaction is predominantly:

$$\frac{1}{2}O_2 + 2H^+ + 2e^- = H_2O \tag{3.56}$$

for which $$E^{\ominus} = 1.228 \text{ V SHE}$$

and in neutral and alkaline water it is predominantly:

$$\frac{1}{2}O_2 + H_2O + 2e^- = 2OH^- \tag{3.57}$$

for which: $$E^{\ominus} = 0.401 \text{ V SHE}$$

The reactions differ kinetically in the requirement of Reaction 3.56 for a
copious supply of hydrogen ions and thermodynamically in the standard
states to which activities of the ions are referred. In reaction 3.56, the standard state is unit activity of hydrogen ions, i.e., for pH = 0 but in
Reaction 3.57 it is unit activity of hydroxyl ions, i.e., for pH = 14 at 25°C,
at which temperature $K_W = (a_{H^+}) \times (a_{OH^-}) = 10^{-14}$.

The reduction of oxygen at metal surfaces is difficult to characterize.
One problem is its sensitivity to concentration polarization due to the low
solubility of oxygen in water. Another is that the conditions of metal surfaces at potentials prevailing during corrosion can differ substantially
from those at equilibrium potentials for Reactions 3.56 and 3.57 as evident
in the Pourbaix diagrams given in Figures 3.2 through 3.8. These effects
are approached quantitatively in Example 9, given in Section 3.2.3, after
introducing the concept of corrosion potentials.

3.2.3   Polarization Characteristics and Corrosion Velocities

The *polarization characteristics* for an electrode refers to the relation
between applied potential and current, including activation, concentration and resistance contributions. These characteristics differ from

electrode to electrode; some are dominated by activation polarization, others by concentration or resistance polarization as illustrated in the contrasting examples of the evolution of hydrogen on platinum and the reduction of oxygen at an iron surface, just described.

3.2.3.1   Corrosion Velocity Diagrams

Polarization characteristics provide the basis for a convenient graphical method of presenting information on corrosion in the form of *corrosion velocity diagrams*, introduced Evans*. The method is to display on the same diagram the polarization characteristics of all electrode processes that contribute anodic and cathodic reactions in a particular corroding system.

The construction is as follows:

1. The equilibrium potentials, E' for possible reactions are determined, applying the Nernst equation for the prevailing activities. Taking the corrosion of iron in neutral aerated water as an example (neglecting the contribution to the cathodic current from the discharge of hydrogen, that is found to be small in Example 9 below), the constituent electrochemical reactions, written in the conventional direction, are:

$$Fe^{2+} + 2e^- = Fe \tag{3.58}$$

$$\tfrac{1}{2}O_2 + H_2O + 2e^- = 2OH^- \tag{3.59}$$

Assuming that the pH of the water is 7, that it is saturated with oxygen from the air and that by the conventional criterion for corrosion $a_{Fe^{2+}} = 10^{-6}$, application of the Nernst equation yields the values $E'_{Fe^{2+}} = -0.62$ V SHE and $E_{O_2} = 1.0$ V SHE.

2. Axes labelled E and i are drawn as in Figure 3.13, using the symbols, $\overset{\rightharpoonup}{i}$ and $\overset{\leftharpoonup}{i}$ to denote anodic and cathodic current densities, respectively.

3. The equilibrium values of E', are marked on the potential axis and the overpotentials are plotted as functions of i for anodic and cathodic polarization. The metal acquires a potential between $E'_{Fe^{2+}}$ and E'_{O_2}, called the *corrosion potential, $E_{CORROSION}$*. Conservation of electrons determines that the corrosion potential has a value at which the total anodic and cathodic currents are equal. The anodic current is called the *corrosion current, $i_{CORROSION}$* and the metal is dissolved at the Faradaically equivalent rate.

---

* Cited in "Further Reading."

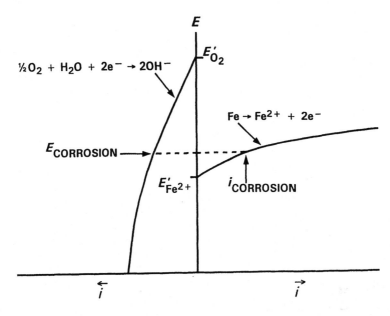

FIGURE 3.13
A corrosion velocity diagram, with iron corroding in aerated neutral water as a tangible example. The curve shapes are arbitrary but indicate high concentration polarization for oxygen absorption. $\vec{i}$ — Current density for anodic reaction. $\overset{\leftarrow}{i}$ — Current density for cathodic reaction.

Sometimes, the anodic reaction by which a metal is dissolved is sustained by not one but two significant cathodic reactions, e.g., in an aerated dilute acid, where both oxygen reduction and hydrogen discharge can contribute to the total cathodic current. The corrosion potential then assumes a value at which the anodic current due to dissolution of the metal equals the sum of the currents from the cathodic reactions.

Example 9
Using the following data, show that the absorption of oxygen is the dominant cathodic reaction sustaining the corrosion of iron in neutral water in equilibrium with air.
Data: For the reaction $Fe^{2+} + 2e^- = Fe$:

$E^{\ominus} = -0.44$ V SHE.
$a_{Fe^{2+}} = 10^{-6}$ (assumed for corrosion by convention)
Tafel constants: b = 0.06 V/decade, $i_0 = 10^{-11}$ A cm^{-2}.

For the reaction $2H^+ + 2e^- = H_2$ on iron:
$E^{\ominus} = 0.00$ V SHE.
Tafel constants: b = 0.12 V/decade, $i_0 = 10^{-10}$ A cm^{-2} at pH 7

Corrosion potential, $E_{CORROSION}$, on iron in aerated water at pH 7: -0.45 V SHE.

SOLUTION:
Anodic Current Density Due to Dissolution of Iron:
Applying the Nernst equation yields the equilibrium potential, $E'_{Fe^{2+}}$:

$$E'_{Fe^{2+}} = -0.44 - \frac{0.0591}{2} \log \frac{1}{10^{-6}} = -0.62 \text{ V SHE}$$

Applying the Tafel equation yields the anodic dissolution current for iron:

$$\eta_{ANODIC} = -b \log i_0 + b \log i_{ANODIC}$$

Substituting for η_{ANODIC}, b and i_0:

$$E_{CORROSION} - E'_{Fe^{2+}} = -0.06 \log 10^{-11} + 0.06 \log i_{ANODIC}$$

$$-0.45 - (-0.62) = -0.06 \log 10^{-11} + 0.06 \log i_{ANODIC}$$

whence, $i_{ANODIC} = 6.8 \times 10^{-9} \text{ A cm}^{-2}$

_____

Cathodic Current Density Due to Evolution of Hydrogen:
Applying the Nernst equation yields the equilibrium potential, E'_{H^+}:

$$E'_{H^+} = 0.00 - \frac{0.0591}{2} \log \frac{1}{\left(a_{H^+}\right)^2}$$

$$= -0.0591 \text{ pH} = -0.0591 \times 7 = -0.414 \text{ V SHE}$$

Applying the Tafel equation yields the current due to the discharge of hydrogen:

$$\eta_{CATHODIC} (H_2) = -b \log i_0 (H_2) + b \log i_{CATHODIC} (H_2)$$

$$E'_{H_2} - E_{CORROSION} = -0.12 \log 10^{-10} + 0.12 \log i_{CATHODIC (H_2)}$$

$$-0.414 - (-0.45) = -0.12 \log 10^{-10} + 0.12 \log i_{CATHODIC (H_2)}$$

whence, $i_{CATHODIC (H_2)} = 2 \times 10^{-10} \text{ A cm}^{-2}$

_____

Cathodic Current Density Due to Oxygen Reduction:
At the corrosion potential, the anodic current density due to the dissolution of iron is equal to the sum of the current densities due to the evolution of hydrogen and the reduction of dissolved oxygen:

$$\tfrac{1}{2}O_2 + H_2O + 2e^- = 2OH^-$$

i.e., $i_{ANODIC}(Fe^{2+}) = i_{CATHODIC}(H_2 \text{ evolution}) + i_{CATHODIC}(O_2 \text{ reduction})$

$$6.8 \times 10^{-9} = 2 \times 10^{-10} + i_{CATHODIC}(O_2 \text{ reduction})$$

whence: $i_{CATHODIC}(O_2 \text{ reduction}) = 6.6 \times 10^{-9} \text{ A cm}^{-2}$

This is more than an order of magnitude greater than the cathodic current density due to evolution of hydrogen, identifying the reduction of oxygen as the dominant cathodic reaction.

Overpotential for Oxygen Reduction:
For the reaction $\tfrac{1}{2}O_2 + H_2O + 2e^- = 2OH^-$:

$$E^{\ominus} = +0.401 \text{ V SHE.}$$

Applying the Nernst equation yields the equilibrium potential:

$$E' = E^{\ominus} - \frac{0.0591}{z} \log \frac{\left(a_{OH^-}\right)^2}{\left(a_{O_2}\right)^{1/2}}$$

$$= +0.401 - \frac{0.0591}{2} \log \frac{\left(10^{-7}\right)^2}{(0.21)^{1/2}}$$

$$= +0.805 \text{ V SHE}$$

At the corrosion potential the cathodic overpotential for oxygen reduction, η is:

$$\eta = E' - E_{CORROSION} = +0.805 - (-0.45) = 1.255 \text{ V}$$

This is a very high overpotential for a cathodic current of only 6.6×10^{-9} A cm^{-2} and implies strong polarization of the reaction. Thus other effects must predominate over activation polarization as predicted by the Tafel equation. These are:

1. Resistance polarization due to a tendency towards film formation.
2. Concentration polarization due to slow diffusion of oxygen, depleting its concentration at the metal surface below the concentration in the bulk of the electrolyte.

These effects conspire to make the corrosion process very sensitive to oxygen concentration.

3.2.3.2   Differential Polarization for Oxygen Reduction and Crevice Corrosion

It is well known that if oxygen dissolved in neutral water cannot reach part of the surface of an active metal but has unrestricted access to the rest, the oxygen-starved surface can suffer enhanced, sometimes intense, corrosion. The effect is called *differential aeration* and if the oxygen starvation is within crevices, the corrosion is known as crevice corrosion.

The enhanced corrosion is evidence that the anodic current density causing metal dissolution is higher on oxygen-starved surfaces and hence is less heavily polarized than it is on surfaces with open access to oxygen. On analysis, the origins of differential aeration turn out to be more complex than at first appears and has exercised many minds.

One factor is that films produced on the metal surface in the course of the oxygen reduction reaction, as deduced in Section 3.2.2.6, influence polarization not only of the cathodic reaction but also the anodic dissolution reaction.

Crevice corrosion is a common problem but forethought can ameliorate its depredations. Crevices can exist through faulty geometric design, lack of full penetration in welds, ill-fitting gaskets and underneath dirt, loose rust, or builders debris. Other differential aeration effects are intensified corrosion at the water-line of partly immersed metals and differential corrosion as a function of depth in stagnant water.

3.2.4   Passivity

In appropriate conditions, some base metals can develop a surface condition that inhibits interaction with aqueous media. The condition is described as *passivity* and its development is called *passivation*. The effect is valuable in conferring corrosion resistance on bare metal surfaces even in aggressive environments.

3.2.4.1   Spontaneous Passivation

Some metals passivate spontaneously in water if the pH is within ranges corresponding to potential-independent domains of stability for oxides or hydroxides. These domains appear in the Pourbaix diagrams, e.g., within the pH ranges 3.9–8.5 for aluminum illustrated in Figure 3.4, and 8.5–10.5 for zinc, illustrated in Figure 3.5. The effect is due to the formation of thin protective oxide or hydroxide films. The existence of a domain of stability for an oxide or hydroxide in the Pourbaix diagram is not the sole condition for passivation. To be protective, the solid product must not only be coherent and adherent to the metal surface but also its integrity must not be impaired by impurities contributed by the metal or the environment. Foreign ions such as chlorides and sulfates can become incorporated in the passive film, reducing its protective power. Impurities and extraneous inclusions in the metals can contribute minority phases that breach the passive surface. With attention to formulation, metal quality and environmental application, corrosion-resistant alloys based on copper, aluminum, zinc, and other metals are widely applied.

3.2.4.2   Anodic Passivation

For some transition metals and their alloys, passivation is both pH and potential dependent. In these circumstances, the metal may corrode actively at low potentials but can be passivated by raising its potential to a more positive value. The phenomenon is described as "anodic passivation".

The principle can be illustrated by the behavior of iron. In aerated neutral water, i.e., with pH 7, the oxygen absorption reaction polarizes the iron anodically to a potential of approximately -0.45 V SHE, as indicated in Example 9. Reference to the iron-water Pourbaix diagram in Figure 3.2 shows that these coordinates for pH and potential define a point in the domain of stability for Fe^{2+}, so that dissolution of iron is expected as happens in practice. If the anodic polarization is increased to impose a potential more positive than -0.15 V SHE, the new coordinates define a point in the domain of stability for Fe_2O_3, indicating a possible passive condition. This is of limited application for corrosion protection, because the passive condition can be implemented only by imposing the potential needed by adding oxidizing agents to the aqueous environment or by impressing anodic current on iron from an external source. The real value of anodic passivation is realized when the potential at which a metal passivates is low enough to be realized by a cathodic reaction naturally available from the environment in which the metal is required to serve. This is especially true of stainless steels, which are formulated, as described in Chapter 8, to passivate even in moderately strong acids when polarized by the oxygen reduction reaction from oxygen solutions naturally in equilibrium with the atmosphere.

3.2.4.3   Theories of Passivation

Passivation is a complex phenomenon and is difficult to explain completely. There are two approaches, the film theory, originated by Faraday, and the more recent absorption theory, associated with the names of Kolotyrkin and Uhlig*.

---

* Cited in "Further Reading."

Film Theory

There is experimental evidence for tangible oxide films on some but not all kinds of passivated metal surfaces. When the film is established, its ability to protect the metal depends on the degree of restraint it imposes on dissolution of the metal. The dissolution is the combined result of three processes:

1. Entry of metal atoms into the film as cations at the metal/film interface.
2. Transport of the metal cations or of oxygen anions through the oxide.
3. Dissolution of metal cations from the film at the film/environment interface.

At ambient temperatures, the processes are driven by the electric field across the film and so the following properties of a film commend it as a passivating agent:

1. Stability over a wide potential range.
2. Mechanical integrity.
3. Low ionic conductivity.
4. Good electron conductivity to reduce the potential difference across the film.
5. Low solubility and slow dissolution in the prevailing aqueous medium.

Adsorption Theory

The theory recognizes that strong anodic passivation is a characteristic of alloys of metals in the transition series, notably iron, chromium, nickel, molybdenum, and cobalt and it is natural to associate it with the partly filled $3d$ or $4d$ shells in their electron configurations, described in Chapter 2. An explanation is that singly occupied or unfilled atomic orbitals offer the facility for the highly polarized water molecules and anions in immediate contact with the metal to bind to the surface metal atoms by sharing electrons, creating the passive condition that protects the metal. A dynamic steady state is envisioned with continuous exchange of molecules between the adsorbed layer and the aqueous environment. Theories of alloying to promote passivating capability by optimizing the bond strength are associated with Uhlig who calculated, for example, that the critical composition for passivation in the iron-chromium binary system should be 13.6 atomic-% Cr, which is close to the observed value of 12.7 atomic-%.

The theory can account for passivation but it must also explain the failure of the metal to passivate at potentials in the active potential range. The difference in behavior has been attributed to potential-sensitive interaction of susceptible metals with the first-row water molecules adsorbed on the surface, described in Chapter 2, Section 2.2.8. At the lower potentials in the active range, surface atoms of the metal, M, are assumed to lose electrons and become solvated by first row water molecules, forming soluble cations:

$$M + n(H_2O) \text{adsorbed} \rightarrow M(H_2O)_n^{z+} + ze^- \qquad (3.60)$$

In the passive range it is assumed that the interaction described by Equation 3.60 is replaced by:

$$M + (H_2O) \text{adsorbed} \rightarrow M(O^{2-}) + 2H^+ \qquad (3.61)$$

in which adsorbed water loses protons to become passivating adsorbed anions. The passivating interaction, Equation 3.61 is assumed to begin at an appreciable rate only at the *passivating potential*, a threshold potential denoting the lower limit of the passive range.

The view that a monolayer of adsorbed anions is sufficient to confer passivity in some circumstances is probably well founded but ideas of the detailed mechanism are uncertain, because they are undoubtedly complex and difficult to verify by experiment.

Compatibility of Film and Adsorption Theories

The main criticism of the adsorption theory is that although it can explain instantaneous passivation by very thin films, the passive state is more often associated with a film thicker than a monolayer, where the film theory is more appropriate. The two theories are not incompatible and it is best to regard each as revealing part of the truth, the film theory dealing with the nature of the film and the adsorption theory drawing attention to the nature of the metal and to the relation between the kinetics of dissolution and the kinetics of passivation.

3.2.5 Breakdown of Passivity

Metals and alloys that rely on passivity for protection are vulnerable to corrosion failure if the passivity breaks down. The nature and distribution of the ensuing corrosion damage often indicates the cause. If the prevailing conditions cannot maintain passivity, the whole surface becomes active and attack is by uniform dissolution. If isolated sites become active in an otherwise passive surface they are selectively attacked. Local dissolution at the active sites is often very intense due to the establishment of *active/passive* cells, i.e., a galvanic effect with a small local anode at the active site stimulated by an extensive cathode provided by the surrounding passive metal surface.

These failure modes are now briefly indicated but they are considered again at greater depth within the contexts of mixed metal systems, stress-corrosion cracking and applications of stainless steels, where their effects are more clearly relevant.

3.2.5.1   General Breakdown of Passivity

Failure to establish or maintain the metal in the passive potential range leads to the uniform corrosion of nominally passivating metals. It implies mismatch of the metal or alloy with the environment to which it is exposed. For example, as discussed in Chapter 8, the hydrogen evolution reaction is incapable of imposing a corrosion potential in the passive ranges of stainless steels and the oxygen reduction reaction is essential to establish and maintain passivity. Thus, if a stainless steel is required to resist a non-oxidizing acid it is essential to maintain a sufficient concentration of dissolved oxygen. Assuming correct material selection, general failure by depassivation is often associated with inadvertent failure to replenish the oxygen as it becomes depleted.

For some metals, such as stainless steels, the passive condition is terminated at an upper potential, the *breakdown potential*, and higher potentials are designated *transpassive potentials*. General breakdown of passivity at a transpassive potential is not often a problem because the breakdown potential is usually above corrosion potentials normally encountered in service.

3.2.5.2   Local Breakdown of Passivity

Crevices

Small quantities of stagnant solutions inside crevices and under shielded areas on a passivated metal surface are selectively depleted of dissolved oxygen because of difficulty in replenishing it by diffusion from the bulk of the liquid. If, over a period, the oxygen content falls below a critical concentration in such a crevice, the passivity breaks down, establishing an active/passive cell, throwing intense attack on the small local anode thereby created.

Pitting

Pitting corrosion is intense chemical attack at dispersed points on an unshielded passive metal surface, forming pits that can perforate thin gauge metal. They are due to breakdown of passivity at very small isolated sites distributed over the metal surface, creating local active/passive cells. They can be initiated by heterogeneities in the metal surface due to minority phases in the microstructure of the metal as for aluminum alloys or by environmental agents as for stainless steels, notably halide and hypochlorite ions, in particular the chloride ion, that imposes severe limitations on the uses of passivating metals for service in seawater and chemical and food processing where the solutions to be resisted can contain chloride contents in excess of about 0.01 molar, equivalent to a 0.06 weight% solution sodium chloride. Disinfectants containing hypochlorites, ClO^-, are another hazard. If the chloride content is much above 0.1 molar, the tendency to pit may be overtaken by general depassivation. Susceptible metals include not only stainless steels and aluminum but also some copper-based alloys and mild steels in certain environmental conditions. The highly localized damage can render equipment unserviceable even if attack on the rest of the metal surface is negligible.

3.2.5.3 Mechanical Breakdown of Passivity

Passivity can be broken down by mechanical means to produce damaging effects, by stress-corrosion cracking, corrosion fatigue and erosion-corrosion dealt with in Chapter 5, but mentioned here for completeness. These effects apply to metals generally, whether passivating or not, but the presence of an initial passive condition is an influential factor.

Stress-corrosion cracking is the result of synergy between sustained mechanical stress and specific agents in the environment, producing cracks causing fracture. For passive metals, applied stress facilitates passivity breakdown at crack sites and prevents repassivation.

Fatigue cracking is caused by application of cyclic stress that first initiates and then propagates a crack. Both initiation and propagation stages are sensitive to environmental intervention, especially in aqueous media, in which event the phenomenon is said to be corrosion fatigue. The mechanical disturbance enhances localized anodic dissolution that concentrates the cyclic stress which in turn amplifies the mechanical disturbance and so on, setting up a synergistic cyclic mechanism. For passivating metals, the role of mechanical disturbance in stimulating anodic dissolution is to break down passivity during the crack initiation stage and to prevent it reforming at the crack root during propagation.

In erosion-corrosion, damage to passivated surfaces by high velocity or turbulent liquid flow enhances corrosion, especially if particulates are present.

3.2.6 Corrosion Inhibitors

Very low concentrations of solutes with particular characteristics can intervene with corrosion kinetics and thereby protect metals from corrosion and are described by the general term, *inhibitors*. Some occur naturally and others are introduced artificially as a strategy for corrosion control. Examples of their applications include:

1. Preserving existing iron or steel delivery systems for municipal water supplies.

2. Maintaining clean thermal transfer surfaces in closed heating and cooling circuits.

3. Protecting mixed metal systems.

4. Enhancing the protective value of paints.

All of these matters are considered where appropriate in Chapters 6 and 9 through 13.

Inhibitors intervene in corrosion kinetics in various ways. Some inhibit cathodic reactions, others inhibit anodic reactions and yet others, *mixed inhibitors*, do both. The detailed mechanisms by which the individual substances produce their effects can be quite complex and is the subject of extensive on-going research. Nevertheless, information on certain well-established principles is needed to apply inhibitors correctly.

3.2.6.1   Cathodic Inhibitors

Cathodic inhibitors produce adherent compact deposits on metal surfaces that suppress the cathodic reaction,

$$\tfrac{1}{2}O_2 + H_2O + 2e^- = 2OH^- \tag{3.3}$$

by creating a barrier to oxygen diffusion and preventing transfer of electrons from the metal.

The most familiar cathodic inhibitors are the solutes in hard water supplies that yield scale. In the natural environment, water containing carbon dioxide derived from the atmosphere can dissolve calcium carbonate from calcareous geological material over which it flows, yielding soluble calcium bicarbonate by the reaction described in Section 2.2.9:

$$CaCO_3\,(s) + CO_2(g) + H_2O(l) = Ca^{2+}_{(aq)} + 2HCO^-_{3(aq)} \tag{2.25}$$

Magnesium carbonate is often associated with calcium carbonate in nature and so some magnesium bicarbonate can be present, produced by an analogous reaction.

In the natural environment, these reactions do not adjust to equilibrium because of difficulty in precipitating the sparingly soluble carbonates, $CaCO_3$ and $MgCO_3$, in open waters. In consequence, hard waters are usually supersaturated by up to an order of magnitude with respect to the carbonates. When the water is abstracted, the surfaces of the systems into which it is transferred can assist nucleation, allowing the reactions to approach equilibrium by precipitating carbonates as a scale on the walls of pipes, tanks and ancillary equipment. The carbonates of certain other divalent cations, e.g., Zn^{2+} and Mn^{2+} are chemically equivalent and have similar calcite structures so that solutions of their bicarbonates can behave similarly as cathodic inhibitors. Since these cathodic inhibitors depend only on the chemistry of the water, they are applicable to all metals.

Soft waters do not naturally form protective scales but they can be induced to do so by treatment with lime and polyphosphates or silicates that can yield alternative deposits on iron and steel that may include iron oxides, calcium phosphates such as hydroxylapatite, $Ca_5(PO_4)_3OH$ or siliceous material. These materials are mixed inhibitors and are considered again in Section 3.2.6.2 below, in the context of anodic inhibitors.

3.2.6.2   Anodic Inhibitors

Anodic inhibitors suppress anodic reactions by assisting the natural passivation tendencies of metal surfaces or forming deposits that are impermeable to the metal ions. The attenuation of anodic activity can be confirmed experimentally by a comparison of polarization characteristics for metals in aqueous solutions with and without one of these substances. For convenience, anodic inhibitors can be considered in two groups, highly oxidizing, and less oxidizing.

Self-Sufficient Oxidizing Inhibitors

Sodium nitrite and sodium chromate are two of the most effective of all inhibtors. They are oxidizing to iron and are applied as inhibitors for ferrous metals in near neutral solutions. Nitrite, NO_2^- and chromate, CrO_4^{2-}, ions can impose redox potentials on the metal by reactions such as:

$$2NO_2^- + 6H^+ + 4e^- = N_2O(g) + 3H_2O \qquad (3.62)$$

for which:
$$E^{\ominus} = 1.29 \text{ V (SHE)}$$

and
$$2Cr^{VI}O_4^{2-} + 10H^+ + 6e^- = Cr^{III}{}_2O_3 + 5H_2O \qquad (3.63)$$

for which:
$$E^{\ominus} = 1.33 \text{ V (SHE)}$$

Even in small concentrations, the available potentials are sufficient to passivate iron by producing a coherent film of γ- Fe_2O_3. This is only a partial explanation for the action of chromates because they are also effective for metals such as aluminum that do not exhibit anodic passivation. Chromium can be identified in the surface film, indicating reinforcement of the existing passive film by the insoluble chromic oxide, $Cr^{III}{}_2O_3$, produced in Reaction 3.63. This topic is considered again in Section 6.4.3.1 in the context of chromate conversion coatings, especially when they are used as underlays to inhibit corrosion by moisture permeating through paints. The additions of sodium nitrite or sodium chromate used depend on water purity and are typically in the range 1 to 10 g dm^{-3} of solution.

Contamination with depassivating ions, e.g., chlorides increases the quantities needed.

Inhibitors Assisted by Dissolved Oxygen

Sodium polyphosphates, sodium silicates, sodium tetraborate (borax) and sodium benzoate are non-oxidizing inhibitors used to protect iron and steels in neutral and slightly alkaline waters; they are effective only when the solution contains dissolved oxygen. All of these substances ionize in aqueous solution yielding anions that are the active species at the iron surface with Na^+ counterions. Polyphosphate ions are poly-ions containing multiple phosphorous atoms, typically $P_3O_{10}^{5-}$. The tetraborate ion is also a poly-ion containing four boron atoms, $B_4O_7^{2-}$. The benzoate ion, $(C_6H_5)COO^-$ is a benzene ring with a carboxylic acid group. A feature that these apparently disparate substances have in common is that whereas their sodium salts are very soluble in water, they have the potential to form very insoluble iron(III) phases and hence to lay down deposits on an iron surface. The requirement for dissolved oxygen is probably associated with a need to promote some iron(II) to iron(III) in these phases. Polyphosphates, silicates, and tetraborates are used as inhibitors for very large volumes of water such as municipal supplies and concentrations in the range .000005 to 0.00002 g dm^{-3} are recommended. Sodium benzoate can be used only for small volumes, as in recirculating systems, because as much as 15 g dm^{-3} may be needed.

Certain other ions, less widely used, e.g., vanadates, tungstates and molybdates require dissolved oxygen to act as inhibitors, even though some of them, especially vanadates are associated with high redox potentials. They exist in aqueous solution as large polyions and the chemistry of equilibria between these ions and insoluble solids that they may produce on metal surfaces, such as iron(III) phases, is obscure, not well established and beyond the scope of the present text.

The benefits from many of the anodic inhibitors found to be effective on iron and steel are transferable to other metals such as aluminum, although the mechanisms may differ, but there are some reservations, e.g., sodium nitrite can attack lead solders and copper in mixed metal systems, such as automobile cooling systems.

Safe and Dangerous Inhibitors

The concept of *safe* and *dangerous* inhibitors is important in practical applications. Cathodic inhibitors are safe in the sense that if they are present at insufficient concentration, they simply fail to protect completely and do not stimulate corrosion on unprotected areas. In contrast, anodic inhibitors are dangerous because cathodic reactions are not suppressed on protected areas. Therefore, if by inadequate initial additions or subsequent poor maintenance the supply of inhibitor fails locally, corrosion is stimulated on the depleted areas because they become anodes in active/passive cells supported by large cathodic currents collected by the passive area.

3.3 Thermodynamics and Kinetics of Dry Oxidation

3.3.1 Factors Promoting the Formation of Protective Oxides

On first exposing a clean metal surface to air, a thin film of oxide forms within a few minutes, covering the metal and separating it from the hostile gaseous environment. It forms under the influence of an electric field developed by a quantum mechanical effect at the metal surface, explained in the Cabrera-Mott theory introduced briefly in Section 3.3.2, and virtually ceases to grow when it is typically 3–10 nm thick. Although thin, the film is almost impenetrable at ambient temperature and the protection it affords explains why many common metals, e.g., aluminum, zinc, copper, nickel, and even iron are virtually permanent in perfectly dry air at ambient temperature.

On raising the temperature, reaction resumes if the integrity of the film is breached or if metal and/or oxygen atoms can penetrate it by diffusion at significant rates. In general, metals and alloys can be assigned to one or the other of two classes according to whether or not their oxide films are non-protective.

1. Metals forming non-protective films: As the oxide thickens, differences between the volumes of oxide and the metal from which it is formed induce tensile or compressive stresses at the metal/oxide interface, that can crack or buckle the oxide, so that its protective value is diminished or destroyed. Examples are magnesium, with an oxide/metal volume ratio of 0.8, producing an oxide under tension and uranium, with a volume ratio of >3, producing such a voluminous oxide product that it fails to adhere.

2. Metals forming protective films: These metals develop and maintain continuous coherent oxide films adhering to the metal substrate and the oxidation rate is controlled by transport of the metal and/or oxygen through the film itself. The term, protective, is relative because if diffusion is rapid at only moderately elevated temperatures, as it is for metals such as iron and copper, the oxide grows into thick scales, progressively consuming the metal surface. Such metals are unsuitable for high temperature service. Oxidation-resistant metals and alloys, such as stainless

steels and alloys based on the nickel-chromium system are designed *inter alia* for their ability to maintain films sufficiently impermeable to the reacting species to limit the growth of the oxide film in severe high-temperature service conditions, such as those prevailing in electric heating elements or gas turbine engines.

When a metal forms a continuous oxide layer, the oxide protects the metal and the reaction of the metal with air is inhibited. Since the oxide layer separates the reactants, continued growth of the oxide is sustained by two processes:

1. Interface reactions by which the metal enters the oxide at the metal/oxide or oxygen enters at the oxide/atmosphere interface.
2. Diffusion of reactants through the oxide.

The rate of oxide growth is controlled by the slower of these two processes.

3.3.2   Thin Films and the Cabrera-Mott Theory

Thin films are characteristic of the almost imperceptible oxidation of engineering metals in dry air at ambient temperatures. Their essential features are:

1. The film growth is initially very rapid but virtually ceases when the films are only 3 to 10 nm thick, corresponding to oxide layers of 30 to 100 atoms.
2. The time-dependence of film thickening, expressed as *growth laws* is uncharacteristic of diffusion-controlled processes. Examples for different metals include logarithmic, inverse logarithmic and cubic functions of time.

From a practical point of view, these films are vitally important, because they confer initial protection on all new surfaces of engineering metals, pending the application of more permanent protection, if required. Theoretical treatments of how the thin films grow are beset by difficulties in acquiring direct information on the structures of such extraordinary thin materials and inconsistencies are found in the growth laws for thin oxides formed on different metals. They are within the remit of physicists, who have exercised great ingenuity in fitting experimentally observed growth laws to theories that are based on short-range electric fields at the metal surface. The basic ideas, due to Cabrera and Mott[*], are:

---

[*] Cited in "Further Reading."

1. Oxygen atoms are adsorbed at the oxide/atmosphere interface.
2. The film is so thin that electrons from the metal pass through it, either by thermionic emission or tunneling (a quantum mechanical effect).
3. The oxygen atoms capture these electrons, becoming anions.

This produces a very strong electric field across the thin film, e.g., 10^6 V cm^{-1}, if the film is 10 nm thick, that is mainly responsible for driving metal cations through the oxide. Film growth is limited by the short range of electrons available by thermionic emission or tunneling.

3.3.3 Thick Films, Thermal Activation and the Wagner Theory

The growth of thick oxide films at elevated temperatures is more relevant to technological interests in the application and fabrication of metals at high temperatures and it is easier to characterize. Oxidation in these circumstances can be approached by Wagner's theory, which applies to an oxide that is thick enough for equilibrium to be maintained locally at both the oxide/atmosphere and metal/oxide interfaces. According to the theory, the oxide grows by complementary reactions with oxygen at the oxide/atmosphere interface and with the metal at the metal/oxide interface and its rate of growth is controlled by the rate at which reacting species diffuse through the oxide via lattice defects.

The theory must be applied with discretion, because strictly it represents oxidation processes which fulfil the following implicit assumptions:

1. The oxide remains coherent and adherent to the metal substrate.
2. The growing oxide is a stable isotropic phase, uniform in thickness.
3. Diffusion is via lattice defects in the oxide and is not significantly supplemented by diffusion via alternative paths, e.g., crystal boundaries.

With these provisos, the theory can often explain the short-term oxidation behavior of engineering metals and alloys and it is useful even where it fails to do so, because incompatibility between theory and observation can sometimes identify factors which might otherwise be overlooked.

The mechanisms of the reactions are determined by the natures of defects in the oxides, as explained below. The illustrative equations are written using the conventional symbols given in Section 2.3.5.5.

3.3.3.1   Oxidation of Metals Forming Cation-Interstitial (n-type) Oxides

Atomic Mechanisms

The unit step extending the oxide lattice is adsorption of an oxygen atom from the atmosphere, its ionization to an anion, O^{2-}, and its co-ordination with a metal cation at the oxide/atmosphere interface. The electrons ionizing the oxygen atom are supplied from the conduction band of the oxide and the cation is extracted from the existing local population of interstitials. Taking zinc as a tangible example, the reaction is:

$$\tfrac{1}{2}O_2 + Zn^{2+} \bullet + 2e \bullet = ZnO \text{ (i. e. } Zn^{2+} + O^{2-}) \tag{3.64}$$

The complementary reaction at the metal/oxide interface is the entry of atoms from the metal into the oxide as cations and electrons, replenishing the interstitial species and consuming the metal:

$$Zn = Zn^{2+} \bullet + 2e \bullet \tag{3.65}$$

Addition of Equations 3.64 and 3.65 yields the overall oxidation reaction:

$$Zn + \tfrac{1}{2}O_2 = ZnO \tag{3.66}$$

Interface Equilibria

Recalling that the theory assumes local equilibrium at the bounding interfaces of the oxide layer, the equilibrium constant, K, for the reaction given in Equation 3.64 characterizes the defect populations at the oxide/atmosphere and metal/oxide interfaces:

$$K = \frac{1}{\left[a_{O_2}\right]^{1/2} \times a_{(Zn^{2+}\bullet)} \times \left[a_{(e\bullet)}\right]^2} \tag{3.67}$$

Since the defect population is small, $a_{(Zn^{2+}\bullet)} \propto n_{(Zn^{2+}\bullet)}$ and $a_{(e\bullet)} \propto n_{(e\bullet)}$, where $n(Zn^{2+}\bullet)$ and $n(e\bullet)$ are the numbers of interstitial zinc ions and interstitial electrons. Assuming unit activity for ZnO, Henrian activity for the species, $Zn^{2+}\bullet$ and $e\bullet$ and ideal behavior for oxygen gas:

$$K = \frac{1}{\left[p_{O_2}\right]^{1/2} \times n_{(Zn^{2+}\bullet)} \times \left[n_{(e\bullet)}\right]^2} \tag{3.68}$$

From stoichiometric considerations, $n_{(Zn^{2+}\bullet)} = \tfrac{1}{2}n_{(e\bullet)} = n$. Substituting, re-arranging and dropping subscripts, Equation 3.68 becomes:

$$n = p^{-1/6} \cdot k^{-1/3} \qquad (3.69)$$

where k includes K and coefficients relating n to $a_{(Zn^{2+}\bullet)}$ and $a_{(e\bullet)}$.

The number of interstitials at the oxide/atmosphere interface, n_{atm} is given by putting p equal to the prevailing oxygen pressure, p_{atm}:

$$n_{atm} = p_{atm}^{-1/6} \cdot k^{-1/3} \qquad (3.70)$$

and the number at the metal/oxide interface, n_0 by putting p equal to the dissociation pressure of the oxide, p_0:

$$n_0 = p_0^{-1/6} \cdot k^{-1/3} \qquad (3.71)$$

Oxidation proceeds if the oxygen pressure in the environment exceeds the dissociation pressure of the oxide, i.e., if $p_{atm} > p_0$, and since by Equations 3.70 and 3.71 $n_{atm} < n_0$, there is a concentration gradient down which cation interstitials diffuse from the metal/oxide interface, where they enter, to the oxide/atmosphere interface, where they are consumed.

3.3.3.2 Oxidation of Metals Forming Cation Vacancy (p-type) Oxides

Atomic Mechanisms

The unit step extending the oxide lattice is also the adsorption of an oxygen atom from the atmosphere, its ionization to an anion, O^{2-}, and its coordination with a metal cation at the oxide/atmosphere interface but the mechanism is determined by the ability of the oxide to accept cation vacancies. The cation is transferred from a lattice site, adding to the existing local population of lattice vacancies, and the electrons to ionize the oxygen are taken from the metal ion d-shell, creating electron holes. Taking nickel as an example, the reaction is:

$$\tfrac{1}{2}O_2 = NiO + Ni^{2+}\square + 2e\square \qquad (3.72)$$

The complementary partial reaction at the metal/oxide interface is the entry of atoms from the metal into the oxide, dissociating to cations and electrons, which annihilate cation vacancies and electron holes in the oxide:

$$Ni + Ni^{2+}\square + 2e\square(oxide) = 0 \qquad (3.73)$$

Addition of Equations 3.72 and 3.73 yields the overall oxidation reaction:

$$Ni + \tfrac{1}{2}O_2 = NiO \qquad (3.74)$$

Interface Equilibria

Using the same approach as for cation interstitial oxides, the equilibrium constant, K, for the reaction given in Equation 3.72 characterizes the defect populations at the oxide/atmosphere and metal/oxide interfaces:

$$K = \frac{a_{(Ni^{2+}\square)} \times \left[a_{(e\square)}\right]^2}{\left[a_{O_2}\right]^{1/2}} \tag{3.75}$$

whence:
$$n = p^{1/6} \cdot k^{1/3} \tag{3.76}$$

where:
$$n_{(Ni^{2+}\square)} = \tfrac{1}{2}n_{(e\square)} = n$$

and k includes K and coefficients relating n to $n_{(Ni^{2+}\square)}$ and $n_{(e\square)}$.

The number of cation vacancies at the oxide/atmosphere interface is:

$$n_{atm} = p_{atm}^{1/6} \cdot k^{1/3} \tag{3.77}$$

and the number at the metal/oxide interface is:

$$n_0 = p_0^{1/6} \cdot k^{1/3} \tag{3.78}$$

As before, oxidation proceeds if $p_{atm} > p_0$, i.e., if the oxygen pressure in the environment exceeds the dissociation pressure of the oxide and $n_{atm} > n_0$. This means that there is a concentration gradient from the oxide/atmosphere interface to the metal/oxide interface, down which cation vacancies diffuse *inwards* from the oxide/atmosphere interface, where they are created, to the metal surface where they are annihilated. This is equivalent to the transport of nickel ions *outwards* from the metal through the oxide to the oxide/atmosphere interface, replenishing those consumed in the formation of new oxide.

3.3.3.3 Oxidation of Metals Forming Anion Vacancy (n-type) Oxides

Atomic Mechanisms

Because the lattice defects permit the diffusion of oxygen but not the metal, the unit step extending the oxide lattice takes place at the at the metal/oxide interface by the entry of a metal atom, its ionization to a cation and its coordination with an oxygen anion. The anion is transferred from a lattice site, adding to the existing local population of lattice vacancies, and the excess electrons are added to the conduction band of the oxide. Taking titanium as an example, the reaction is:

$$Ti = TiO_2 + 2O^{2-}\square + 4e\bullet \tag{3.79}$$

The complementary process at the oxide/atmosphere interface is the entry of adsorbed oxygen atoms, which annihilate anion vacancies and remove excess electrons:

$$TiO_2 + O_2 + 2O^{2-}\square + 4e\bullet = TiO_2(\text{fewer defects}) \qquad (3.80)$$

Interface Equilibria

Changes in such a small defect population are virtually without effect on the activity of the oxide, so that $a_{(TiO_2)} = a_{(TiO_2)}$ (with fewer defects) ≈ 1 and can be omitted from the activity quotient so that the equilibrium constant for the reaction given in Equation 3.80 is:

$$K = \cfrac{1}{\left[a_{O_2}\right] \times \left[a_{O^{2-}}\right]^2 \times \left[a_{(e\bullet)}\right]^4} \qquad (3.81)$$

Proceeding in the same way as for cation interstitial and cation vacancy oxides:

$$n = k^{-1/6} \cdot p^{-1/6} \qquad (3.82)$$

producing a concentration gradient down which anion vacancies diffuse from the metal/oxide interface to the oxide/atmosphere interface.

3.3.3.4 Oxidation of Metals Forming Stoichiometric Ionic Oxides

Two common metals, magnesium and aluminum form stoichiometric oxides in which the only significant defects are Schottky pairs of cation and anion vacancies. The only sustainable transport mechanism through a homogeneous oxide growing on the hypothetical perfectly pure metals is the slow cooperative diffusion of the pairs, by *intrinsic ionic conductivity*.

However, the oxides are so sensitive to impurities that the oxides forming on even high purity commercially-produced metals exhibit a degree of *extrinsic conductivity*. This arises because impurity atoms can give electron energy levels introducing some electronic conductivity, which can support independent diffusion of one of the ion species. These impurities are usually traces of other metals but MgO is also susceptible to the impurity anions OH⁻, ions injected at the oxide/atmosphere interface by atmospheric water vapor.

Oxygen diffuses via the $O^{2-}\square$ vacancies much more slowly than magnesium does via the $Mg^{2+}\square$ vacancies so that magnesium is the independently diffusing species in the impure oxide. The electron carriers introduced by the impurities permit limited oxidation mechanisms similar to those for cation vacant oxides.

Because the concentrations of impurities introducing electronic conductivity is low, oxidation is very slow. For example, when MgO forms a protective layer, as it can when a very thin layer is formed in dry air on magnesium metal or when it is formed on some aluminum-magnesium alloys for which the oxide/metal volume ratio is favorable, the only significant transport of magnesium is through the disorder at the oxide crystal boundaries.

3.3.3.5 Time and Temperature Dependence of Diffusion-Controlled Oxidation

When oxidation is diffusion-controlled as assumed in the Wagner theory, the rate at which the oxide thickens is proportional to the rate at which the diffusing species, e. g., Zn in ZnO or Ni in NiO, is transported through the oxide, which is inversely proportional to the oxide thickness:

$$\frac{dx}{dt} = \frac{k}{x} \tag{3.83}$$

where x is the instantaneous oxide thickness after the elapse of time, t.
Integration yields a parabolic time law:

$$x = 2kt^{\frac{1}{2}} \tag{3.84}$$

Since it is diffusion-controlled, the dependence of oxidation rate on temperature, T, is derived from that of the diffusion coefficient, D, for the diffusing species that, over a limited temperature range, follows an Arrhenius type relation:

$$\ln D = \frac{-\Delta H^*}{RT} + \text{constant} \tag{3.85}$$

leading to a relation between the parabolic rate constant, k, and T of the form:

$$\log k = \frac{-A}{T} + B \tag{3.86}$$

Ample experimental verification of Equations 3.84 and 3.86 shows that the short-term oxidation of many metals, including iron, nickel, and copper is consistent with the theory.

3.3.3.6 Correlations with Other Observations

Electrical Conductivities of n-type and p-type oxides

n in Equations 3.69, 3.76 and 3.82 is a measure not only of the number of lattice defects but also the number of charge carriers (interstitial electrons or electron holes) that confer semi-conducting properties on the oxide. These equations predict that the electrical conductivity of semi-conducting oxides are functions of the oxygen pressure in the environment. The function is direct for p-type oxides, e.g., $n \propto p^{1/6}$ for NiO and inverse for n-type oxides, e.g., $n \propto p^{-1/6}$ for ZnO. Experimental evidence for one or the other of these relationships can indicate whether a particular oxide is n-type or p-type.

Inert Marker Experiments

When the metal is the diffusing species, the oxide/atmosphere interface can be identified as the growth interface by observing that experimental inert markers placed on the metal surface become enveloped in the growing oxide. Conversely, when it is oxygen that diffuses so that the metal/oxide interface is the growth interface, the markers remain on the oxide surface.

Development of Voids at Metal/Oxide Interfaces

With prolonged oxidation, the entry of metal atoms into a cation excess or cation vacant oxide can inject sufficient vacant lattice sites into the metal to condense as voids that are sometimes observed as at the metal/oxide interface because atomic disorder there provides the easiest site for them to nucleate.

3.3.3.7 Effects of Impurities

In real oxidation, impurities entering the oxides from impurities in the metal or contaminants in the atmosphere can have pronounced effects on the oxidation rate out of all proportion to the quantities present.

Valency Effects in Semi-conducting oxides

The introduction into an oxide of impurity cations with an oxidation state different than that of the host metal disturbs the balance of lattice and electronic defects with effects on the oxidation rate that depend on whether the oxide is n-type or p-type.

1. n-type oxides — Example ZnO:

   The replacement of some host metal cations, Zn^{2+}, by impurity cations in a *lower* oxidation state, e.g., Li^+, produces a deficit of positive charge on the lattice that is compensated by adjustment of the defect structure. The populations of interstitial cations, $n_{(Zn^{2+}\bullet)}$ and electrons, $n_{(e\bullet)}$ are interdependent by the equilibrium described in Equation 3.68, so that the adjustment is by both an increase in the population of interstitial cations and a corresponding reduction in the population of interstitial electrons, $e\bullet$. The additional interstitial cations improve the transport of Zn^{2+} ions, enhancing the oxidation rate. Conversely, the replacement of host cations by impurity cations in a *higher* oxidation state, say Cr^{3+}, introduces an excess of positive charge that is compensated in the opposite sense, with more interstitial electrons and fewer interstitial cations, diminishing the rate of oxidation.

2. p-type oxides — Example NiO:

Similar considerations apply to a p-type oxide but the consequences are reversed. A deficit of positive charge introduced by impurity cations in a *lower* oxidation state is compensated by a lower population of cation vacancies, $n_{(Ni^{2+} \square)}$ and a higher population of electron holes, $n_{(e\square)}$, under the control of the equilibrium described in Equation 3.75. This curtails the transport of Ni^{2+} through the oxide, depressing the oxidation rate. Conversely, an excess of positive charge introduced by impurity cations in a *higher* oxidation state is compensated by a higher population of cation vacancies, $n_{(Ni^{2+} \square)}$ and a lower population of electron holes, $n_{(e\square)}$ enhancing the oxidation rate.

The acceleration or retardation can be striking. An addition of 0.5% of chromium to nickel or of lithium to zinc can increase the oxidation rates by orders of magnitude. These effects are collectively known as *Hauffe's valency rules*, for their originator.

Catastrophic Oxidation Induced by Impurities

This aptly named effect describes accelerated oxidation associated with the destruction of protection by oxides when liquid phases are introduced by particular impurities contributed from the atmosphere or the metal. The oxides of several metals and other impurities are liquid at relatively low temperatures, especially vanadium, V_2O_5, boron, B_2O_3 and molybdenum, MoO_3, that have melting points of 690, 450, and 795 °C respectively. The melting points of liquid phases may be lowered further by the formation of eutectics with oxides contributed by the metal or with other substances such as sodium sulfate formed from sulfur in oxidizing atmospheres. A potent source of these undesirable impurities is the ash of fossil fuels in products of combustion that impinge on metals. Most coal, fuel oils, and gas contain sulfur and the ash of some fuel oils, is particularly rich in vanadium oxide. Another source is vanadium and molybdenum from the alloy content of special steels but this can be avoided by correct material selection.

3.3.3.8 Loss of Integrity of Protective Oxides

Oxide films formed on some metals may be only partially or temporarily protective, leading to *paralinear oxidation* or *breakaway oxidation*. Both effects are due to the loss of integrity of an initially protective oxide layer.

Paralinear Oxidation

As the name suggests, the oxidation rate at first diminishes with time by an apparent parabolic rate law until it reaches a critical value, after which it remains constant. It can be explained as the consequence of two simultaneous processes:

1. The oxide develops as a compact protective *barrier layer* with a diminishing growth rate consistent with the Wagner theory.
2. The outer surface of the barrier layer transforms at a constant rate into an unprotective, e.g., cracked or porous, form.

When the oxide is thin, the first process dominates but eventually a steady state is reached at which a linear rate-control is exercised by the second process, which restrains growth of the barrier layer to a limiting thickness.

Breakaway Oxidation

In breakaway oxidation, the oxide begins to grow at a rate diminishing with time, as if to establish a protective layer, but at critical oxide thickness it accelerates. After acceleration, the growth rate may be constant, as for magnesium at 500°C or it may rise to a maximum and then diminish, as for tungsten at 700°C. Among factors which can initiate this behavior is the development of lateral stresses in an oxide, due to unfavorable metal-substrate/oxide volume ratios, causing it to crack or shear, exposing unprotected metal to the atmosphere.

3.3.4 Selective Oxidation of Components in an Alloy

One or more of the components of some alloy systems can oxidize selectively and the effect can be exploited in alloy formulation to provide protective films.

3.3.4.1 Principles

The expected oxidation product is the oxide of whichever component can reduce the oxygen activity at the alloy surface to the lowest value. This depends on the activities of the components and therefore on the alloy composition. Consider two metals A and B, both of which can oxidize, in a hypothetical alloy system with the following simplifying characteristics:

1. The metal system is a complete series of solid solutions.
2. Only the simple oxides of the metals, AO and BO can form.
3. The oxides are immiscible.

The oxidation reaction for metal A is:

$$2A + O_2 = 2AO \tag{3.87}$$

and the activities of the reactants are related by the equilibrium constant:

$$K_A = \frac{\left(a_{AO}\right)^2}{\left(a_A\right)^2 \times \left(a_{O_2}\right)} \tag{3.88}$$

For the pure metal in equilibrium with its oxide, a_A and a_{AO} are both unity, giving a_{O_2} a unique equilibrium value, $[a_{O_2}]_{AO}$, where:

$$[a_{O_2}]_{AO} = 1/K_A \tag{3.89}$$

but when metal A is diluted in an alloy with metal B, a_A is reduced to a value depending on composition:

$$a_{A\ IN\ ALLOY} = \gamma_A \cdot X_A \tag{3.90}$$

where: X_A = mole fraction of metal A in the alloy
          γ_A = activity coefficient for metal A (note $\gamma_A \rightarrow 1$ as $X_A \rightarrow 1$)
and the value of a_{O_2} in equilibrium with the metal and its oxide is given by:

$$K_A = \frac{\left(a_{AO}\right)^2}{\left(a_{A\ IN\ ALLOY}\right)^2 \times \left(a_{O_2}\right)} \tag{3.91}$$

$$= \frac{1}{\left(\gamma_A \cdot X_A\right)^2 \cdot a_{O_2}} \tag{3.92}$$

Substituting for K_A from Equation 3.89 and rearranging:

$$a_{O_2} = \frac{[a_{O_2}]_{AO}}{\left(\gamma_A \cdot X_A\right)^2} \tag{3.93}$$

Equation 3.93 gives the minimum oxygen activity to form AO at the metal surface.

Applying the same argument to the oxidation reaction for metal B:

$$2B + O_2 = 2BO \tag{3.94}$$

$$a_{O_2} = \frac{\left[a_{O_2}\right]_{BO}}{\left(\gamma_B \cdot X_B\right)^2} \tag{3.95}$$

Equation 3.95 gives the minimum oxygen activity to form BO at the metal surface.

Equations 3.93 and 3.95 show that as the composition of a binary alloy changes progressively from $(X_A = 1, X_B = 0)$ to $(X_A = 0, X_B = 1)$:

a_{O_2} for AO formation rises from $[a_{O_2}]_{AO}$ to infinity and
a_{O_2} for BO formation falls from infinity to $[a_{O_2}]_{BO}$.

This is shown in Figure 3.14. Theoretically, for compositions between pure metal A and a critical composition, $X_{CRITICAL}$, AO is the exclusive oxidation product because it buffers a_{O_2} at a value below that needed to form BO. Similarly, for compositions between $X_{CRITICAL}$ and pure metal B, BO is the exclusive product because it buffers a_{O_2} at a value below that needed to form AO.

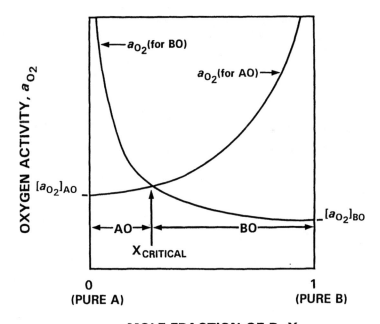

FIGURE 3.14
Oxygen activities in equilibrium with the oxides AO and BO formed on binary alloys in the system A-B, as functions of composition.
$[a_{O_2}]_{AO}$, Oxygen activity in equilibrium with oxide AO and pure metal A.
$[a_{O_2}]_{BO}$, Oxygen activity in equilibrium with oxide BO and pure metal B.

3.3.4.2   *Oxidation of Alloys Forming Complex Oxides*

The possible oxidation products for some technically important alloy systems include spinel-type oxides containing two metals in addition to the oxides of the component metals individually. This happens when there is a composition range within which the lowest oxygen activity is produced by two metals in combination. Examples of systems with solid solutions exhibiting this behavior include:

1. Aluminum-magnesium alloys.
2. Nickel-chromium alloys, from which nickel-base superalloys were developed.
3. Iron-chromium alloys, which form the basis for stainless steels.

The Aluminum-Magnesium-Oxygen system

As an example consider a particular system, the aluminum-magnesium system. The possible oxidation products are MgO, Al_2O_3, and the spinel, $MgAl_2O_4$. The same principles apply as described Section 3.3.4.1, extended to include the spinel. Assuming that the oxides are pure, $a_{Al_2O_3}$, a_{MgO}, and $a_{MgAl_2O_4}$ are all unity and can be neglected in expressions for the equilibrium constants. Writing the equations conventionally for one mole of oxygen:

For the reaction: $4/3[Al]_{\text{SOLUTION IN ALLOY}} + O_2 = 2/3Al_2O_3$

$$K_{Al_2O_3} = \frac{1}{\left(a_{Al}\right)^{4/3} \times a_{O_2}} \qquad (3.96)$$

For the reaction: $2[Mg]_{\text{SOLUTION IN ALLOY}} + O_2 = 2MgO$

$$K_{MgO} = \frac{1}{\left(a_{MgO}\right)^{2} \times a_{O_2}} \qquad (3.97)$$

For the reaction: $\tfrac{1}{2}[Mg]_{\text{SOLUTION IN ALLOY}} + [Al]_{\text{SOLUTION IN ALLOY}} + O_2$
$$= \tfrac{1}{2}MgAl_2O_4$$

$$K_{MgAl_2O_4} = \frac{1}{\left(a_{Mg}\right)^{1/2} \times \left(a_{Al}\right) \times a_{O_2}} \qquad (3.98)$$

Example 10

Show which of the oxides, Al_2O_3, MgO or $MgAl_2O_4$ is most stable for a binary alloy of aluminum with 4 weight % magnesium in solid solution at 500°C.

SOLUTION:
The most stable oxide is the one that is in equilibrium with the lowest oxygen activity. The equilibrium oxygen activities are calculated using Equations 3.96, 3.97, and 3.98. The equilibrium constants, $K_{Al_2O_3}$, K_{MgO}, and $K_{MgAl_2O_4}$ are obtained by application of the Van't Hoff isobar to the Gibbs free energies of formation, $\Delta G^{\ominus}$, for the oxides:

$$\ln K = \frac{-\Delta G^{\ominus}}{RT} \tag{3.99}$$

Standard Gibbs free energies are given as functions of temperature in standard references, e.g., Kubaschewski and Alcock*. Bhatt and Garg* give values for the activities of aluminum and magnesium in aluminum - magnesium binary alloys; for an aluminum - 4 weight % magnesium binary alloy at 500°C, the values are:

$$a_{Mg} = 0.088 \qquad a_{Al} = 0.96$$

The calculations are as follows:

FOR Al$_2$O$_3$ AT 500 °C (773 K)

$$\Delta G^{\ominus} = -968529 \text{ J mol}^{-1} \text{ of oxygen}$$

Applying the Van't Hoff isobar, $\ln K = -(-968529)/(8.314 \times 773) = 150.7$

whence: $\qquad K = 2.8 \times 10^{65}$

From Equation 3.96: $\qquad K = 1/[(a_{Al})^{4/3} \times a_{O_2}]$

$$a_{O_2} = 1/[K \times (a_{Al})^{4/3}] = 1/(2.8 \times 10^{65} \times 0.947) = \mathbf{3.83 \times 10^{-66}}$$

FOR MgO AT 500°C (773 K)

$$\Delta G^{\ominus} = -1043135 \text{ J mol}^{-1} \text{ of oxygen}$$

Applying the Van't Hoff isobar, $\ln K = -(-1043135)/(8.314 \times 773) = 162.3$

whence: $\qquad K = 3.1 \times 10^{70}$

From Equation 3.97: $\quad K = 1/[(a_{Mg})^2 \times a_{O_2}]$

$$a_{O_2} = 1/[K \times (a_{Mg})^2] = 1/(3.1 \times 10^{70} \times 7.744 \times 10^{-3}) = \mathbf{4.2 \times 10^{-69}}$$

---

* Kubaschewski, O. and Alcock, C. B., *Metallurgical Thermochemistry*, Pergamon Press, New York, 1979; Bhatt and Garg, *Met. Trans.*, B, 7, p. 227, 1976.

FOR MgAl₂O₄ AT 500 °C (773 K)

$$\Delta G^{\ominus} = -1007094 \text{ J mol}^{-1} \text{ of oxygen}$$

Applying the Van't Hoff isobar, $\ln K = -(-1007094)/(8.314 \times 773) = 156.7$

whence: $K = 1.14 \times 10^{68}$

From Equation 3.98: $K = 1/[a_{O_2} \times (a_{Al}) \times (a_{Mg})^{\frac{1}{2}}]$

$a_{O_2} = 1/[K \times (a_{Al}) \times (a_{Mg})^{\frac{1}{2}}] = 1/(1.14 \times 10^{68} \times 0.96 \times 0.297) = \mathbf{3.08 \times 10^{-68}}$

Hence MgO is the most stable oxide for the conditions considered beause it is in equilibrium with the lowest oxygen activity.

Figure 3.15 gives plots of oxygen activities at 500°C in equilibrium with the three oxides for all compositions of aluminum-magnesium alloys in the range 0 to 10 weight % magnesium (equivalent to $X_{Mg} = 0.1$), calculated as in Example 10, using the appropriate values of a_{Mg} and a_{Al} given by Bhatt and Garg.

The plots for $[a_{O_2}]_{MgAl_2O_4}$, $[a_{O_2}]_{MgO}$, and $[a_{O_2}]_{Al_2O_3}$ intersect, defining *two* critical compositions, 0.16 % magnesium and 1.05 % magnesium, dividing the composition axis into three ranges:

% Magnesium	Expected oxidation product
1. < 0.16 %	Al₂O₃
2. 0.16 % to 1.05 %	MgAl₂O₄
3. > 1.05 %	MgO

This result is consistent with the practical observation that magnesium oxide is formed in preference to alumina on aluminum-magnesium alloys with high magnesium contents during hot-working and heat-treatments in the course of commercial manufacture.

The Nickel-Chromium-Oxygen System

The nickel-chromium-oxygen system also includes two single oxides and a spinel, i.e., NiO, Cr₂O₃ and NiCr₂O₄. The theoretical treatment is essentially similar to that described for the aluminum-magnesium-oxygen system.

The Iron-Chromium-Oxygen System

The iron-chromium-oxygen system resembles the aluminum-magnesium-oxygen and nickel-chromium-oxygen systems but is complicated by five oxides:

1. A series of single oxides from the iron-oxygen system, FeO(wüstite)/Fe$_3$O$_4$(magnetite)/Fe$_2$O$_3$(hematite) described in Chapter 7.
2. The single oxide from the chromium-oxygen system: Cr$_2$O$_3$.
3. The spinel, FeCr$_2$O$_4$ (chromite).

If attention is confined to the oxides that form adjacent to the metal surface, i.e., FeO, FeCr$_2$O$_4$ and Cr$_2$O$_3$, similar conclusions are reached for the expected oxidation products as for the other systems, but kinetic factors can influence the compositions of the scales.

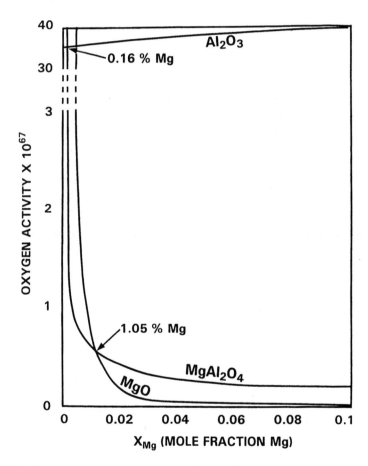

FIGURE 3.15
Oxygen activity in equilibrium with Al$_2$O$_3$, MgO, and MgAl$_2$O$_4$ on aluminum-magnesium alloys as functions of magnesium content. Note change of scale for oxygen activity in equilibrium with Al$_2$O$_3$. Critical compositions expressed as magnesium %.

3.3.4.3   Kinetic Considerations

Diffusion of Oxidizing Species in the Metal

The foregoing considerations apply to the metal composition immediately adjacent to the oxide, i.e., the composition at the extreme surface of the metal. However, the selective oxidation of one component of a binary alloy reduces its concentration at the metal surface, shifting the local alloy composition towards the critical composition at which another component oxidizes. Whether or not the critical composition is reached depends on how far the initial composition of the alloy is from the critical value and how rapidly the depleted composition can be restored by diffusion from the bulk of the metal. In practice, there is a range of composition around the critical composition within which both oxides form, giving a two-phase oxide structure.

Nucleation

The most stable oxide is sometimes difficult to nucleate at the metal surface, allowing the preferential nucleation of a less stable oxide. This can happen for an oxide with a structure that is epitaxially incompatible with the underlying metal or for a spinel with a stoichiometric ratio different from the ratio of different metal atoms in the alloy on which it is favored. This is considered in Chapter 8 within the context of the practical oxidation resistance of stainless steels. These non-equilibrium structures are often temporary and revert to equilibrium structures, given time. Meanwhile, the effects on oxidation rates can be very significant.

Further Reading

Glasstone, S., *Thermodynamics for Chemists*, Van Nostrand, New York, 1947.

Lewis, G. N., Randall, M., Pitzer, K. S. and Brewer, L., *Thermodynamics*, McGraw-Hill, New York, 1961.

Bodsworth, C., *The Extraction and Refining of Metals*, CRC Press, Boca Raton, Fl, 1994.

Davies, C. W., *Electrochemistry*, George Newnes, London, 1967, Chap. 12.

Ives, D. J. G., and Janz, G. J., *Reference Electrodes*, Academic Press, London, 1961.

Pourbaix, M., *An Atlas of Electrochemical Equilibria*, Cebelcor, Brussels, 1965.

Hinshelwood, C. N., *Kinetics of Chemical Change*, Clarendon Press, Oxford, 1940.

Bokris, J. O'M. and Reddy, *Modern Electrochemistry*, Plenum Press, New York, 1970, Chap. 8.

Evans, U. R., *The Corrosion and Oxidation of Metals*, Edward Arnold, London, 1971.

Betts, A. J. and Boulton, L. H., Crevice corrosion: review of mechanisms, modelling and mitigation, *Br. Corros. J.*, 28, 279, 1993.

France, W. D., *Crevice Corrosion of Metals*, ASTM-STP516, American Society for Testing and Materials, p. 164, 1972.

Kolotyrkin, Y. M., Electrochemical behaviour of metals during anodic and chemical passivation in electrolytic solutions, *Proc. 1st Inter. Symp. Corrosion*, Butterworths, London, 1962.

Uhlig, H. H., Electron configuration and passivity in alloys, *Z. Electrochem.*, 62, 700, 1958.

Rosenfeld, I. L., *Localized Corrosion*, National Association of Corrosion Engineers, Houston, 1974, p. 373.

Fontana, M. G. and Green, N. D., *Corrosion Engineering*, McGraw-Hill, New York, 1967.

Sendriks, A. J., *Corrosion of Stainless Steels*, John Wiley, New York: 1979, Chaps. 4 and 5.

Hatch, J. E. (Ed.), *Aluminum: Properties and Physical Metallurgy*, American Society for Metals, 1984, Chap. 7.

Proc. 8th European Syposium on Corrosion Inhibitors, The Institute of Materials, London, 1996.

Kubaschewski, O. and Hopkins, B. E., *Oxidation of Metals and Alloys*, Butterworths, London, 1962, Chap. 2.

Wagner, C., *Seminar on Atom Movements*, ASM, Cleveland, 153, 1951.

Hauffe, K. and Vierk, A. L., *Z. Phys Chem.*, 196, 160, 1950.

Problems for Chapter 3

1. Using the standard approximation, $e^x = 1 + x$ for small values of x, show that for small overpotentials, η, the Butler-Volmer equation simplifies to $i = i_0 F \eta / RT$. If $\beta = 0.5$, $z = 1$, and $T = 298$ K, show that this approximation introduces an error of $< 1\%$ for $\eta < 0.01$ V.

2. The standard electrode potentials for the following reactions in alkaline solutions are:

$$2H_2O + 2e^- = H_2 + 2OH^- \qquad E^{\ominus} = -0.827 \text{ V (SHE)} \qquad (1)$$

$$\tfrac{1}{2}O_2 + H_2O + 2e^- = 2OH \qquad E^{\ominus} = +0.401 \text{ V (SHE)} \qquad (2)$$

Show that these values are compatible with the standard electrode potentials for the corresponding reactions in acid solutions:

$$2H^+ + 2e^- = H_2 \qquad E^{\ominus} = 0.000 \text{ V SHE.} \qquad (3)$$

$$\tfrac{1}{2}O_2 + 2H^+ + 2e^- = H_2O \qquad E^{\ominus} = +1.228 \text{ V SHE.} \qquad (4)$$

3. This problem illustrates that information given in Pourbaix diagrams applies only to the species considered and can be misleading if other information is ignored.

   Using the following data for 25°C:

$$Cd^{2+} + 2e^- = Cd \qquad\qquad E^{\ominus} = -0.403 \text{ V (SHE)} \qquad (1)$$

$$Cd(OH)_2 + 2H^+ + 2e^- = Cd + 2H_2O \quad E^{\ominus} = +0.017 \text{ V (SHE)} \qquad (2)$$

$$Cd(OH)_2 + 2H^+ = Cd^{2+} + 2H_2O \quad (a_{Cd^{2+}})/(a_{H^+})^2 = 10^1 \qquad (3)$$

Construct a partial Pourbaix diagrams for the cadmium-water system for $a_{Cd^{2+}} = 10^{-6}$. Use it to show that the information it gives is inconsistent with the excellent corrosion resistance of the metal in neutral media. (Note: the information not available from the diagram is that the metal probably passivates by interaction with atmospheric carbon dioxide, forming the stable insoluble carbonate, $CdCO_3$).

4. Domestic bleach is produced by treating sodium hydroxide with chlorine gas:

$$2NaOH + Cl_2(g) = NaClO + NaCl + H_2O \qquad \text{(Reaction 1)}$$

The active agent is the hypochlorite ion, ClO^- that exerts its effect by the reaction:

$$ClO^- + H_2O + 2e^- = Cl^- + 2OH^- \qquad \text{(Reaction 2)}$$

for which: $\qquad\qquad E^{\ominus} = 0.89 \text{ V (SHE)}$

Calculate the pH of a particular solution containing 0.01 M quantities of each of ClO^-, Cl^-, and OH^- ions and the potential it imposes on a metal. Referring these values to the diagrams given in Figures 3.2 to 3.8, which of the following metals would you eliminate immediately for use in storage vessels for the solution (a) iron (or steel) (b) aluminum (c) zinc (d) copper (e) tin (f) nickel?

5. Copper oxidizes at 700°C in oxygen at 1 atmosphere pressure, yielding an adherent layer of the metal deficit p-type oxide, Cu_2O. Derive expressions for the lattice and electronic defect populations in the oxide at the oxide/oxygen and metal/oxide

interfaces. Hence estimate by how much an increase in the oxygen pressure to 100 atmospheres influences (a) the rate of oxidation of the metal and (b) the electrical conductivity of the oxide.

6. Determine conditions for which iron is theoretically stable in water at 25°C with respect to $a_{Fe2+} = 10^{-6}$. Suggestion: Consult the iron-water Pourbaix diagram, given in Figure 3.2 and consider applying a high pressure to the system.

Solutions to Problems for Chapter 3

Solution to Problem 1.
The solution yields the low-field approximation for the Butler-Volmer equation:

$$i = i_0 \left[\exp \frac{(1 - \alpha)zF\eta}{RT} - \exp \frac{-\alpha zF\eta}{RT} \right]$$

If η is small enough, the standard substitution of the form $e^x = 1 + x$ can be made for both exponentials, yielding:

$$i \approx i_0 \left[1 + \frac{(1 - \alpha)zF\eta}{RT} - 1 - \frac{-\alpha zF\eta}{RT} \right] = \frac{i_0 F\eta}{RT}$$

Applying the Butler-Volmer Equation for $\eta = 0.01$ V:

$$i = i_0 \left[\exp \frac{(1 - 0.5) \times 1 \times 96400 \times 0.01}{8.314 \times 298} - \exp \frac{-(0.5 \times 1 \times 96400 \times 0.01)}{8.314 \times 298} \right]$$

$$= i_0 \left[\exp 0.195 - \exp -(0.195) \right] = 0.392 \, i_0$$

Applying the low field approximation:

$$i = \frac{i_0 \times 96400 \times 0.01}{8.314 \times 298} = 0.389 i_0$$

The difference between the two values is $< 1\%$ and diminishes as $\eta \to 0$.

Solution to Problem 2.

The standard state for Reaction 1, to which $E^{\ominus}$ applies is $a_{OH^-} = 1$. The standard state for Reaction 3 is $a_{H^+} = 1$, for which $a_{OH^-} = 10^{-14}$, because $K_w = 10^{-14}$ at 298 K. Applying the Nernst Equation to Reaction 1 in the standard state for Reaction 3 yields:

$$E = E^{\ominus} - \frac{0.0591}{z} \log J$$

$$= -0.827 - \frac{0.0591}{2} \log\left(10^{-14}\right)^2$$

$$= 0.000 \text{ V}\,(\text{SHE})$$

which is the value of $E^{\ominus}$ for Reaction 3. Thus the standard electrode potentials for Reactions 1 and 3 are compatible.

Similarly, applying the Nernst Equation to Reaction 2 in the standard state for Reaction 4 yields:

$$E = E^{\ominus} - \frac{0.0591}{z} \log J$$

$$= +0.401 - \frac{0.0591}{2} \log\left(10^{-14}\right)^2$$

$$= 1.228 \text{ V}\,(\text{SHE})$$

which is the value of $E^{\ominus}$ for Reaction 4. Thus the standard electrode potentials for Reactions 2 and 4 are also compatible.

Solution to Problem 3.

Proceeding as for the calculation in Example 3, given in the Appendix to Chapter 3:

Reaction 1: $Cd^{2+} + 2e^- = Cd$ $E^{\ominus} = -0.403$ V SHE.

Applying the Nernst Equation: $E = -0.403 - \dfrac{0.0591}{2} \cdot \log\dfrac{1}{a_{Cd^{2+}}}$

$$E = -0.403 + 0.0296 \log\left(a_{Cd^{2+}}\right)$$

For: $a_{Cd^{2+}} = 10^{-6}$: $E = -0.581$

Reaction 2: $Cd(OH)_2 + 2H^+ + 2e^- = Cd + 2H_2O$ $\quad E^\ominus = +0.017$ V SHE.

Applying the Nernst Equation: $\quad E = +0.017 - \dfrac{0.0591}{2} \cdot \log \dfrac{1}{\left(a_{H^+}\right)^2}$

$$E = +0.017 - 0.0591 \text{ pH}$$

Reaction 3: $\qquad\qquad Cd(OH)_2 + 2H^+ = Cd^{2+} + 2H_2O$

$$(a_{Cd^{2+}})/(a_{H^+})^2 = 10^{14}$$

Taking logarithms: $\quad \log(a_{Cd^{2+}}) - 2\log(a_{H^+}) = 14$

Re-arranging: $\quad \text{pH} = \{-\log(a_{Cd^{2+}}) + 14\}/2$

For: $\qquad\qquad a_{Cd^{2+}} = 10^{-6}: \text{pH} = 10$

Plotting lines on a partial Pourbaix diagram using the equations obtained for Reactions 1, 2, and 3 reveals a domain of stability for Cd^{2+} ions for $E > 0.558$ V (SHE) and pH < 10. Hence, by the conventional criterion for corrosion, i.e., $a_{Cd^{2+}} > 10^{-6}$, Cd^{2+} ions are stable with respect to cadmium metal within the domain of stability for water for all pH < 10. The metal might therefore be expected to corrode in near neutral waters but this is inconsistent with experience that cadmium resists humid air and natural waters well enough to serve as a protective coating for steels. An explanation is that in constructing the diagram, no information was included for species derived from carbon dioxide, that is a ubiquitous component of air and natural waters; the good performance of cadmium is usually attributed to passivation by the very insoluble compound, cadmium carbonate.

Solution to Problem 4.

0.01 M solutions are fairly dilute and a_{ClO^-}, a_{Cl^-}, and a_{OH^-} can all be assumed to be 0.01, i.e., 10^{-2}, without serious error.

Since $K_w = 10^{-14}$ at 298 K, $a_{H^+} = 10^{-12}$ and so pH = 12.
Applying the Nernst equation to Reaction 2:

$$E = E^\ominus - \frac{0.0591}{z} \log \frac{a_{Cl^-} \times \left(a_{OH^-}\right)^2}{a_{ClO^-}}$$

$$= +0.89 - \frac{0.0591}{2} \log \frac{(0.01) \times (0.01)^2}{0.01}$$

$$= +1.01 \text{ V (SHE)}$$

The coordinates, pH = 12, E = 1.01 V (SHE), can be referred to the Pourbaix diagrams given in Figures 3.2 and 3.4 through 3.8. They lie well inside domains of stability for soluble ions for the zinc- and aluminum- systems, so that these metals can probably be eliminated from consideration. The same coordinates lie in domains of stability of solid phases for the iron-, copper-, tin-, and nickel-systems, so that these metals offer prospects for resisting the solution. As always, such conclusions are tentative, pending confirmation.

Solution to Problem 5.

The procedure is similar to that used in Section 3.3.3.2 for the formation of NiO on nickel, taking account of the different stoichiometric ratio. The reaction is:

$$\tfrac{1}{2}O_2 = Cu_2O + 2Cu^+ \square + 2e \square$$

The equilibrium constant, K, characterizes the defect populations at the oxide/atmosphere and metal/oxide interfaces:

$$K = \frac{\left\{a_{(Cu^+\square)}\right\}^2 \times \left\{a_{(e\square)}\right\}^2}{\left\{a_{O_2}\right\}^{1/2}}$$

whence:

$$n = p^{1/8} \cdot k^{1/4}$$

where:

$$n_{(Cu^+\square)} = n_{(e\square)} = n$$

and k includes K and coefficients relating n to $n_{(Cu^+\square)}$ and $n_{(e\square)}$.

The number of cation vacancies at the oxide/atmosphere interface, n_{atm}, is:

$$n_{atm} = p_{atm}^{1/8} \cdot k^{1/4}$$

where p_{atm} is the external oxygen pressure and the number at the metal/oxide interface, n_o, is:

$$n_0 = p_0^{1/8} \cdot k^{1/4}$$

where p_0 is the oxygen pressure in equilibrium with copper and Cu_2O.

The concentration gradient of vacancies through the oxide layer is virtually determined by p_{atm}, because $p_{atm} \gg p_0$. With the assumptions implicit in the Wagner theory, the oxidation rate is a linear function of this concentration gradient, so that if the oxygen pressure is raised from

1 to 100 atmospheres the rate of oxidation of copper is expected to increase by a factor of $(100)^{1/8} = 1.8$. Since the number of electron holes is also equal to η_0, the electrical conductivity of Cu_2O in equilibrium with oxygen at a pressure of 100 atmospheres is also expected to be about a factor of 1.8 times greater than that of the oxide in equilibrium with oxygen at a pressure of 1 atmosphere.

Solution to Problem 6.

Iron is stable in oxygen-free water if there is a common domain of stability. Inspection of Figure 3.2 shows that, by the convention that iron is stable when $a_{Fe^{2+}} < 10^{-6}$, there is only a small gap between the domains of stability of water and of iron at the Fe/Fe_3O_4 boundary. This gap can be closed by increasing the pressure on the system, so that $a_{H_2} > 1$, to extend the domain of stability of water to more negative potentials.

For pH > 9, the upper boundary of the domain of stability for iron is determined by the reaction:

$$Fe_3O_4 + 8H^+ + 8e^- = 3Fe + 4H_2O$$

and is described by Equation 3.20 in Section 3.1.5.2:

$$E = -0.085 - 0.0591 \text{ pH}$$

The lower boundary of the domain of stability for water is determined by the reaction:

$$2H^+ + 2e^- = H_2$$

and is given by the Nernst equation:

$$E = 0.000 - \frac{0.0591}{2} \log \frac{p_{H_2}}{\left(a_{H^+}\right)^2}$$

$$= \frac{0.0591}{2} \log p_{H_2} - 0.0591 \text{ pH}$$

The critical condition is when the two boundaries coincide, i.e:

$$-0.085 - 0.0591 \text{ pH} = \frac{0.0591}{2} \log p_{H_2} - 0.0591 \text{ pH}$$

Hence $\log p_{H_2} = 2.876$ and so $p_{H_2} = 752$ atm and the conditions for a common domain of stability are $p_{H_2} = 752$ atm and pH > 9.

Appendix

Examples 2 to 6 — Construction of Some Pourbaix Diagrams

Example 2.　The Aluminum–Water System
Species considered: Al, Al^{3+}, AlO_2^-, Al_2O_3

Reaction 1:　　　　　　　　　　　$Al^{3-} + 3e^- = Al$

Information needed:　　　　　　　　$E^{\ominus} = -1.663$ V SHE.

Applying the Nernst Equation:　　$E = -1.663 - \dfrac{0.0591}{3} \cdot \log\dfrac{1}{a_{Al^{3+}}}$

$$E = -1.663 + 0.0197 \log\left(a_{Al^{3+}}\right)$$

For:　　$a_{Al^{3+}} = 10^{-6}: E = -1.780$　　　$a_{Al^{3+}} = 10^{-2}: E = -1.701$

　　　　$a_{Al^{3+}} = 10^{-4}: E = -1.741$　　　$a_{Al^{3+}} = 1: \quad E = -1.662$

Reaction 2:　　　$Al_2O_3 + 6H^+ + 6e^- = 2Al + 3H_2O$

Information needed:　　　　　　　　$E^{\ominus} = -1.550$ V SHE.

Applying the Nernst Equation:　　$E = -1.550 - \dfrac{0.0591}{6} \cdot \log\dfrac{1}{\left(a_{H^+}\right)^6}$

$$E = -1.550 - 0.0591 \text{ pH}$$

Reaction 3:　　　$AlO_2^- + 4H^+ + 3e^- = Al + 2H_2O$

Information needed:　　　　　　　　$E^{\ominus} = -1.262$ V SHE.

Applying the Nernst Equation: $E = -1.262 - \dfrac{0.0591}{3} \cdot \log\dfrac{1}{\left(a_{AlO_2^-}\right) \times \left(a_{H^+}\right)^4}$

$$E = -1.262 + 0.0197 \log\left(a_{AlO_2^-}\right) - 0.0788 \text{ pH}$$

For:
$$a_{AlO_2^-} = 10^{-6}: E = -1.380 - 0.0788 \text{ pH}$$

$$a_{AlO_2^-} = 10^{-4}: E = -1.341 - 0.0788 \text{ pH}$$

$$a_{AlO_2^-} = 10^{-2}: E = -1.301 - 0.0788 \text{ pH}$$

$$a_{AlO_2^-} = 1: \quad E = -1.262 - 0.0788 \text{ pH}$$

Reaction 4: $\qquad Al_2O_3 + 6H^+ = 2Al^{3+} + 3H_2O$

Information needed: $\qquad (a_{Al^{3+}})/(a_{H^+})^3 = 5 \times 10^5.$

Taking logarithms: $\quad \log(a_{Al^{3+}}) - 3\log(a_{H^+}) = 5.70$

Re-arranging: $\qquad$ **pH = {(– log ($a_{Al^{3+}}$) + 5.70)}/3**

For: $\quad a_{Al^{3+}} = 10^{-6}:$ pH = 3.9 $\qquad a_{Al^{3+}} = 10^{-2}:$ pH = 2.6

$\qquad a_{Al^{3+}} = 10^{-4}:$ pH = 3.2 $\qquad a_{Al^{3+}} = 1:$ $\quad$ pH = 1.9

Reaction 5: $\qquad 2AlO_2^- + 2H^+ = Al_2O_3 + H_2O$

Information needed: $\qquad (a_{AlO_2^-}) \times (a_{H^+}) = 2.51 \times 10^{-15}.$

Taking logarithms: $\quad \log(a_{AlO_2^-}) + \log(a_{H^+}) = -14.6$

Re-arranging: $\qquad$ **pH = log ($a_{AlO_2^-}$) + 14.6**

For: $\quad a_{AlO_2^-} = 10^{-6}:$ pH = 8.6 $\qquad a_{AlO_2^-} = 10^{-2}:$ pH = 12.6

$\qquad a_{AlO_2^-} = 10^{-4}:$ pH = 10.6 $\qquad a_{AlO_2^-} = 1:$ $\quad$ pH = 14.6

Diagram: Figure 3.4 gives a diagram for $a_{Al^{3+}} = a_{AlO_2^-} = 10^{-6}.$

Example 3. The Zinc-Water System
Species considered: Zn, Zn^{2+}, $HZnO_2^-$, ZnO_2^{2-}, $Zn(OH)_2$

Reaction 1: $\qquad Zn^{2+} + 2e^- = Zn$

Information needed: $\qquad E^{\ominus} = -0.763 \text{ V SHE}$

Applying the Nernst Equation: $E = -0.763 - \dfrac{0.0591}{2} \cdot \log \dfrac{1}{a_{Zn^{2+}}}$

$$E = -0.763 + 0.0296 \log\left(a_{Zn^{2+}}\right)$$

For: $a_{Zn^{2+}} = 10^{-6}: E = -0.940$ $a_{Zn^{2+}} = 10^{-2}: E = -0.822$

            $a_{Zn^{2+}} = 10^{-4}: E = -0.881$ $a_{Zn^{2+}} = 1:$ $E = -0.763$

Reaction 2: $Zn(OH)_2 + 2H^+ + 2e^- = Zn + 2H_2O$

Information needed: $E^{\ominus} = -0.439$ V SHE

Applying the Nernst Equation: $E = -0.439 - \dfrac{0.0591}{2} \cdot \log \dfrac{1}{\left(a_{H^+}\right)^2}$

$$E = -0.439 - 0.0591\,pH$$

Reaction 3: $HZnO_2^- + 3H^+ + 2e^- = Zn + 2H_2O$

Information needed: $E^{\ominus} = +0.054$ V SHE

Applying the Nernst Equation: $E = +0.054 - \dfrac{0.0591}{2} \cdot \log \dfrac{1}{\left(a_{HZnO_2^-}\right) \times \left(a_{H^+}\right)^3}$

$$E = 0.054 + 0.0296 \log\left(a_{HZnO_2^-}\right) - 0.0887\,pH$$

For: $a_{HZnO_2^-} = 10^{-6}:$ $E = -0.124 - 0.0887\,pH$

            $a_{HZnO_2^-} = 10^{-4}:$ $E = -0.064 - 0.0887\,pH$

            $a_{HZnO_2^-} = 10^{-2}:$ $E = -0.005 - 0.0887\,pH$

            $a_{HZnO_2^-} = 1:$ $E = +0.054 - 0.0887\,pH$

Reaction 4: $ZnO_2^{2-} + 4H^+ + 2e^- = Zn + 2H_2O$

Information needed: $E^{\ominus} = +0.441$ V SHE

Applying the Nernst Equation: $E = +0.441 - \dfrac{0.0591}{2} \cdot \log \dfrac{1}{\left(a_{ZnO_2^{2-}}\right) \times \left(a_{H^+}\right)^4}$

$$E = +0.441 + 0.0296 \log\left(a_{ZnO_2^{2-}}\right) - 0.118 \text{ pH}$$

For: $a_{ZnO_2^{2-}} = 10^{-6}$: $E = +0.264 - 0.118 \text{ pH}$

       $a_{ZnO_2^{2-}} = 10^{-2}$: $E = +0.323 - 0.118 \text{ pH}$

       $a_{ZnO_2^{2-}} = 10^{-4}$: $E = +0.382 - 0.118 \text{ pH}$

       $a_{ZnO_2^{2-}} = 1$: $E = +0.441 - 0.118 \text{ pH}$

Reaction 5: $Zn(OH)_2 + 2H^+ = Zn^{2+} + 2H_2O$

Information needed: $(a_{Zn^{2+}})/(a_{H^+})^2 = 10^{11}$

Taking logarithms: $\log(a_{Zn^{2+}}) - 2\log(a_{H^+}) = 11$

Re-arranging: **pH = {−log($a_{Zn^{2+}}$) + 11}/2**

For: $a_{Zn^{2+}} = 10^{-6}$: pH = 8.5 $a_{Zn^{2+}} = 10^{-2}$: pH = 6.5

       $a_{Zn^{2+}} = 10^{-4}$: pH = 7.5 $a_{Zn^{2+}} = 1$:   pH = 5.5

Reaction 6: $Zn(OH)_2 = H^+ + HZnO_2^-$

Information needed: $(a_{H^+}) \times (a_{HZnO_2^-}) = 2.09 \times 10^{-17}$

Taking logarithms: $\log(a_{H^+}) + \log(a_{HZnO_2^-}) = -16.7$

Re-arranging: **pH = log($a_{HZnO_2^-}$) + 16.7**

For: $a_{HZnO_2^-} = 10^{-6}$: pH = 10.7 $a_{HZnO_2^-} = 10^{-2}$: pH = 14.7

       $a_{HZnO_2^-} = 10^{-4}$: pH = 12.7 $a_{HZnO_2^-} = 1$:   pH = 16.7

Reaction 7: $\qquad\qquad\qquad\qquad$ $HZnO_2^- = H^+ + ZnO_2^{2-}$

Information needed: $\quad (a_{H^+}) \times (a_{ZnO_2^{2-}})/(a_{HZnO_2^-}) = 7.8 \times 10^{-14}$

Taking logarithms: $\quad \log(a_{H^+}) + \log(a_{ZnO_2^{2-}}) - \log(a_{HZnO_2^-}) = -13.1$

Re-arranging: $\qquad$ **pH = 13.1** (since $a_{ZnO_2^{2-}} = a_{HZnO_2^-}$ by definition)

Diagram: Figure 3.5 gives the diagram for $a_{Zn^{2+}} = a_{HZnO_2^-} = a_{ZnO_2^{2-}} = 10^{-6}$.

Example 4. The Copper-Water System

Species considered: Cu, Cu^{2+}, Cu^+, CuO, Cu_2O

Reaction 1: $\qquad\qquad$ $Cu^{2+} + 2e^- = Cu$

Information needed: $\qquad\qquad$ $E^{\ominus} = +0.337$ V SHE.

Applying the Nernst Equation: $\quad E = +0.337 - \dfrac{0.0591}{2} \cdot \log\dfrac{1}{a_{Cu^{2+}}}$

$$E = +0.337 + 0.0296 \log\left(a_{Cu^{2+}}\right)$$

For: $\qquad a_{Cu^{2+}} = 10^{-6}: E = 0.159 \qquad\qquad a_{Cu^{2+}} = 10^{-2}: E = 0.278$

$\qquad\qquad a_{Cu^{2+}} = 10^{-4}: E = 0.219 \qquad\qquad a_{Cu^{2+}} = 1: \quad E = 0.337$

Reaction 2: $\qquad\qquad$ $Cu^+ + e^- = Cu$

Information needed: $\qquad\qquad$ $E^{\ominus} = +0.520$ V SHE.

Applying the Nernst Equation: $\quad E = +0.520 - \dfrac{0.0591}{1} \cdot \log\dfrac{1}{a_{Cu^+}}$

$$E = +0.520 + 0.0591 \log\left(a_{Cu^+}\right)$$

For: $\qquad a_{Cu^+} = 10^{-6}: E = 0.165 \qquad\qquad a_{Cu^+} = 10^{-2}: E = 0.402$

$\qquad\qquad a_{Cu^+} = 10^{-4}: E = 0.284 \qquad\qquad a_{Cu^+} = 1: \quad E = 0.520$

Reaction 3: $2Cu^{2+} + H_2O + 2e^- = Cu_2O + 2H^+$

Information needed: $E^{\ominus} = +0.203$ V SHE.

Applying the Nernst Equation: $E = +0.203 - \dfrac{0.0591}{2} \cdot \log \dfrac{\left(a_{H^+}\right)^2}{\left(a_{Cu^{2+}}\right)^2}$

$$E = 0.203 + 0.0591\,pH + 0.0591 \log\left(a_{Cu^{2+}}\right)$$

For: $a_{Cu^{2+}} = 10^{-6}$: $E = -0.152 - 0.0591\,pH$

$a_{Cu^{2+}} = 10^{-2}$: $E = +0.085 - 0.0591\,pH$

$a_{Cu^{2+}} = 10^{-4}$: $E = -0.033 - 0.0591\,pH$

$a_{Cu^{2+}} = 1$: $E = +0.203 - 0.0591\,pH$

Reaction 4: $Cu^{2+} + H_2O = CuO + 2H^+$

Information needed: $(a_{H^+})^2/(a_{Cu^{2+}}) = 1.29 \times 10^{-8}$

Taking logarithms: $2\log(a_{H^+}) - \log(a_{Cu^{2+}}) = -7.89$

Re-arranging: $pH = \{-\log(a_{Cu^{2+}}) + 7.89\}/2$

For: $a_{Cu^{2+}} = 10^{-6}$: pH = 6.95 $a_{Cu^{2+}} = 10^{-2}$: pH = 4.95

$a_{Cu^{2+}} = 10^{-4}$: pH = 5.95 $a_{Cu^{2+}} = 1$: pH = 3.95

Reaction 5: $Cu_2O + 2H^+ + 2e^- = 2Cu + H_2O$

Information needed: $E^{\ominus} = +0.471$ V SHE.

Applying the Nernst Equation: $E = +0.471 - \dfrac{0.0591}{2} \cdot \log \dfrac{1}{\left(a_{H^+}\right)^2}$

$$E = +0.471 - 0.0591\,pH$$

Reaction 6: $2CuO + 2H^+ + 2e^- = Cu_2O + H_2O$

Information needed: $E^{\ominus} = +0.669$ V SHE.

Applying the Nernst Equation: $E = +0.669 - \dfrac{0.0591}{2} \cdot \log \dfrac{1}{\left(a_{H^+}\right)^2}$

$$E = 0.669 - 0.0591 \, pH$$

Diagram: Figure 3.6 gives the diagram for $a_{Cu^{2+}} = a_{Cu^+} = 10^{-6}$.

Example 5.   The Tin-Water System
Species considered: Sn, Sn^{2+}, Sn^{4+}, SnO_3^{2-}, SnO_2

Reaction 1: $Sn^{2+} + 2e = Sn$

Information needed: $E^{\ominus} = -0.136$ V SHE.

Applying the Nernst Equation: $E = -0.136 - \dfrac{0.0591}{2} \cdot \log \dfrac{1}{a_{Sn^{2+}}}$

$$E = -0.136 + 0.0296 \log\left(a_{Sn^{2+}}\right)$$

For: $a_{Sn^{2+}} = 10^{-6}: E = -0.313$ $a_{Sn^{2+}} = 10^{-2}: E = -0.195$

$a_{Sn^{2+}} = 10^{-4}: E = -0.254$ $a_{Sn^{2+}} = 1:$ $E = -0.136$

Reaction 2: $SnO_2 + 4H^+ + 4e^- = Sn + 2H_2O$

Information needed: $E^{\ominus} = -0.106$ V SHE.

Applying the Nernst Equation: $E = -0.106 - \dfrac{0.0591}{4} \cdot \log \dfrac{1}{\left(a_{H^+}\right)^4}$

$$E = -0.106 - 0.0591 \, pH$$

Reaction 3: $SnO_2 + 4H^+ + 2e^- = Sn^{2+} + 2H_2O$

Information needed: $E^{\ominus} = -0.077$ V SHE.

Applying the Nernst Equation: $E = -0.077 - \dfrac{0.0591}{2} \cdot \log \dfrac{\left(a_{Sn^{2+}}\right)}{\left(a_{H^+}\right)^4}$

$$E = -0.077 - 0.118 \, pH - 0.0296 \, \log\left(a_{Sn^{2+}}\right)$$

For: $a_{Sn^{2+}} = 10^{-6}$: $E = +0.100 - 0.118 \, pH$

$a_{Sn^{2+}} = 10^{-4}$: $E = +0.041 - 0.118 \, pH$

$a_{Sn^{2+}} = 10^{-2}$: $E = -0.018 - 0.118 \, pH$

$a_{Sn^{2+}} = 1$: $E = -0.077 - 0.118 \, pH$

Reaction 4: $Sn^{4+} + 2H_2O = SnO_2 + 4H^+$

Information needed: $(a_{Sn^{4+}})/(a_{H^+})^4 = 2.08 \times 10^{-8}$

Taking logarithms: $\log(a_{Sn^{4+}}) - 4\log(a_{H^+}) = -7.68$

Re-arranging: $pH = -\{\log(a_{Sn^{4+}}) + 7.68\}/4$

For: $a_{Sn^{4+}} = 10^{-6}$: $pH = -0.4$ $a_{Sn^{4+}} = 10^{-2}$: $pH = -1.4$

$a_{Sn^{4+}} = 10^{-4}$: $pH = -0.9$ $a_{Sn^{4+}} = 1$: $pH = -1.9$

Reaction 5: $Sn^{4+} + 2e^- = Sn^{2+}$

Information needed: $E^{\ominus} = +0.151$ V SHE.

Applying the Nernst Equation: $E = +0.151 - \dfrac{0.0591}{2} \cdot \log \dfrac{a_{Sn^{2+}}}{a_{Sn^{4+}}}$

$$= +0.151 - 0.0296 \, \log\left\{\left(a_{Sn^{2+}}\right)/\left(a_{Sn^{4+}}\right)\right\}$$

$$E = +0.151 \text{ since } \left(a_{Sn^{2+}}\right)/\left(a_{Sn^{4+}}\right) \text{ is } 1$$

Reaction 6:
$$SnO_3^{2-} + 2H^+ = SnO_2 + H_2O$$

Information needed:
$$(a_{SnO_3^{2-}}) \times (a_{H^+})^2 = 6.92 \times 10^{-32}$$

Taking logarithms:
$$\log(a_{SnO_3^{2-}}) + 2\log(a_{H^+}) = -31.16$$

Re-arranging:
$$\mathbf{pH = \{log\,(a_{SnO_3^{2-}}) + 31.16\}/2}$$

For:

$a_{SnO_3^{2-}} = 10^{-6}$: pH = 12.6 $a_{SnO_3^{2-}} = 10^{-2}$: pH = 14.6

$a_{SnO_3^{2-}} = 10^{-4}$: pH = 13.6 $a_{SnO_3^{2-}} = 1$:   pH = 15.6

Diagram: Figure 3.7 gives the diagram for $a_{Sn^{2+}} = a_{Sn^{4+}} = a_{SnO_3^{2-}} = 10^{-6}$.

Example 6.   The Nickel-Water System
Species considered: Ni, Ni^{2+}, $Ni(OH)_2$

Reaction 1: $Ni^{2+} + 2e^- = Ni$

Information needed: $E^{\ominus} = -0.250$ V SHE.

Applying the Nernst Equation: $E = -0.250 - \dfrac{0.0591}{2} \cdot \log\dfrac{1}{a_{Ni^{2+}}}$

$$E = -0.250 + 0.0296 \log\left(a_{Ni^{2+}}\right)$$

For: $a_{Ni^{2+}} = 10^{-6}$: $E = -0.428$ $a_{Ni^{2+}} = 10^{-2}$: $E = -0.309$

$a_{Ni^{2+}} = 10^{-4}$: $E = -0.368$ $a_{Ni^{2+}} = 1$: $E = -0.250$

Reaction 2: $Ni(OH)_2 + 2H^+ + 2e = Ni + 2H_2O$

Information needed: $E^{\ominus} = +0.110$ V SHE.

Applying the Nernst Equation: $E = +0.110 - \dfrac{0.0591}{2} \cdot \log\dfrac{1}{\left(a_{H^+}\right)^2}$

$$E = +0.110 - 0.0591\ \text{pH}$$

Reaction 3: $\qquad$ $Ni(OH)_2 + 2H^+ = Ni^{2+} + 2H_2O$

Information needed: $\qquad$ $(a_{Ni^{2+}})/(a_{H^+})^2 = 1.58 \times 10^{12}$

Taking logarithms: $\quad$ $\log(a_{Ni^{2+}}) - 2\log(a_{H^+}) = 12.20$

Re-arranging: $\qquad$ $\mathbf{pH = \{-\log(a_{Ni^{2+}}) + 12.2\}/2}$

For: $\qquad$ $a_{Ni^{2+}} = 10^{-6}: pH = 9.1$ $\qquad\qquad$ $a_{Ni^{2+}} = 10^{-2}: pH = 7.1$

$\qquad\qquad$ $a_{Ni^{2+}} = 10^{-4}: pH = 8.1$ $\qquad\qquad$ $a_{Ni^{2+}} = 1: \quad pH = 6.1$

Diagram: Figures 3.8 and 6.2 give partial diagrams for $a_{Ni^{2+}} = 10^{-6}$ and $a_{Ni^{2+}} = 1$, respectively.

4

Mixed Metal Systems
and Cathodic Protection

4.1 Galvanic Stimulation

Contact of dissimilar metals in aqueous media is a frequent cause of premature corrosion yet, with appropriate foreknowledge, it can be controlled and with proper design and materials selection it should seldom be encountered. Nevertheless, there are several reasons why it prevails:

1. In good faith, a designer may combine incompatible metals that separately have good corrosion resistance, unaware that the combination constitutes a corrosion hazard. This can happen, for example, in incorrectly specifying steel or stainless steel fasteners for joining a relatively soft metal such as aluminum.

2. Contractors may assemble systems empirically, using standardized components made from incompatible metals. Such systems may have combinations from among copper and steel pipe, brass fittings, galvanized steel and aluminum bodies.

3. A component or system purposefully constructed from a single metal or from combinations of compatible metals may yet suffer damage stimulated by traces of incompatible metals introduced by contamination in various ways.

4. In areas with hard water supplies, metals in supply systems are usually protected by coatings of deposited lime-scale, as explained in Section 2.2.9.2. Besides slowing corrosion generally, it can mask the effect of incompatible metals so that the possible effects are not appreciated. Such experience leads to problems if transferred to soft-water areas.

In what follows, the nature and origin of the effect is first explained and then some common practical implications are discussed.

4.1.1 Bimetallic Couples

When two dissimilar metals in electrical contact are exposed to an aqueous medium, the less noble metal in general suffers more corrosion and the more noble metal suffers less than if they were isolated in the same medium. The increased attack on the less noble metal is called *galvanic stimulation*. The combination of metals producing the effect is a *bimetallic couple*.

The effect can be very intense and is a potent cause of the premature failure of the less noble metal. It does not correlate quantitatively either with differences between the standard electrode potentials of a pair of metals, $(E^{\ominus}_{\text{FIRST METAL}}) - (E^{\ominus}_{\text{SECOND METAL}})$, or with differences between the corrosion potentials that each would have separately in the same environment. Hence other factors modify the relationship between the metals in a bimetallic couple and the phenomenon is more complex than might appear at first sight.

4.1.2 The Origin of the Bimetallic Effect

The origin of the effect is as follows. In isolation in an environment in which they attract the same cathodic reaction(s), e.g., in aerated neutral water:

$$\tfrac{1}{2}O_2 + H_2O + 2e^- = 2OH^- \tag{4.1}$$

the metals would acquire different corrosion potentials, $E_{\text{CORROSION}}$, but when they are in electrical contact, they must adopt a common potential and hence the potential of the less noble metal is raised and that of the more noble metal is depressed from their respective normal values. This is illustrated in Figure 4.1, which shows the current distribution in a system with two metals, A (more noble) and B (less noble) when brought to a common potential in the same aqueous environment, the *mixed potential*, E_{MIXED}. As a consequence, the corrosion current on metal B is increased from $i_{\text{CORROSION}}$ (B) to $i'_{\text{CORROSION}}$ (B) and that on metal A is reduced from $i_{\text{CORROSION}}$ (A) to $i'_{\text{CORROSION}}$ (A)

The total anodic and cathodic currents are balanced in the system as a whole, but the anodic and cathodic current densities are unequal on either of the two metals. The anodic current exceeds the cathodic current on the less noble metal and the cathodic current exceeds the anodic current on the more noble metal. The proportion of the *total* anodic current each carries and hence the relative intensity of attack depends on their relative areas. If the noble metal area greatly exceeds the base metal area, the attack on the base metal is intense, e.g., steel fasteners in copper sheet suffer rapid attack. Conversely, if the base metal area greatly exceeds the noble metal area, the enhanced attack on the base metal is more widely spread and less severe, e.g., brass fasteners in steel sheet.

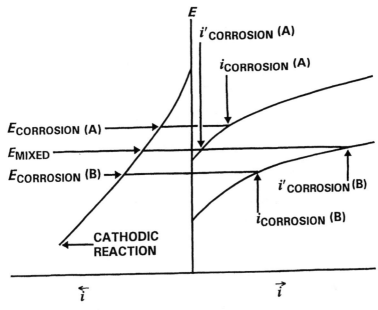

FIGURE 4.1
Current distribution on metals A (more noble) and B (less noble) in a corroding bimetallic system. If the metals are insulated, their corrosion potentials are $E_{CORROSION}(A)$ and $E_{CORROSION}(B)$ and their corrosion currents are $i_{CORROSION}$ (A) and $i_{CORROSION}$ (B). If connected, they acquire a common potential, E_{MIXED}, and the anodic current on A falls to $i'_{CORROSION}$ (A) and that on B rises to $i'_{CORROSION}$ (B).

4.1.3 Design Implications

4.1.3.1 Active and Weakly Passive Metals

The indiscriminate mixing of metals with large differences in standard electrode potentials in structures or components exposed to aqueous environments, rain or condensation can be a bad design fault. As examples, aluminum or galvanized steel fail rapidly in contact with copper, brass or stainless steels. Contact between such incompatible metals in plumbing, roofing, chemical plant, cooling systems, etc. must be strictly avoided.

The photograph in Figure 4.2 shows a rather extreme but interesting example. The component illustrated was part of a pressure-sensing device exposed to an external marine environment and used to control a flow of liquid in a large pipe. The photograph shows a painted aluminum cap which was secured to the main body of the device by stainless steel bolts. The cap was further fitted with brass unions for connection to small-bore tubing forming part of the device. The immediate cause of failure was the production of a such a pressure from aluminum corrosion product formed locally within the screw threads that pieces of the aluminum cap cracked away. General corrosion of the component was insignificant. Chemical analysis revealed significant chloride contamination in the corrosion

FIGURE 4.2
Aluminum cap cracked by pressure of the product of corrosion galvanically stimulated in the threads of stainless-steel bolts.

product. Recalling the discussion of the integrity of passivity in Section 3.2.5 it is probable that although stainless steel and aluminum are inherently incompatible, exposure to chloride from the marine environment exacerbates the effect by breaking down the passivity of aluminum more readily than the stronger passivity of the stainless steel.

The remedy was to re-specify the materials of construction. Since it would have been inconvenient to replace aluminum as the material for the cap, the bolt material was changed to heavily cadmium-plated plain carbon steel. This alone would have been an incomplete remedy because the stainless steel/aluminum bi-metallic couple had masked the latent bi-metallic couple presented by the brass union screwed into the cap. It was expected that eliminating the stainless steel/aluminum couple would activate the brass/aluminum couple so that an essential part of the remedy was to replace this too by an equivalent cadmium-plated item.

4.1.3.2 Active/Passive Couples

Certain combinations of metals and alloys with nominally similar electrode potentials and which might on first consideration seem acceptable can in practice be electrochemically dissimilar and form very active couples. This can happen if one of a pair of metals is strongly passivated in the environment of interest. Such strong passivation can promote the nobility of a base metal, so that in appropriate circumstances, strongly passivating base metals such as titanium and chromium and alloys in which they are components can act as noble partners in a bimetallic pairs with base metals such as iron, which remain active, or aluminum, which is much less strongly passive.

An example is the combination of a plain carbon steel with a stainless steel in a mildly oxidizing environment, e.g., aerated water; a couple forms between the active surface of the plain carbon steel and the passive surface of the stainless steel, throwing attack on the plain steel.

4.1.3.3 Cathodic Collectors

An electrically conducting material which does not itself support an anodic dissolution current can accelerate the corrosion of a metal with which it is in electrical contact. It does so because it supplements the area on which cathodic reactions can occur and the complementary anodic current is superimposed on the normal anodic current on the metal with which it is in contact. Such a material thus simulates a noble metal in bimetallic couples, and it is referred to as a *cathodic collector*. Examples of such materials are graphite, certain interstitial metal carbides and borides, e.g., titanium diboride TiB_2 and sufficiently conducting p-type metal oxides, e.g., copper(I) oxide, Cu_2O.

4.1.3.4 Compatibility Groups

In many designs, it is impracticable to avoid mixing metals because different properties or characteristics are required for components. A simple example is the use of mechanical fasteners, such as bolts in constructions with aluminum alloys, where the stresses imposed on screw threads may require a stronger material, e.g., a steel. If metals must be mixed, the arrangement must be designed to eliminate or at least minimize bimetallic effects.

One option is to insulate incompatible metals from each other but this is not always effective because of transport through the aqueous medium as described in Section 4.1.3.5. It is better to select metal combinations which experience has shown to be acceptable. To aid selection, metals and alloys can be assigned on the basis of experience to several *compatibility groups* as follows:

Group 1 — Strongly electronegative metals:
    Magnesium and its alloys.

Group 2 — Base metals:
    Aluminum, cadmium, zinc, and their alloys.

Group 3 — Intermediate metals:
    Lead, tin, iron, and their alloys, *except stainless steels*

Group 4 — Noble metals:
    Copper, silver, gold, platinum, and their alloys.
Group 5 — Strongly passivating metals:
    Titanium, chromium, nickel, cobalt, their alloys, and
    stainless steels.
Group 6 — Cathodic collectors:
    Graphite and electrically conducting carbides, borides, and
    oxides

Cathodic collectors are included because although they cannot sustain *anodic* reactions, they extend the electrical conducting surface available for *cathodic* reactions supporting the anodic reaction of metals with which they are in contact.

Great care is needed in applying this information because the compatibility groups do not imply that it is always safe to use metals together from within the same compatibility group. It should be used rather to warn of potentially disastrous situations if metals in different groups are mixed. Generally, the greater the separation of the groups in the above sequence from which metals are combined, the greater is the risk of galvanic stimulation. The groups would, for example, indicate that cadmium-plated or zinc-plated steel bolts are probably safe to use for an aluminum alloy structure exposed outside but that stainless steel bolts could promote catastrophic corrosion of the adjacent aluminum. Nevertheless, the predictive value of the compatibility grouping is sometimes uncertain, as the following example illustrates. Contact with a small area of graphite may have only a small effect on copper, as the compatibility grouping suggests, but there are circumstances in which a carbon/copper couple is dangerous. If a copper surface is *nearly completely* covered with copper oxide or with a thin film of carbon from carbonization of lubricant residues during the heat-treatment of fabricated products which is normally carried out in reducing atmospheres, small areas of copper exposed, e.g., at abrasions, can suffer intense attack, leading to perforation by pitting. This is because a high anodic current density on the very small copper anodes is sustained by the total cathodic current collected from the very large area of the oxide or carbon-covered surface. An example of corrosion initiated in this way is illustrated by the photograph of the inside of a copper water main, given in Figure 4.3.

4.1.3.5   Indirect Stimulation

Even when two incompatible metals are not in direct contact, it is still possible for metals to be damaged by bimetallic cells set up indirectly. A common example is perforation of aluminum utensils regularly filled through copper piping with soft water containing a small quantity of dissolved carbon dioxide, such as that supplied from acidic strata or that delivered by

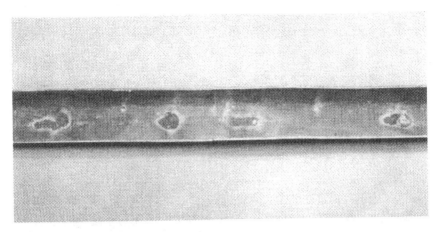

FIGURE 4.3
Corrosion of copper stimulated galvanically by carbonized lubricant residues on the inside of a copper water main.

domestic water softeners. Since the water is soft, the pipework and utensils are unprotected by chalky deposits, and traces of copper are dissolved from the piping as copper bicarbonate, from which copper metal is subsequently deposited very thinly on the aluminum at weak points in the aluminum oxide film by a replacement reaction:

$$3Cu(HCO_3)_2 + 2Al = 3Cu + 2Al(OH)_3 + 6CO_2 \qquad (4.2)$$

In this way, multiple small aluminum/copper bimetallic cells are set up *in situ* on the aluminum surface. To illustrate how catastrophic the effect can be, domestic utensils fabricated from commercial pure aluminum, which can have life expectancy of ten or more years, can fail by perforation in a few weeks use, if subjected to indirect stimulation as just described. The photograph of part of the inside base of an aluminum electric kettle in Figure 4.4 illustrates the effect.

Other examples are found in domestic and industrial plumbing, where incompatible components bought separately are indiscriminately assembled. Such components can include copper pipe, steel pipe, galvanized (zinc-coated) steel tanks and fittings, and brass compression convenience fittings.

Systems incorporating incompatible metals appear to be trouble-free in particular situations, e.g.:

1. If the metals are isolated and water flow is from base to noble metals.

2. In regions with hard water supplies, where lime-scale deposits mask bimetallic couples and provide cathodic inhibition.

FIGURE 4.4
Part of the inside base of an aluminum kettle showing white corrosion product around multiple galvanic cells stimulated by copper deposition from soft cuprosolvent water. One cell has perforated the aluminum sheet.

3. In recirculating systems, e. g. central heating systems or automobile cooling systems, where the dissolved oxygen is consumed and not replenished, thereby denying access to cathodic reactions.

Nevertheless, such situations are unsatisfactory because a change in conditions, e.g., a change in the source of water, may activate latent bimetallic effects.

4.1.3.6   Variable Polarity

The complexities of the bimetallic effect were noted in Section 4.1.1. In some systems, the intervention of environmental factors can determine which metal of a bimetallic couple is anodic and which is cathodic. The phenomenon is called *polarity reversal*. The tin coating on food cans is quoted as an example in Section 4.2.2.

Example

Following rationalization of the water supply industry, the hard water supply to a certain city is replaced by soft water. Within a short time, there is an alarming increase in the corrosion of metals in contact with the water.

One particularly prevalent problem is rapid perforation of pure aluminum domestic cooking utensils. Using electron probe microanalysis, a local laboratory has identified traces of copper in rings of corrosion product surrounding the perforations. Assess the probable cause and recommend remedies that could be implemented by (1) householders, (2) plumbers, and (3) the water supply company.

SOLUTION:
Domestic water supply systems are often installed by builders on the basis of local experience, as implied later in Chapter 13. The change of water supply has probably revealed mixed metal plumbing systems whose potential deleterious effects had previously been suppressed by limescale deposits from hard water. The particular problem described is a common example of indirect stimulation due to cuprosolvent soft water conveyed by domestic copper plumbing, producing *in situ* copper/aluminum bimetallic cells by replacement reaction, depositing copper at weak points in the oxide film protecting aluminum surfaces.

1. Householders could counter the effect by replacing aluminum cooking utensils with copper, stainless steel or vitreous enamelled steel alternatives. The strong passive condition of stainless steel surfaces, explained further in Chapter 8, resists the replacement reactions. Enamelled steel is insulated.

2. Plumbers could remove the problem by replacing all of the copper plumbing with an innocuous material, e.g., polyethylene.

3. The water supply company could inhibit the water e.g., by polyphosphate treatment.

4.2 Protection by Sacrificial Anodes

4.2.1 Principle

Bimetallic effects can be turned to a useful purpose by using the corrosion of one metal to protect another. The protecting metal, B in Figure 4.1, is the less noble of the pair and is selected so that the mixed potential lies below the equilibrium potential of the protected metal, A. It therefore takes the whole of the anodic dissolution current of the system. Since it is progressively consumed in fulfilling this role, it is called a *sacrificial anode*.

Sacrificial protection is mostly applied to steel. The protecting metals used for sacrificial anodes are magnesium, aluminum, and zinc, that are inherently more expensive than the metal they protect and the economic benefit is the preservation of the added value of expensive structures and inaccessible systems.

By suitable choice of the protecting metal, the ratio of the exposed area of the protecting metal to that of the protected metal and the number and disposition of sacrificial anodes, a structure or system can be completely protected. The choice of the metal is determined by the nature and severity of the environment. The metal must corrode fast enough to apply the protection and yet not be so active that it is wastefully consumed. For example, magnesium is too reactive in seawater but is more effective than most other suitable metals in buried structures.

4.2.2 Application

Buried Steel Pipework

Zinc or magnesium anodes are connect to the pipe at recorded intervals along it. The anodes are regularly lifted for inspection and replaced when necessary. To minimize anode consumption, the pipework is given primary protection by a coating such as a tarred wrap. During installation, there is an opportunity to backfill the excavation with material that allows soil water to maintain electrolytic contact between the protected system and the anodes.

Ships Hulls

Bare aluminum anodes are attached to the hull at intervals below the water-line. The hull is protected by painting, for the main protection, leaving the aluminum anodes unpainted. Alloying additions may be made to the aluminum to control the rate of dissolution in seawater. Extra anodes may be needed at the stern to counter the bimetallic effect of propellers, which are often cast from a copper-base alloy, manganese bronze.

Galvanic Protection of Steel by Coatings

One of the most effective and inexpensive methods of protecting steel sheet in near neutral aqueous environments is to coat it with a thin layer of zinc, either by hot-dipping, i.e., *galvanizing*, or more usually by electrodeposition. In near-neutral aqueous media, the zinc coating resists corrosion by forming a passive surface of $Zn(OH)_2$. At defects in the coating caused by abrasion and at cut edges, the exposed iron is cathodically protected by corrosion of the newly exposed zinc, producing $Zn(OH)_2$, which covers the exposed iron, stifling further attack.

It may seem surprising that tin coatings can galvanically protect steel food cans, in view of the fact that standard electrode potential of the reaction:

$$Sn^{2+} + 2e^- = Sn \tag{4.3}$$

for which: $\qquad$ $E^{\ominus} = -0.13$ V SHE

is more noble than that for the reaction:

$$Fe^{2+} + 2e^- = Fe \qquad\qquad (4.4)$$

for which: $\qquad$ $E^{\ominus} = -0.44$ V SHE

From this information, it might be expected that the extensive tin surface would stimulate corrosion of steel exposed at breaks in the coating. This is so in humid atmospheres and in hot water, but the polarity is reversed in the environment of a sealed food can, which is a de-aerated aqueous solution containing organic acids, such as fruit juices. Thus, provided that the pH is above that needed to passivate tin, the coating inside a can protects the steel exposed at imperfections.

These and other examples of galvanic protection are re-examined within wider contexts, in later chapters.

4.3   Cathodic Protection by Impressed Current

A metal in an aqueous environment, usually fresh or sea water, can be protected by impressing an electric current at its surface, delivered by a DC generator or rectifier, to polarize the potential to a value more negative than the equilibrium potential. The metal to be protected is almost invariably steel and the objective is to preserve the added value of expensive structures and installations. The current is applied through the aqueous medium between the structure or system to be protected and multiple strategically placed inert anodes, insulated from the metal and mounted close to its surface.

The choice between this approach and the alternative, the use of sacrificial anodes, is determined mainly by which of them is the most cost effective for particular projects. Sacrificial protection incurs the initial and replacement costs of the anodes, the costs of regular inspection and the costs of labor and facilities needed to replace them, e. g. dry docks for marine applications. Impressed current protection incurs capital costs for power sources and semi-permanent anodes and running costs to supply the electric power. The decision on which to use in a particular situation is based not only on technical considerations but also on the usual input from accountants, taking account of matters such as discounted cash flow and tax deductible expenditure. Generally, large and particularly inaccessible systems such as the hulls of ocean going ships and long pipelines favor impressed current protection. It is also applied in food and chemical plants where contamination from the corrosion of sacrificial anodes cannot be tolerated.

The selection of power requirements, wiring systems and disposition of the anodes for a particular project is an engineering concern, taking account of water or soil conductivity, exposed area of metal and ohmic potential drops along or within the protected system. The principal concern is to ensure that all parts of the system are depressed to a protective potential economically. The cost of power to protect a large bare steel system by impressed current alone would be prohibitive, so that impressed current protection is used to complement protective coatings. For a ship's hull it is applied together with the paint system to take care of local paint failure, bare patches, abrasions etc; a buried pipeline is usually protected with tarred wrapping and when wet, the groundbed of soil around the pipe must have sufficient conductivity to sustain the protecting current.

The inert anodes are not consumable items but good ones are expensive and are intended to be both efficient and unattacked. The most popular, especially for marine use, are of titanium with a very thin electrodeposited coating of platinum. The oxide on bare titanium has properties that protect it from corrosion in an electrolyte but resist the passage of the protecting current into an aqueous medium; it will however freely pass electrons into another metal, so that application of the platinum coating produces an efficient anode with a substrate that is protected in the event of local failure of the platinum film. Less expensive anode materials are graphite and cast irons with high silicon contents; graphite is fragile and tends to pulverize in service and silicon irons are brittle and can be shaped only by casting.

The cathodic current stimulates the reaction:

$$\tfrac{1}{2}O_2 + H_2O + 2e^- = 2OH^- \tag{4.5}$$

and the generation of OH^- ions raises the pH of the electrolyte at the metal surface, producing some important incidental effects. The rise in alkalinity is protective to the metal because it tends to passivate the steel and also promotes precipitation of a protective carbonate scale on the metal surface if the water is hard but it tends to degrade adjacent paint coatings.

In ensuring that the potential is sufficiently depressed for protection at parts of a structure remote from the anodes of a large complex system, the potential at other parts can be excessively negative, promoting hydrogen evolution by the cathodic reaction:

$$2H^+ + 2e^- = H_2 \tag{4.6}$$

Iron and steel are unique among engineering metals in that they are permeable to hydrogen at ambient temperatures and some of the cathodically-produced hydrogen enters the metal:

$$2H^+ + 2e^- = 2H_{(\text{SOLUTION IN IRON})} \tag{4.7}$$

Hydrogen absorbed from Reaction 4.7 accumulates at points of triaxial stress, where it can initiate cracks by the well-known phenomenon of *delayed failure* if the stress is sufficient. Iron or steel rendered susceptible to embrittlement in this way is said to be *overprotected*. An important part of the design of a protection system is to dispose the anodes to give a reasonably uniform potential over the surface, so that excessive cathodic polarization is not needed and the hazard need not arise.

Further Reading

Brown, R. H., *Galvanic Corrosion*, American Society for Testing and Materials, Bulletin 126, 1944.

Fontana, M. G. and Greene, N. D., *Corrosion Engineering*, McGraw-Hill, New York, 1978, Chap. 10.

Martin, B. A., *Cathodic Protection Theory and Practice*, Ellis Horwood, Chichester, U.K., 1986

5

The Intervention of Stress

Stresses applied to a metal structure can interact synergistically with environmentally induced electrochemical effects in aqueous media causing failure of structures by cracking at loads far below those indicated by the normal mechanical properties of the metal. The effects are confined to the crack sites and there is usually no significant damage by general corrosion. The generic term environmentally sensitive cracking describes these destructive effects. Two principal forms are recognized:

1. Stress-corrosion cracking that describes delayed failure under static loading.
2. Corrosion-fatigue that describes failure induced by cyclic loading.

A further form is embrittlement by hydrogen generated electrochemically that dissolves at the metal surface and migrates to stressed locations, where it can reduce the bond strength between metal atoms so that they separate under the applied stress, It is sometimes treated as a separate effect but is often dealt with as a special case of stress-corrosion cracking.

Failure by environmental cracking is unpremeditated and it is sometimes catastrophic. It can inflict substantial economic loss and at worst it can be life-threatening. For example, it can damage the integrity of aluminum alloy airframes in flight, cause explosions of pressurized stainless steel chemical plant, and initiate collapse of steel civil engineering structures. Naturally, such hazards attract close attention and susceptibility to cracking is often the limiting factor in a design rather than the ultimate strength of metals.

5.1 Stress-Corrosion Cracking (SCC)

5.1.1 Characteristic Features

Stress-corrosion cracking is a system property influenced by factors contributed by the metal and by the environment. The following features are characteristic:

1. **Conjoint Action** — Cracking is caused by the synergistic combination of stress and a specific environmental agent, usually in aqueous solution. Separate or alternate application of stress and exposure to the agent is insufficient.

2. **Stress** — A constant stress intensity in a crack-opening mode, K_1, K_2, or K_3, is sufficient. It may be applied externally or by internal strains imparted by fabrication, contraction after welding or mechanically-corrected mismatch.

3. **Environment** — Conditions for SCC are highly specific and for a given metal or alloy, cracking is induced only if one or another of particular species, the *specific agents* is present. General corrosion or the presence of other environmental species is insufficient.

4. **Crack Morphology** — The cracks appear brittle with no deformation of adjacent metal, even if the mode of failure under stress alone is ductile.

5. **Life-to-Failure** — The life decreases with increasing stress and is the sum of two parts (a) an induction period which determines most of the life, e.g., weeks or years and (b) a rapid crack propagation period, typically hours or minutes.

6. **Crack Path** — The crack path is a characteristic of particular metals or alloys For some it is *intergranular*, i.e., along the grain boundaries between the metal crystals; for others it is *transgranular*, i.e., through the crystals, avoiding grain boundaries; for yet others it is indiscriminate.

Features of some important alloy systems exhibiting SCC are summarized in Table 5.1. This illustrates some of the factors which influence susceptibility. They include:

1. Nature and the composition of the metal.
2. Crystal structure of the metal.
3. Thermal and mechanical treatments given to the metal.
4. Species present in the environment.
5. Temperature.
6. Magnitude and state of stress.

The relative importance of the factors listed above differs from metal to metal, adding to the complexity of the phenomenon and, for practical reasons, separate bodies of evidence are built up by interests in the safe use of particular metal product groups, such as aluminum alloys, stainless steels or plain carbon steels. For example, users of high-strength aluminum alloys pay particular attention to the sensitivity of the alloys to heat-treatments

TABLE 5.1

Characteristics of Stress-corrosion Cracking in Some Alloy Systems

System	Specific Agent	Crack Path	Remarks
Aluminum alloys	Cl⁻ ions	Grain boundaries	Age-hardened alloys and alloys with > 3% magnesium.
Stainless steels	Cl⁻ ions	Through grains	Serious for austenitic steels at high temperatures.
Plain carbon steels	OH⁻ or NO₃⁻ ions, hydrogen sulfide	Grain boundaries	The terms *caustic* and *nitrate* cracking are sometimes used.
Brasses	Ammonia	Indiscriminate	Formerly called *season cracking*

with special reference to the structure near grain boundaries which provide the crack paths. In contrast, users of stainless steels are less concerned with such aspects but take account *inter alia* of the fact that the cracking is especially associated with the face-centered cubic austenite phase, in which it is transgranular. Hence, to gain an impression of the problems encountered it is more useful to consider some selected important alloy groups rather than to attempt to generalize.

5.1.2 Stress-Corrosion Cracking in Aluminum Alloys

5.1.2.1 Susceptible Alloys

Stress-corrosion cracking can afflict some but not all aluminum alloys. Unfortunately the alloys at risk include high-strength alloys for critical applications such as airframes, especially since runway de-icing materials and marine atmospheres are potential sources of the chloride ion, Cl^-, that is the specific agent. The susceptible alloys, with compositions specified by the Aluminum Association codes given in Table 9.2, Chapter 9, are:

1. Alloys that develop strength by quenching from high temperature and reheating at lower temperature to produce a dispersion of sub-microscopic precipitate particles within their structures (age-hardening):

   Aluminum-copper-magnesium-silicon alloys in the AA 2000 series

   Aluminum-zinc-magnesium-copper alloys in the AA 7000 series

   Aluminum-magnesium-silicon alloys in the AA 6000 series.

   Aluminum alloys containing lithium.

2. Alloys that develop strength by mechanically working them:

   Aluminum-magnesium based alloys in the AA 5000 series.

The age-hardening alloys suffer SCC only when hardened. The aluminum-magnesium alloys are susceptible only for magnesium contents over 3%.

Commercial grades of pure aluminum, aluminum-manganese alloys, and aluminum-silicon alloys are immune. The crack path is intergranular and is associated with the characteristics of the grain boundary area.

5.1.2.2   Probable Causes

Aluminum alloys, in general, derive protection by passivation within the pH range 3.9 to 8.6, as might be inferred from the Pourbaix diagram in Figure 3.4. and effects related to localized corrosion such as SCC are associated with local passivity breakdown. The main theories are based on this assumption.

Crack propagation during SCC is discontinuous and is generally attributed to alternating mechanical and electrochemical actions. Attention is drawn to the capacity of the depassivating chloride ions to stimulate localized corrosion at the grain boundaries, producing fissures which act as stress-raisers. Stress concentrated at the fissures opens them, exposing new surfaces sensitive to further corrosion, extending the cracks and increasing the stress concentration. The alternating electrochemical and mechanical effects constitute an iterative synergistic process. The following section considers how the grain boundaries become selective sites for the electrochemical aspect of SCC.

5.1.2.3   Mechanisms

The intergranular crack path naturally focusses attention on the metal structure in the immediate vicinity of the grain boundaries and explanations are sought by two different approaches that are not necessarily mutually exclusive. One is based on anodic activity due to selective precipitation of intermetallic phases at the boundaries and the other is based on the entry of cathodically produced hydrogen along grain boundaries.

Grain Boundary Precipitation

A feature common to aluminum alloys susceptible to SCC, i.e., age-hardening alloys and alloys with high magnesium contents, is precipitation of intermetallic compounds at the grain boundaries.

The age-hardening alloys are formulated to promote, by a suitable sequence of heat-treatments, a sub-microscopic precipitate of intermetallic compound particles highly dispersed within the crystals. Its function is to obstruct the movement of dislocations in the lattice and thereby harden and strengthen the metal. There is, however, a pre-disposition for some larger precipitate particles to nucleate and grow at the grain boundaries, with the result that precipitate particles are aligned along the grain boundaries and surrounded on both sides by very narrow bands of metal that are depleted in the alloy components contributing to the precipitate, known as the *grain boundary denuded zone*. There is thus established a continuous network of three closely adjacent bands of metal with different compositions, the precipitate, the denuded zone and the matrix, between which there are possibilities for electrochemical interactions. The role of the specific agent, the chloride ion is to initiate these interactions by assisting in local depassivation.

In aluminum-copper-magnesium-silicon alloys of the AA 2000 series, the grain boundary precipitate includes the compound, $CuAl_2$, and the denuded zone is anodic to both the precipitate and the matrix, creating possibilities for local action electrochemical cells, somewhat similar in action to the bi-metallic cells described in Chapter 4. In high strength aluminum-zinc-magnesium-copper alloys of the AA 7000 series, the precipitate, $MgZn_2$, is anodic to the matrix, creating local action cells of reverse polarity. Other age-hardening alloys can be considered similarly but with differences in detail.

Heat-treatment programs for the age-hardening alloys have a pronounced effect on the susceptibility to SCC and must be carefully designed and controlled to minimize it. In practice, the alloys are strengthened by two sequential heat treatments. The first is a high-temperature *solution treatment*, at a temperature in the range 500-510°C, according to the alloy, in which the alloy components responsible for the hardening are dissolved and the metal is quenched in water to retain them in solid solution. The second is a low-temperature *artificial aging treatment* at temperatures of the order of 170°C, according to the alloy, in which the alloy components in enforced solution go through a complex sequence of clustering, coalescence and precipitation. During the aging treatment, the strength and hardness of the metal rise to a maximum peak hardness and then as the precipitation process nears completion, they decline and the metal is then described as *overaged*. From a purely mechanical viewpoint, the best combination of properties is reached at peak hardness but unfortunately, this coincides with maximum susceptibility to SCC. In critical applications, reduction of sensitivity to SCC is more important than fully exploiting the mechanical properties available, and a suitable compromise is attained by controlled over-aging. There are alternative approaches, include two-stage aging and modified solution-treatment or quenching. Alloys containing copper can be *naturally aged* by holding them at ambient temperature after quenching but materials hardened in this way are very susceptible to SCC.

Plain aluminum–magnesium alloys derive their strength from work-hardening and not age-hardening but because the solubility of magnesium falls rapidly with falling temperature, alloys with >3% magnesium, tend to precipitate the compound Mg_5Al_8, that is anodic to the matrix, at the grain boundaries.

Hydrogen Embrittlement

An alternative or supplementary theory of stress-corrosion cracking is based on experimental observations that although hydrogen is virtually immobile in the matrices of aluminum alloys at ambient temperature, it can permeate the grain boundaries in quenched aluminum alloys containing magnesium. The idea is that hydrogen generated by a complementary cathodic reaction at the metal surface can enter the metal via the grain boundaries and disrupt it by *decohesion,* an effect well known in other contexts.

Structural Factors

Intergranular stress-corrosion cracking is expected to be sensitive to the metal structure. In rolled sheet and plate and sections fabricated by extrusion, the metal crystals or grains are elongated in the direction of working. Therefore a much greater grain boundary area is presented to a stress applied normal to the working direction, the *short transverse* direction, than to stresses parallel to or across it, the *longitudinal* and *transverse* directions and the sensitivity to SCC is correspondingly greatest in the short transverse direction. Problems due to sensitivity in this direction are confined mainly to thick sections, such as rolled plate and forgings. In thinner sheet, where the stresses are longitudinal or transverse, the manganese and chromium contents, often present in alloy formulations, are said to assist resistance to SCC by promoting elongated grain structures. The chromium content is also credited with improving resistance to SCC by inducing general precipitation of aging precipitates throughout the grains in preference to grain boundaries.

5.1.3 Stress-Corrosion Cracking in Stainless Steels

5.1.3.1 Brief Description of Phenomena

Stress-corrosion cracking in stainless steels exemplifies the uncertainties and controversy which surround the topic in general. For a long time it was assumed that SCC was confined to austenitic stainless steels formulated to be composed almost entirely of the face-centered *austenite* phase, as described in Chapter 8, and that cracking was always transgranular, i.e., through the grains. Ferritic stainless steels, composed of the body-centered *ferrite* phase, also described in Chapter 8, were assumed to be immune, introducing a crystallographic consideration. It is now known that SCC can afflict ferritic steels and that the crack paths can be intergranular in both ferritic and austenitic steels in special circumstances but the detail is uncertain, due mainly to less experience of the effects in ferritic steels because, in general, they are not used for such critical applications as austenitic steels. Further complexity is sometimes introduced by including cracking in stainless steels in which carbides are present at the grain boundaries, due to incorrect heat-treatment or to unsuitable selection of steels for welding, as described in Chapter 8. The following discussion is restricted to transgranular cracking in correctly heat-treated austenitic stainless steels which constitutes the main body of the practical problems encountered.

5.1.3.2   Environmental Influences

The specific agent is chloride and the problem is serious only at elevated temperatures, especially for environments of boiling or superheated aqueous solutions. The time-to-failure is reduced as the temperature is raised and as the chloride content in the environment increases. The effective chloride content may be much higher than the nominal content if there is a concentration mechanism, by, for example, cyclic evaporation of condensates or evaporation of water from dilute chloride solutions leaking from or dripping on to a heated pressure vessel. The susceptibility to cracking is not particularly sensitive to pH, but in general it is more severe in more acidic solutions. In principle, the dissolved oxygen content is an important factor because it provides the oxygen absorption cathodic reaction to complement the electrochemical contribution to SCC but it is difficult to divorce from the temperature-sensitivity of the effect because the solubility of oxygen in aqueous media diminishes rapidly as the temperature is raised and, in any case, since SCC is a phenomenon associated with solutions at high temperatures, the oxygen content is inevitably low.

5.1.3.3   Sensitivity to Steel Structure and Composition

The relationships between the susceptibility to SCC and steel compositions, condition, and microstructures are complex, often known only empirically and sometimes not reproducible.

The cracks are transgranular and can exhibit multiple branching. The presence of some ferrite in nominally austenitic stainless steels can block the cracks suggesting that ferrite is more resistant to SCC. Worked austenitic steels do not have sufficient ferrite for the effect to have any practical value but castings of equivalent steels have non-equilibrium structures as described in Chapter 8 and can contain as much as 13% ferrite, persisting from an uncompleted peritectic reaction, which affords a useful degree of protection. For the same reason, *duplex steels*, that are formulated to contain both austenite and ferrite whether cast or wrought are less susceptible than austenitic steels.

The influence of the composition of an austenitic stainless steel on its susceptibility to SCC is not well-defined. High nickel and chromium contents improve resistance to SCC but, judging from published information, the influences of other alloy components, including molybdenum and carbon, which are often components of austenic stainless steels, do not seem to be reproducible, perhaps because of uncertainties in interactions between the components of the alloys.

5.1.3.4   Mechanisms

The electrochemical contribution to stress-corrosion cracking is confirmed by the academic observation that it can be suppressed by cathodic protection and enhanced by galvanic stimulation but there is no consensus on how stress and environmental attack interact. Two stages need explanation, i.e., induction and crack propagation.

Induction

Induction must be associated with local breakdown of the passive film and active/passive cells are established to initiate the electrochemical contribution. Since cracking is transgranular, intrinsic features of the austenite grains or their surfaces must be invoked and not features of their boundaries, as for aluminum alloys. One idea is that local perforation of the passive layer occurs by penetration of microscopic slip steps; this is supported by a relationship between susceptibility to SCC and the energy of faults in the sequence of atom layers, the *stacking fault energy*. Another is that chloride-induced pits can act both as stress-raisers and sites for active/passive cells. Dissolution is driven by the potential difference between the large passive metal surface and the active metal at the crack tips.

Propagation

The cracks can follow crystallographic planes but do not do so consistently and this must be reflected in any proposed explanation. There are two approaches, depending on the view taken on whether it is the electrochemical or mechanical contribution that propagates the cracks. One view asserts that the cracks are propagated by electrochemical dissolution of a narrow strip of metal and the role assigned to the mechanical contribution is to maintain the metal at the crack in an active condition tip by causing it to yield, preventing re-passivation. The other view considers that the metal is separated mechanically under tension and the electrochemical contribution is to resharpen the crack tip when it is blunted by yielding. There is probably some truth in both ideas.

5.1.4 Stress-Corrosion Cracking in Plain Carbon Steels

5.1.4.1 Brief General Description of Phenomena

Plain carbon steels are susceptible to stress-corrosion cracking in the presence of one or another of specific agents that include nitrates, bicarbonates and alkalis and also hydrogen sulfide. Cracking in alkalis is sometimes called *caustic cracking*. All of these agents have technological significance. Additions of nitrates and *pH correction*, which means raising the pH, are sometimes used in moderation to inhibit general corrosion of steel and hydrogen sulfide is encountered in oilfield sour liquors. The crack paths are characteristically intergranular.

Stress-corrosion cracking promoted by nitrates can occur at ambient temperatures but it is more often experienced with superheated solutions under pressure where the temperature exceeds 100 oC. Caustic cracking is associated with hot solutions with high pH values but mildly alkaline solutions also pose risks because the pH can be raised locally, e.g., by evaporation of water from leaks in heated vessels, such as boilers.

5.1.4.2 Mechanisms

Unlike aluminum alloys and stainless steels, plain carbon steels are active in neutral aqueous solutions, but increasing the pH or raising the potential by nitrate additions induces passivity due to the formation of a surface film of magnetite, Fe_3O_4, as suggested by the Pourbaix diagram in Figure 3.3. At any faults in the film, the exposed iron is the anodic partner in an active/passive cell and consequent intense local attack stimulated by the large surrounding passive surface can provide the electrochemical component for SCC. Explanations for the synergism between electrochemical and mechanical components of SCC can be sought along the same lines as for the aluminum and stainless steel systems described above. The intergranular crack path suggests that structural features at the grain boundaries, such as the presence of carbides, facilitate the depassivation that can initiate and sustain cracking.

Cracking in hydrogen sulfide solutions is a form of hydrogen embrittlement in which the sulfide stimulates catalytic activity assisting the entry of cathodically produced hydrogen into the metal where it can promote decohesion. Special steels with exceptionally low sulfur contents < 0.002% are produced for application to structures at risk because they are more resistant.

5.2 Corrosion Fatigue

5.2.1 Characteristic Features

Corrosion fatigue cracking differs from stress-corrosion cracking in two respects:

1. **Stress** — The cracking is induced by a cyclic applied stress in a crack-opening mode.
2. **Environment** — Combinations of environmental conditions and metal compositions for corrosion fatigue are not specific.

Fatigue failure of a metal is characterized by delayed fracture associated with cracking induced by cyclic stresses well below the maximum constant stress that the material can bear. A good general account is given in Reed-Hill's standard text on physical metallurgy cited in "Further Reading."

The life expectancy depends not only on the properties of the metal and on the stress system but also on the nature of the environment. All environments influence the fatigue life but the distinction of most concern is the marked reduction in life expectancy when an air environment is replaced by aqueous media, for which the term *corrosion fatigue* implies interaction between mechanical and electrochemical factors. The overall effect is usually quantified on a comparative basis by laboratory fatigue tests, usually by rotating bending or reverse bending tests using internationally standardized samples. By convention, the results are presented graphically as *S-N* curves, in which the number of stress cycles needed to cause failure are expressed as a function of the stress amplitude. The logarithmic plot of cycles-to-failure is used only for convenient presentation. The life-to-failure becomes progressively shorter as the cyclic stress amplitude is raised. A typical example of effect of an aqueous environment is illustrated in Figure 5.1, which compares the fatigue life of an austenitic stainless steel in 0.5 M sodium chloride solution with its fatigue life in air. For steels cyclically loaded in air, there is an *endurance limit*, that is a safe limiting stress below which failure does not occur but for non-ferrous metals and for metals cyclically loaded in aqueous media, there is no safe limit and on the basis of empirical information and experience, stresses are restricted to values ensuring survival for a prescribed design life, typically 10^7 or 10^8 cycles.

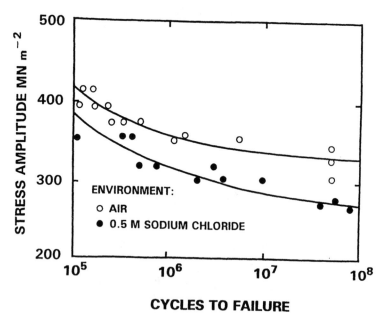

Figure 5.1

Comparison of fatigue lives of AISI 316 stainless steel in air and in 0.5 M sodium chloride solution. Results of tests on 3 mm thick plate in reverse bending at 24 Hz. Nominal composition of steel 17% Cr, 12% Ni, 2.5% Mo, 0.06% C. (Reproduced with permission of the Institute of Materials, London.)

5.2.2 Mechanisms

5.2.2.1 *Mechanical Events Causing Failure*

Fatigue failure of a metal is the culmination of a complex sequence of events, summarized briefly as follows:

1. **Crack Initiation** — An incubation period, that can constitute a significant part of the fatigue life, precedes crack initiation. In this period, the characteristic event is the formation of *persistent slip bands* PSBs, a term describing highly localized regions of cyclically yielding metal at the surface, that are the precursors of cracks.

2. **Crack Propagation** — Cracks grow initially along crystallographic planes within the PSBs, a process called *Stage 1 cracking* but usually then change direction to follow planes normal to the maximum applied stress, a process called *Stage 2 cracking*, often leaving microscopic striations on the fracture surface, marking successive increments of crack advance.

3. **Overload Fracture** — Ultimately, the load-bearing section is reduced to an area which is unable to sustain the maximum stress and the metal separates by fast fracture.

5.2.2.2 *Intervention of the Environment*

The events listed above provide considerable scope for intervention by the environment to assist the surface events associated with the PSBs. Several environment/stress interactions are possible in principle, as follows:

1. Local corrosion damage, e.g., pitting, intensifying the surface stress.

2. Stimulated dissolution of active metal by the yielding within PSBs.

3. Dissolution of metal exposed by the rupture of passive films over PSBs.

4. Localized embrittlement by hydrogen absorbed from cathodic reactions.

According to circumstances, all are possible contributing factors, founded on well-established concepts. Local corrosion damage can act in the same way as other surface imperfections in raising the local surface stress above the nominal applied stress, thereby reducing fatigue life but in many other examples of environment-accelerated failure, there is no local corrosion damage and one of the other factors must be invoked. These imply various forms of electrochemical activity stimulated by the applied cyclic stress. They have been identified experimentally by monitoring corrosion current transients synchronized with stress cycles. For a passivating metal, such as a stainless steel, their amplitudes increase if depassivating ions are present and for an active metal, such as steel, their development depends on the accumulation of plastic strain in the PSBs. These observations provide evidence for the electrochemical activity envisioned.

5.3   Erosion-Corrosion and Cavitation

Erosion-corrosion and cavitation effects are induced by rapid relative movement between a flowing electrolyte. e.g., water, and metal parts, pipes, or containers. All metals can be affected to a greater or lesser degree.

5.3.1   Erosion-Corrosion

Erosion-corrosion accelerates corrosion by dispersing the mechanisms that protect metals in static or slow-moving contact with an aqueous environment. As the name implies, one aspect of the effect is to scour the metal surface, interfering with the formation of films that would otherwise offer protection. This applies not only to passivating metals but also to other metals that, although not normally protected by passivity, derive at least some protection from surface films. If the moving liquid carries solid particles in suspension, the scouring effect is so much the greater.

The relative movement also tends to sweep away the boundary layer of static liquid present at the metal/liquid interface. This further stimulates corrosion by dispersing concentration polarization, especially for the oxygen reduction reaction, discussed in Section 3.2.2.5, and for anodic reactions yielding soluble products.

5.3.2   Cavitation

Cavitation is produced by the impingement of a liquid on a metal at high velocity and is particularly associated with rapidly moving metal parts in water, such as propellers and pump impellers. The relative movement induces a hydrodynamic condition that creates streams of small cavities in the liquid which collapse, delivering multiple sharp blows at the metal surface. The disturbance disrupts protective films, leading to a corroded surface with a characteristic rough and pitted appearance.

Material properties which militate against cavitation and also erosion-corrosion are good general corrosion resistance, strong passivating characteristics and hardness of the metal.

5.4   Precautions Against Stress-Induced Failures

Irrespective of the particular characteristics of systems susceptible to stress-corrosion cracking, corrosion-fatigue or other stress/environment interactions, it is prudent to take fairly obvious precautions related to the materials, the specific agents and the magnitude of applied stresses.

Materials

Probably the greatest hazard is failure to appreciate the problems. It is essential to be aware of the metal/environment systems that are suscepti-ble to stress-corrosion cracking and take the available information into account when specifying materials. A few examples suffice to make the point. If a stainless steel must be used in an application in which there is a possibility of exposure to chlorides, it is wise to consider one of the steels with lower susceptibility to stress-corrosion cracking such as a duplex steel, even if there is a cost or technical incentive to consider a more sus-ceptible steel, such as a fully austenitic steel. It has already been noted that with age-hardening aluminum alloys, it is prudent to forego the maximum mechanical properties available by controlled overaging to reduce suscep-tibility to stress-corrosion cracking.

Environments

In all load-bearing applications of metals susceptible to stress-corrosion, it is essential to be sure that neither obvious nor latent sources of the spe-cific agent(s) are present. Examples of obvious sources are process liquors, marine atmospheres and de-icing salts containing chlorides and oil instal-lations contributing sour liquors. Examples of less obvious sources are spills, leaks, condensates and salts leached from insulation by drips from above.

It is not always appreciated that fatigue is a system property and is environmentally-sensitive. Information given in the context of material characterization alone is usually based on fatigue tests conducted in air but it is essential that even comparative data is evaluated and matched to the actual stress system and environment of the application envisioned.

Stresses

If metals must be used in a system in which environmentally sensitive cracking or corrosion-fatigue is a possible hazard, viable designs are still possible if stresses in all parts of a structure are kept below known safe limits, using information from experience or reliable sources. A designer would not deliberately exceed nominal safe stresses, discounted by the usual factors of safety but, in practice, design stresses may be inadvert-ently supplemented by other stresses. Stress-corrosion cracking and corrosion fatigue are surface-sensitive and stress-raisers can markedly increase nominal surface stresses. For this reason, abrupt changes in section and artifacts due to poor surface finishing should be eliminated. Metals can, and frequently do, carry internal stresses which are additive to the external loads. Some of these stresses are imparted by the metal supplier, especially stresses locked in by mechanical working, and by differential contraction following heat-treatments. Others are imparted by contraction after welding and by careless assembly. These stresses can and should be minimized by careful working and heat-treatment practices and if possible eliminated by stress-relieving heat-treatments. Surface compressive stress is considered beneficial and shot-peening is sometimes advocated to introduce it.

Geometry

Geometric considerations apply to stress reduction to lessen the risks of environmentally-sensitive cracking but they are also important in ameliorating erosion-corrosion and cavitation corrosion. Systems containing flowing liquids should be designed for minimum relative movement and avoid profiles that contribute to turbulence.

Monitoring

In particularly critical situations, where environmentally sensitive cracking is possible, it may be advisable to regularly monitor the condition of the metal and that of the environment during service. This information can be used to forestall impending damage and adverse environmental changes.

Further Reading

Gangloff, R. P. and Ives, M. B. (Eds.), *Environment-Induced Cracking of Metals*, National Association of Corrosion Engineers, Houston, TX, 1990.

Hydrogen Embrittlement and Stress-Corrosion Cracking, American Society for Metals, Metals Park, OH, 1984.

Staehle, R. W., Forty, A. J., and Van Rooyen (Eds.), *Fundamental Aspects of Stress-Corrosion Cracking*, National Association of Corrosion Engineers, Houston, TX, 1969.

Newman, R. C. and Procter, R. P. M., Stress Corrosion Cracking, 1965-1990, *Br. Corros. J.*, 25, 259, 1990.

Hatch, J. E. (Ed), *Aluminum Properties and Physical Metallurgy*, American Society for Metals, Metals Park, Ohio, 1984, Chap. 7.

Scamens, G. M., *Hydrogen-Induced Fracture of Aluminium Alloys*, Symposium on Hydrogen Effects in Metals, TMS AIME, Warrendale, PA, 1980, p.467 and *Aluminum*, 59, 332, 1982.

Staehle, R. W. (Ed.), *Stress-Corrosion Cracking and Hydrogen Embrittlement of Iron Base Alloys*, National Association of Corrosion Engineers, Houston, TX, 1977.

Parkins, R. N., *Corrosion Sci.*, 20, 147, 1980.

Proc. Symp. on Effect of Hydrogen Sulfide on Steel, Canadian Institute of Mining and Metallurgy, Edmonton, 1983.

Reed-Hill, R. E., *Physical Metallurgy Principles*, Van Nostrand, Princeton, NJ, 1964, p. 559.

Talbot, D. E. J., Martin, J. W., Chandler, C., and Sanderson, M. I., Assessment of crack initiation in corrosion fatigue by oscilloscope display of corrosion current transients, *Metals Technology*, 9, 130, 1982.

6

Protective Coatings

Often, the best strategy to control corrosion of an active metal is to apply a protective surface coating. A further advantage is that it is usually possible to combine the protective function with aesthetic appeal that is valuable for vehicles, domestic and business consumer durables, architecture, and even for ephemeral packaging where sales potential is enhanced by attractive appearance. The principal applications of surface coatings on metals is for the protection of iron and steel products and structures because of the sheer volume of production, estimated at 7.5×10^8 tonnes per annum worldwide. Equivalent coatings are applied to other metals but sometimes as much for appearance as for protection.

Casual observation shows that most steel products and structures are coated either with other metals or with paints. Most metal coats are produced by electrodeposition but the older practice of *hot-dipping*, in which the substrate is passed through the liquified coating metal is still important for producing zinc-coated, *galvanized* steel. The term, *paints*, describes filled polymer binding media, including air-drying and stove-drying formulations, irrespective of trade descriptions. Other less conspicuous but important, coatings are generated by reactions in which the metal surfaces themselves participate. Three are of particular interest (1) phosphate coatings that provide a key for paints on steel, (2) anodic films that extend the applications of aluminum and its alloys, and (3) chromate coatings that can protect or provide a key for paints on non-ferrous metals.

6.1 Surface Preparation

6.1.1 Surface Conditions of Manufactured Metal Forms

The surfaces of manufactured metal forms, including sheet, plate and sections are seldom suitable for the application of coatings. Poor surface quality, detritus and contamination undermine the adhesion of electrodeposits and detract from the protection afforded by paints. They must be prepared before coatings are applied. The preparation needed depends on the metal and the manufacturing and fabricating processes by which they are produced.

Rolled Surfaces

The surfaces of rolled products, plates sheets and sections, can be and usually are heavily contaminated. The kind of contamination depends on the metal, as described below for steels and aluminum alloys.

Steels are hot-rolled at high temperatures, of the order of 1100°C and the steel surface oxidizes to form thick *mill scale*, that spalls away from plain carbon steels under the roll pressure. It is more difficult to detach from some alloy steels and patches may remain on the finished hot-rolled product. When hot-rolling is finished, the metal temperature is still high enough to form a final scale as it cools. If the hot-rolled product is to be cold-rolled, the scale must first be removed. During cold-rolling, the roll gap is lubricated with mineral oils and traces remain on the surface of the finished metal sheet. Any intermediate heat-treatments in the schedule provide further opportunities for scale formation.

Different criteria apply to aluminum alloys that are hot-rolled at temperatures of the order of 500°C. They do not form thick scales like steels but are subject to a different surface effect. The deformation in the roll gap introduces differential velocities between the periphery of the roll and the metal being rolled, so that the metal slips backwards on the roll face on entry and forwards on exit. The relative motion induces transfer of aluminum particles to the roll face that are carried round and transferred back to the metal as the roll revolves. As successive ingots are rolled, a steady state is established in which the roll surface carries a constant thin coating of the metal and the extreme surface of the emerging stock is a layer of compacted flake, a few µm thick. The roll is lubricated and cooled with a recirculated soluble oil/water emulsion that interacts with the metal surface yielding metal soaps, degrading and contaminating the lubricant. The flake is compacted on the metal surface in this unsavory environment, producing a surface condition illustrated for pure aluminum in Figure 6.1. Subsequent cold-rolling further compacts and burnishes the flake surface and also impresses its own characteristic artifact on the surface, i.e., reticulation, a network of very fine, shallow surface cracks, offering traps for further contamination.

Extruded Surfaces

Extruded aluminum sections may be scored but they are reasonably clean because they are abraded against the hard dies through which they are forced.

Surfaces of Press-Formed Products

Aluminum alloys, plain carbon steels and austenitic stainless steels have extensive applications for hollow-ware produced by press-forming, usually by deep-drawing in which rolled sheet is deformed by a punch into a shaped cavity, The sheet is restrained by pressure pads to prevent wrinkling and lubricated with mineral oil to control friction. The metal is

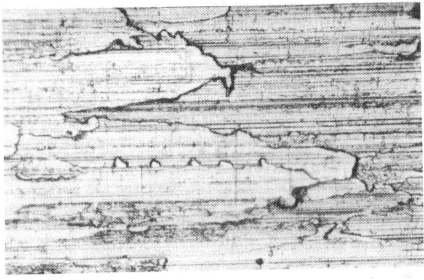

20 μm

Figure 6.1
Compacted flake on hot-rolled pure aluminum (AA 1100) surface.

heated adiabatically by the heavy deformation and the transient surface temperature can be high enough to carbonize some of the lubricant. The resulting contamination is loose to the touch but very resistant to removal by physically and chemically active solutions.

6.1.2 The Cleaning and Preparation of Metal Surfaces

6.1.2.1 Descaling Steels

Manual Methods

Manual methods such as grinding, shot and sand blasting and flame cleaning are slow, subjective and costly. Recourse to them is justified only if chemical descaling or weathering are inappropriate. Shot and sand blasting remove scale and roughen the surface providing a useful key for paint. Flame cleaning detaches scale by differential expansion and burns off oil and grease, leaving a dry surface that can be painted.

Weathering

Exposing steel to the weather for about six months exploits atmospheric corrosion to undermine mill scale so that it is easily removed. Weathering is often an economic way of preparing structural steelwork for painting. Subsequent cleaning must be conscientious, because residual scale stimulates corrosion under paint. Regions shielded from rain may not be adequately cleaned and need remedial attention. When correctly practiced, weathering and subsequent phosphate treatment yields substrates suitable for painting.

Acid Pickling

Mill scale on hot-rolled steel products is removed by pickling in sulfuric acid. It is effective, economical and is adaptable to on-line use in steel mills.

Dissolution of isolated iron oxides in acids is very slow, but oxides in contact with the metal can be attacked by *reductive dissolution*, meaning that the oxide is dissolved by a cathodic reaction. This exploits the structure of scale formed on iron and steel explained more fully in Chapter 7. Scale formed at temperatures < 575°C comprises an inner layer of magnetite, Fe_3O_4 against the metal, overlaid by an outer layer of hematite,Fe_2O_3. Scale on hot-worked steel is cracked due to differential contraction on cooling. Iron is anodic to magnetite and acid percolating through the cracks to the metal establishes local cells in which cathodic dissolution of magnetite:

$$Fe_3O_4 + 2e^- + 8H^+ \rightarrow 3Fe^{2+} + 4H_2O \qquad (6.1)$$

is stimulated by the anodic dissolution of iron:

$$Fe \rightarrow Fe^{2+} + 2e^- \qquad (6.2)$$

The dissolution reactions proceed along the magnetite/iron interface, loosening the scale so that it falls away.

Scale formed at temperatures > 575°C has a third oxide layer, wüstite, FeO, interposed between the metal and the magnetite layer, that is particularly amenable to reductive dissolution, because it is unstable at temperatures < 575°C and on cooling it tends to decompose yielding a eutectoid mixture of iron and magnetite:

$$4FeO \rightarrow Fe + Fe_3O_4 \qquad (6.3)$$

In pickling, multiple dissolution cells rapidly consume the former wüstite layer, detaching the overlying scale. Hot-working finishing temperatures above 575°C are therefore conducive to subsequent easy descaling.

The simplest procedure is to immerse the material as pieces or in coils in static 0.05–0.1 molar sulfuric acid at 60–80°C, treated with restrainers that are organic additives to inhibit dissolution of the descaled metal. The acid is replenished and replaced as necessary. High-speed pickling is needed to descale moving continuous rolled steel strip. The strip is flexed to crack the scale and passed through 0.25 molar acid at 95°C flowing counter-current to it in a sequence of tanks, after which the metal is washed and dried on-line. Looping pits are strategically placed to enable the end of a strip to stop momentarily for the following strip to be joined by welding, permitting continuous operation.

A hazard in pickling is cathodic hydrogen generated on descaled metal:

$$2H^+ + 2e^- = H_2$$

At ambient and slightly elevated temperatures hydrogen can diffuse into iron and steels, raising blisters at subcutaneous flaws. In thick sections, it can be occluded and cause embrittlement by decohesion, as referred to in Chapter 5 in the context of stress-corrosion cracking. These problems can be averted by attention to the pickling program and to the metal quality.

There are various modifications to the basic pickling process, including the application of externally impressed cathodic current to protect exposed descaled metal or impressed anodic current to suppress hydrogen embrittlement.

6.1.2.2 Cleaning Aluminum Surfaces

Many applications of aluminum depend on a combination of corrosion resistance and attractive appearance that can be imparted by metal surface finishing treatments. Whatever treatment is applied, the cleanliness of the metal surface is critically important in determining both the initial quality of the finish and its durability.

The surface of the metal as it emerges from casting or fabrication operations is both physically and chemically contaminated. Physical contamination can comprise lubricant residues from working operations, adherent dirt and particles embedded from mechanical polishing. Chemical contamination includes oxidation and corrosion products.

The principles used to remove contamination are straightforward but acceptable procedures are more difficult to devise. The constraints are:

1. Costs of cleaning.
2. Ability of cleaning agents to wet the surfaces.
3. Dispersion and trapping detritus to prevent re-deposition.
4. Quality of cleaned surfaces.
5. Toxicity and environmental impact of cleaning agents.

Solvent Degreasing

Physical detritus and oily residues are loosened by organic solvents. The choice and application of a solvent is restricted by cost, flammability, low flash points and the toxicity of chlorinated hydrocarbons and of benzene; kerosene is reasonably safe but because its volatility is low it evaporates slowly and it can leave oily residues.

The most effective application of solvents is in vapor degreasing in which articles are first immersed in hot solvent to remove the worst contamination and then cooled and held in its vapor to attract condensate that washes off oils and detritus. There are many variants of the process in both the selection of solvent and physical arrangement of the plant. The choice of solvent is not easy. Trichloroethane, $C_2H_3Cl_3$ is effective, but although safe in many respects, it is a candidate for depleting the ozone layer. Trichloroethylene, C_2HCl_3 is less likely to cause ozone depletion but it is more toxic and there is a small risk of dangerous exothermic polymerization catalyzed by traces of aluminum chloride formed from hydrogen chloride produced by photolytic decomposion of trichloroethylene; to counter this risk, commercial trichloroethylene is *stabilized* by adding a soluble amine as a scavenger for hydrochloric acid. For this reason, the more stable solvent, tetrachloroethylene is preferable.

A limitation of degreasing with organic solvents is that they are inefficient in removing water soluble contaminants such as soaps and soluble oil residues from lubricants used in machining and metal forming. This can be ountered by using a two-phase cleaning fluid in which an aqueous phase and an organic solvent are used together and are made compatible by incorporating an emulsifying agent, such as triethanolamine with oleic acid that produces a soap with both hydrophilic and oleophilic properties. Such a system dissolves or loosens both oil soluble and water soluble matters so that they can both be washed away by water. Arrangements can be made to apply the fluids by immersion in tanks or by spray.

Alkaline Cleaning

Physical cleaning by solvent degreasing is followed by alkaline cleaning, dissolving a surface layer of metal to release firmly adherent oxidation products and subsurface contamination e.g., metal soaps compacted under flake as illustrated in Figure 6.1.

Reference to the aluminum Pourbaix diagram in Figure 3.4 shows that it is easier to attack aluminum in alkaline than in acidic solutions. A pH value in the range 9 to 11 is sufficient. Sodium hydroxide attacks the metal too readily and damages the surface by etching it. Cleaning solutions are therefore based on sodium carbonate, Na_2CO_3, and sodium tertiary phosphate, Na_3PO_4, whose etching action can be suppressed by the addition of sodium metasilicate, Na_2SiO_3, assumed to be due to the deposition of a very thin layer of hydrated silica on the freshly exposed aluminum surface. In practice, solutions containing all three salts together with surfactants to promote wetting are used. The concentrations and relative proportions of the components vary within the range 10 to 60 g dm^{-3} and are determined by experience. Operation is by immersion of the metal for a few minutes in the solution at a temperature in the range 75 to 95°C.

The suitability of nominally cleaned metal for further surface treatments cannot be judged by appearance but may be assessed by spraying it with water; a tendency for the water falling on the metal to break into globules indicates inadequate cleaning.

6.1.2.3 Preparation of Aluminum Substrates for Electrodeposits

Electrodeposits do not adhere well to aluminum and aluminum alloy substrates unless they are specially prepared because of interference from the air-formed oxide film and the reactivity of the metal. Although alternative, more complicated, techniques are sometimes used, the most widely applied surface preparation is to deposit a thin layer of zinc on the aluminum surface by chemical replacement. The process is reliable, inexpensive and suitable for most aluminum alloys.

Cleaned aluminum alloy articles are immersed in an alkaline solution of sodium zincate, Na_2ZnO_2, that dissolves the oxide film and replaces it with the zinc coating. Solutions are prepared by dissolving zinc oxide in sodium hydroxide solution. The relative proportions are fairly critical to produce adherent zinc deposits. The replacement reaction is represented by coupled anodic and cathodic reactions, such as:

$$2Al + 8OH^- \rightarrow 2AlO_2^- + 4H_2O + 6e^- \qquad (6.4a)$$

$$3ZnO_2^{2-} + 12H^+ + 6e^- \rightarrow 3Zn + 6H_2O \qquad (6.4b)$$

Typical solutions contain 40 to 50 g dm^{-3} of zinc oxide and 400 to 450 g dm^{-3} of sodium hydroxide but there are many proprietary variants. A representative treatment is 1 to 3 minutes immersion at ambient temperature but it may be varied to suit alloy compositions and pre-treatments given to the metal.

6.1.2.4 Chemical and Electrochemical Polishing of Aluminum Alloys

Several metals can be brightened by controlled chemical or electrochemical dissolution in specially formulated solutions. There are niche applications for other metals but the principal industrial use is for aluminum alloys for the following reasons:

1. For many decorative uses, value added to the product justifies the costs of treatment.

2. The bright surface can be protected by anodizing, as described in Section 6.4.2.

3. The color of the bright surface matches nickel/chromium coating on steel or plastics and mixtures of them in, e.g., automobile trim presenting an integrated appearance.

4. Chemically and electrochemically polished aluminum can be not only bright but specular, i.e., yielding undistorted image clarity, a quality needed for reflectors.

Evidence from extensive investigation indicates that whether a solution etches, polishes or passivates an aluminum surface depends on which of the following responses it elicits:

1. Direct anodic dissolution of the metal

$$Al \rightarrow Al^{3+} + 3e^- \tag{6.5}$$

2. Formation of a very thin oxide or hydrated oxide film on the metal surface

$$2Al + 3H_2O \rightarrow Al_2O_3 + 6H^+ + 6e^- \tag{6.6}$$

   that does not suppress Reaction 6.5 but exercises diffusion control over it.

3. Formation of a thicker or qualitatively different film that suppresses Reaction 6.5, passivating the metal.

Polishing corresponds to the second of these responses because it is well established that brightening is associated with a tangible film on the metal, qualitatively similar to but thinner than that formed in anodizing, *qv*. The role of the film is to control dissolution, overriding differential dissolution due to the effects of crystallographic features of the metal surface which would otherwise be responsible for etching.

Several proprietary solutions produce the required brightening. They are mainly, but not exclusively, based on mixtures of concentrated phosphoric, sulfuric, and nitric acids. Phosphoric acid is the preferred acid for dissolution because it attacks aluminum reasonably uniformly. The nitric acid is the oxidizing agent responsible for formation of the film. The less expensive sulfuric acid is sometimes used for partial replacement of phosphoric acid as an economy. A representative formulation is 80.5% phosphoric acid + 3.5% nitric acid + 16% water by volume; the addition of 0.01 to 0.2 wt% of copper improves the brightness obtained. The solution is used hot, e.g., at 90°C and the work is immersed in it for a time between 15 seconds and five minutes. The quality of the finish depends on skill and experience in working the solution since the best results are obtained when optimum concentrations of aluminum phosphate (6 to 12 mass%) and free phosphoric acid (65 to 75 mass%) have been established during use. Some patented brand names of solutions in widespread use are ALCOA's *R5* and Albright and Wilson's *Phosbrite 159*.

As concerns with the environment and safety at work increase, polishing baths containing nitric acid that evolve nitric oxides may give way to solutions free from nitric acid and based on phosphoric acid and sulfuric acid, such as Albright and Wilson's *Phosbrite 156*, that yield bright but non-specular finishes, suitable for less critical decorative applications such as anodized domestic and bathroom fittings.

The very best specular finishes are given by electrochemical polishing. One of the more successful is the British Aluminium Company's process used under the registered trademark *Brytal*. The process is expensive and works well only for high-purity aluminum, i.e., > 99.8% preferably 99.99%, and aluminum–magnesium alloys based on these purities but the specular reflectivity of Brytal treated and anodized reflectors for use in e.g., laser cavities is unrivalled except by silver, which tarnishes easily.

The electrolyte is a solution of 20 mass % sodium carbonate, Na_2CO_3, and 6 mass % sodium tertiary phosphate, Na_3PO_4, both of high purity, in de-ionized water yielding a pH of 11.0 to 11.6. The bath is heated by steam coils to 90°C and operated with an anodic potential of + 7 to 12 V applied to the work against an aluminum cathode. It is agitated to disperse concentration polarization, with care to avoid disturbing a sludge deposited in the tank. Manipulation is expensive through the jigging and provision of electrical connections to the work.

Like chemical polishing, electrochemical polishing is associated with diffusion control of the dissolution process by a film on the metal surface; it is assisted by an overlying viscous liquid phosphate complex of unknown constitution. The condition correlates with a *polishing range* in the anodic polarization characteristics for the process in which current density is independent of potential.

6.2 Electrodeposition

6.2.1 Application and Principles

Electrodeposition provides a convenient means of applying a protective coating of one metal on another. A particular attraction is the ability to control the thickness and uniformity of the coatings fairly accurately through the Faradaic equivalent of the total electric charge passed, allowing for any inefficiencies. This permits the surface qualities of expensive metals such as nickel and tin to be imparted to metals of lower value, such as steels, by applying them as very thin coatings. Often a coating is required to have both protective value and aesthetic appeal and in that case there is a further economy in using electrodeposition because it is usually possible to produce bright decorative finishes that require no subsequent treatment. Furthermore, the versatility of electrodeposition commends it for a wide variety of applications ranging from batch plating of small parts to continuous plating of strip emerging at high speed from steel mills.

6.2.1.1 *Cathodic and Anodic Reactions*

Provided that hydrogen evolution is suppressed, metals can be deposited from aqueous solution by application of a sufficient cathodic potential. The species from which the metal is extracted can be cations, e.g.:

$$Zn^{2+} + 2e^- \rightarrow Zn \tag{6.7}$$

oxy-anions, e.g.:

$$Cr_2O_7^{2-} + 14\,H^+ + 12e^- = 2Cr + 7H_2O \tag{6.8}$$

or complex cyanide anions, e.g.:

$$[Cu(CN)_2]^- + e^- = Cu + 2CN^- \tag{6.9}$$

Deposition from cations is relatively simple and is usually the option with lowest cost but deposition from anions is sometimes essential to obtain deposits that match particular applications. The cathodic potential needed is the sum of the equilibrium potential for the reaction and the total polarization for deposition at the required rate. Activation, concentration and resistance polarization can all affect the nature of the deposit.

The return current is led into the plating bath by anodes, that are preferably made of the same metal as that deposited, because the depletion of metal ions at the cathode is continuously replenished by matching dissolution at the anode, e.g.:

$$Zn(anode) \rightarrow Zn^{2+} = 2e^- \tag{6.10}$$

For some metals, notably chromium, it is impracticable to use anodes of the metal deposited and inert anodes must be used, and the anodic reaction by which the return current enters the bath is oxygen evolution:

$$H_2O \rightarrow \tfrac{1}{2}O_2 + 2H^+ + 2e^- \tag{6.11}$$

The depleted metal ion content of the bath is then replenished by intermittent additions of metal-bearing salts.

6.2.1.2 Hydrogen Discharge

Metal deposition reactions compete with the discharge of hydrogen:

$$2H^+ + 2e^- = H_2(gas) \tag{6.12}$$

for which, as shown in Chapter 3:

$$E_{HYDROGEN} = -0.0591\ pH \tag{6.13}$$

Plating solutions are concentrated and the metal activity is of the order of unity so that the potential at which the metal is deposited, $E_{DEPOSITION}$ is close to the standard potential for the electrode process, $E^{\ominus}$. A particular metal can be deposited from aqueous solution only if the cathodic reaction by which it is deposited is thermodynamically favored or kinetically easier than Reaction 6.12. This depends on the following circumstances:

1. $E_{DEPOSITION}$ is less negative than $E_{HYDROGEN}$ and the discharge of hydrogen is not thermodynamically favored. An example is copper deposited from Cu^{2+}.

2. $E_{DEPOSITION}$ is more negative than $E_{HYDROGEN}$ but the hydrogen overpotential on the metal is so high that the discharge of hydrogen is either suppressed as for zinc deposited from Zn^{2+} and tin deposited from Sn^{2+} or limited to a tolerable degree as for nickel deposited from Ni^{2+}.

3. $E_{DEPOSITION}$ is so much less negative than $E_{HYDROGEN}$ that the hydrogen discharge reaction cannot be polarized sufficiently to reach the potential needed to deposit the metal. This is why aluminum and magnesium cannot be electrodeposited from aqueous solution.

6.2.1.3 Throwing Power

Except for materials with the simplest geometry, such as flat sheets, it is seldom possible for all parts of the substrate to be equidistant from anode surfaces. Therefore the current density and hence the thickness of the deposit is not uniform over the substrate surface but varies from place to place due to differences in the ohmic resistance of the path through the electrolyte to an anode. The distribution of deposition current due to this factor is the *primary current distribution*. Anodes are shaped and disposed around the substrate so that it is as uniform as possible.

The actual current distribution is usually more uniform than the primary current distribution suggests because of the effects of polarization. The total potential across the electrodeposition cell is the sum of three components:

1. Any difference between the equilibrium potentials for the reactions at the anode and cathode.
2. The potential drop across the ohmic resistance of the electrolyte.
3. Polarization at the cathode and at the anode.

The potential applied across the cell is constant so that variations in the potential drop due to variations in ohmic resistance are offset by opposite variations in the potential associated with polarization. The effect is to deflect deposition current away from places where the current density is high, so that the deposit is more uniform than would be predicted from the primary current distribution. The quality of a plating bath that promotes this uniformity is called its *throwing power*. A high throwing power is essential for deposits applied to complex shapes.

TABLE 6.1

Representative Solutions for Electrodeposition

Metal	Solution	Solutes	g dm^{-3}	Temperature °C	Cathode Current Density A dm^{-2}
Nickel	Watts solution and derivatives	$Ni_2SO_4 \cdot 6H_2O$ $NiCl \cdot 6H_2O$ H_3BO_3 additives in derivatives	240 45 30	0–50	0.5–2.0
Copper	Cyanide strike solution	$CuCN$ $NaCN$ Na_2CO_3	23 34 15	40	0.5
Copper	Concentrated cyanide solution	$CuCN$ KCN Na_2CO_3 Brightening agents	55 100 7.5	80	3
Chromium	Chromate solution	CrO_3 H_2SO_4	400 4	45	20
Tin	Acid sulfate solution	$SnSO_4$ H_2SO_4 $C_6H_5(OH)SO_3H$ β-naphthol Gelatin	55 100 100 1 2	45	30

Tin	Acid fluoborate solution	$SnSO_4$	54	35	10–50
		HBF_4	120		
		Proprietary additives			
Tin	Halide solution	$SnCl_2$	75		
		NaF	25		
		KF	50		
		NaCl	45		
		β-naphthol			
Tin	Alkaline stannate solution	$Na_2SnO_3 \cdot 3H_2O$	90	80	1.0–2.5
		NaOH	7.5		
		$NaC_2H_3O_2 \cdot 3H_2O$	15		
Zinc	Acid sulfate solution	$ZnSO_4 \cdot 7H_2O$	350	30	3.0
		NH_4Cl	30		
		$NaC_2H_3O_2 \cdot 3H_2O$	15		
		H_2SO_4 to pH 4			
		Proprietary additives			
Zinc	Alkaline cyanide solution	$Zn(CN)_2$	60	50	2.0
		NaCN	25		
		NaOH	50		

6.2.1.4   *Illustrative Selection of Deposition Processes*

Processes yielding useful metal deposits share essential principles and a comprehensive review would involve unnecessary repetition. Detailed empirical descriptions are readily available in handbooks and trade journals. Sections 6.2.2 through 6.2.6 describe the deposition of selected metals, chosen for their wide commercial applications and because as a group they illustrate sufficient of the electrochemical background and operating procedures to characterize electrodeposition generally. The selection comprises nickel, copper and chromium that are the components of a common protective and decorative coating system, tin that is used extensively in contact with foodstuffs and zinc that is applied to steels for protection in external atmospheres. Typical formulations for solutions used in depositing these metals are collected for reference in Table 6.1.

Pourbaix diagrams are sometimes useful in interpreting the chemistry of electrodeposition of particular metals and the appropriate versions are those for ion activities of 1, that apply to metal concentrations close to those used in practice.

6.2.2   Electrodeposition of Nickel

6.2.2.1   *General Considerations*

Electrodeposited nickel is widely used and one of the main applications is as part of a corrosion and tarnish resistant coating system for steel in which the nickel is deposited on a copper undercoat and overlaid by a bright chromium deposit, a finish, usually known as "bright chromium plating".

Nickel cannot be plated successfully from complex ions and commercial baths are based on simple nickel salts, using consumable anodes. The character of the deposit is sensitive to the pH of the plating solution and serviceable coatings can be produced only if the pH is controlled at a value close to pH 4. The object is to avoid co-depositing nickel hydroxide without risking embrittlement by hydrogen occlusion. This can be explained conveniently, using the nickel/water Pourbaix diagram for $a_{Ni^{2+}} = 1$, given in Figure 6.2.

The diagram shows that sufficient nickel is soluble for viable electrodeposition only if pH < 6. The metal is electrodeposited by depressing the potential of the substrate to lie in the domain of stability for nickel (Ni). The line representing the equilibrium potential for hydrogen evolution lies below the equilibrium potential for nickel deposition if pH > 4.2 and this would suggest that it might be possible to deposit nickel without discharging hydrogen in the pH range 4.2 to 6. However, practical processes operate with a cathodic overpotential and some hydrogen is always evolved. Without countermeasures the evolution of hydrogen poses the following problem:

1. If the initial pH > 5, the depletion of H^+ ions at the cathode causes the local pH to drift into the $Ni(OH)_2$ domain and nickel hydroxide is incorporated in the deposit.

2. If the pH < 3 hydrogen is evolved so rapidly that some is occluded in the deposit.

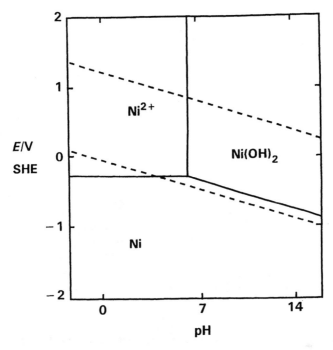

Figure 6.2
Partial Pourbaix diagram for the nickel/water system at 25°C. $a_{Ni2+} = 1$. Domain for the stability of water shown by dotted lines.

Boric acid is added to the solution as a *pH buffer* that maintains a characteristic constant hydrogen ion activity, corresponding to pH 4, by the incomplete dissociation:

$$H_3BO_3 + H_2O \leftrightarrows B(OH)_4^- + H^+ \qquad (6.14)$$

Practical problems must also be addressed in formulating a plating bath:

1. Hydrogen bubbles nucleating at preferred sites on the cathode divert the deposition current, leaving *pits* in the deposit. This is countered with anti-pitting agents, that are wetting agents to detach bubbles and oxidizing agents, such as hydrogen peroxide, that modify the cathodic reaction to produce water instead of hydrogen:

$$2H^+ + H_2O_2 + 2e^- = 2H_2O \qquad (6.15)$$

2. Impurities embrittle and discolor the plate and interfere with subsequent chromium plating and both the salts and the consumable anodes must be pure. Unfortunately, pure nickel anodes passivate easily, interrupting the current and this is averted by adding depassivating Cl^- ions to the solution.

A formulation meeting these requirements is the *Watts solution,* named for its originator, using nickel sulfate as the main source of nickel with a typical composition given in Table 6.1. The Watts solution is suitable for industrial applications but because the deposits are soft, coarse-grained, and dull, it is modified for bright decorative deposits, hard deposits and high-speed plating.

6.2.2.2 Bright Nickel Plating

Bright-plating solutions are Watts solutions with organic additives in empirical proprietary formulations. They contain brighteners that poison easy nucleation sites refining the grain and levellers that even out undulations. The structure of a bright deposit is featureless except for alternating light and dark bands, parallel to the surface, visible in microsections.

The buffered value of pH can be adjusted in the range pH 3 to pH 5 as required by additions of sulfuric acid or nickel carbonate. The high-purity nickel anodes may be dosed with oxygen or sulfur to assist dissolution and are enclosed in fabric bags to contain the residues.

Steel, iron, aluminum, and zinc alloys are attacked by the acidic nickel plating solution and are prepared to receive nickel by depositing an acid-resistant copper undercoat from alkaline solution. Aluminum and its alloys are first depassivated by sodium zincate treatment.

6.2.2.3 Other Nickel Plating Processes

Increasing the proportion of nickel chloride in the Watts solution increases the cost but confers two advantages:

1. The greater ionic mobility of Cl^-, allows higher current densities.
2. The deposits are harder.

The concentration of boric acid is raised and the pH is reduced to allow for the greater pH drift accompanying the high current density.

There are other nickel electrodeposition processes, less frequently used, but based on the principles described, e.g., using nickel fluoborate and nickel sulfamate as the nickel source.

6.2.3 Electrodeposition of Copper

Acceptable deposits of copper can be obtained from acidic copper(II) sulfate solution, using organic additives to refine the structure that is otherwise coarse-grained. By themselves, these deposits have little application as protective coatings because copper is a strong galvanic stimulant to base substrates such as steel exposed at defects in the coating. Because it is easy to build up thick deposits, copper deposition from acidic baths has applications in electroforming, e.g., it was used for type-face in printing techniques now obsolete. The most important use of electrode-posited copper in protective coatings is as an undercoat, especially on steel, forming part of comprehensive coating systems and for this purpose copper is plated from alkaline copper cyanide solutions.

Alkaline solutions have the advantage over acidic solutions that copper can be plated directly on base metals including steel, zinc and aluminum with much less risk of undermining the adhesion of the coating by chemical interaction with the substrate. Even so, the solution from which copper is first applied to steel, the copper *strike solution* is relatively dilute, to discourage deposition of copper by chemical displacement. A typical strike solution, containing CuCN and NaCN, using copper plates as consumable anodes is given in Table 6.1. The balance of solutes is to obtain a pH in the range 11.5–12.5. The coat is struck with a low current density at moderate temperature.

The two cynanides interact and dissociate, yielding cuprocyanide anions:

$$CuCN + NaCN = Na^+ + [Cu(CN)_2]^- \tag{6.16}$$

and the cathodic reaction depositing copper is:

$$[Cu(CN)_2]^- + e^- = Cu + 2CN^- \tag{6.17}$$

Deposition of the complex, negatively charged anion at a cathode leads to high activation polarization, diminishing the effect of easy nucleation sites and yielding a bright deposit. The coating is built up by depositing more copper on the strike coat with a higher anode current density, from a more concentrated solution, with a typical composition given in Table 6.1.

6.2.4 Electrodeposition of Chromium

6.2.4.1 Applications

Electrodeposited chromium is used extensively to give a brilliant, decorative and tarnish-resistant finish. The deposits are very thin, usually < 1 μm, and by themselves they offer little protection because they are passive but porous and galvanic corrosion stimulated at a substrate such as steel exposed by the pores can undermine and detach the coat. For this reason, the chromium is deposited over an initial thicker nickel deposit that provides the main protection. The nickel underlay must itself be bright to benefit from the brilliance imparted by the chromium.

Industrial applications exploit the hardness of chromium deposits that can approach 900 Brinell, and for this purpose, thick deposits, e.g., 25μm. are applied directly to steel for wear resistance. Examples include bearing surfaces and surfaces mating to close tolerances such as the contacts of measuring gauges.

6.2.4.2   Principles of Deposition

It is difficult to produce metallic electrodeposits of chromium from aqueous solution and its successful development owes as much to empirical craft as to electrochemical science. Chromium deposited from Cr^{3+} cations is unsuitable for decorative or industrial use, because it contains entrained oxide and a viable electrodeposition process is based on solutions of chromate ions that yield deposits with the required characteristics.

The chromium source is > 99% pure chromic acid anhydride, CrO_3 that yields several chromium-bearing species in mutual equilibrium in aqueous solution:

$$CrO_3 + H_2O = H_2CrO_4 \tag{6.18 a}$$

$$CrO_3 + H_2O = 2H^+ + CrO_4^{2-} \tag{6.18 b}$$

$$CrO_3 + H_2O = H^+ + HCrO_4^- \tag{6.18 c}$$

$$2CrO_3 + H_2O = 2H^+ + Cr_2O_7^{2-} \tag{6.18 d}$$

in all of which the chromium is hexavalent. The relative abundance of these species is pH-dependent and, even in undisturbed solution, it is complex but in the dynamic situation at a cathode with the high current density used in electrodeposition it is uncertain. Selecting one of these species, $Cr_2O_7^{2-}$, to characterize cathodic reactions that deposit the metal:

$$Cr_2O_7^{2-} + 14\ H^+ + 12e^- = 2Cr + 7H_2O \tag{6.19}$$

To add to the complexity, the process does not work properly unless a small quantity of one of certain other acids is added to the solution. In practice, sulfuric acid is added in sufficient quantity to give a CrO_3/SO_4 ratio between 50 and 150.

6.2.4.3   Operation of Chromic Acid Baths

It is impracticable to use consumable chromium anodes because they are a more expensive source of the metal than chromic acid anhydride. A further disadvantage is that anodic dissolution of chromium is so much more efficient than cathodic deposition that it produces a surplus of chromium in the bath, partly in the form of the unwanted Cr^{3+} ion. Insoluble anodes of a lead-tin or lead-antimony alloy are used and the depletion of the chromium content is replenished by additions to the solution as required.

The concentration of chromic acid anhydride in the bath is 250 to 400 g dm^{-3}. The cathode efficiency is very low, about 15%, and this, together with the high valency state of the chromium, imposes a heavy consumption of electric power for deposition. Raising the chromium content of the electrolyte raises its conductivity but this is offset by reduced cathode efficiency. From experience, the concentration and current densities are optimized to secure the most economic operation. This turns out to be 400 g dm^{-3} with lower current densities for thin bright deposits and 250 g dm^{-3} with high current densities for industrial hard deposits.

Some hexavalent chromium ions are reduced at the cathode to the trivalent ions, Cr^{3+}. These are re-oxidized at the cathode, catalyzed by lead dioxide, PbO_2 that forms on the anode surface and the proportion of chromium in the form of Cr^{3+} is kept in check by using anodes with a large surface area. A passive layer of lead chromate formed on anode surfaces during idle periods must be cleaned off before resuming deposition.

6.2.4.4 *Quality of the Deposit*

To produce a bright deposit, the temperature and current density must be optimized to lie within a restricted *bright plating range* as indicated in Figure 6.3. Outside of this range, the coating has an unacceptable appearance.

The throwing power of the chromate bath is poor and sometimes negative because of the inverse relation between cathode efficiency and current density. However for decorative coatings, it is less important than the ability of the system to yield bright deposits over the whole substrate.

These effects cause difficulty in controlling the process and its success is due to careful jigging, the use of auxiliary anodes, dummy cathodes and shields to supplement or deflect local currents and strict temperature control e.g., 45 ± 1°C.

The existence of the bright plating range, the role of SO_4^{2-} in assisting deposition and the very low cathode efficiency are all difficult to explain. It is known that a thin membrane is formed on the cathode during deposition, perhaps by an supplementary reaction, yielding an oxide species, e.g.:

$$Cr_2O_7^{2-} + 8H^+ + 6e^- = Cr_2O_3 + 4H_2O \qquad (6.20)$$

Resistance polarization for metal deposition reactions introduced by the film is probably one factor contributing to the brightness of the deposit.

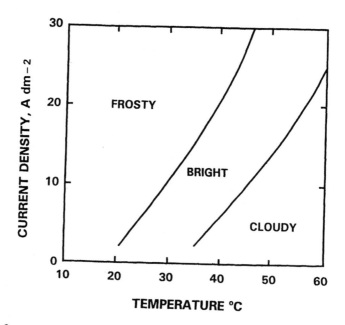

Figure 6.3
Influences of temperature and current density on the appearance of chromium electrode-posited on nickel substrates.

6.2.5   Electrodeposition of Tin

6.2.5.1   General Principles

The Pourbaix diagram for $a_{Sn^{2+}} = 1$, given in Figure 6.4 indicates that tin is sufficiently soluble for electrodeposition if the pH of the solution is either <0.5 or >14; this applies to ions formed by interaction of tin with water, i.e., Sn^{2+}, Sn^{4+}, SnO_3^{2-}, and $HSnO_2^-$. The diagram cannot give the further information that the halide ions, Cl^- and F^- extend the solubility of tin in the intervening pH range by forming complex anions, $SnCl_3^-$ and SnF_3^-.

   The minimum potentials for tin deposition lie significantly below the corresponding potentials for hydrogen at all values of pH but hydrogen evolution is suppressed, because the hydrogen overpotential on tin is very high.

   Three types of commercial electroplating baths are in use (1) strongly acidic solutions, (2) strongly alkaline solutions, (3) halide solutions. Acidic and halide solutions contain divalent tin as Sn^{2+}, $SnCl_3^-$ or SnF_3^-, whereas alkaline solutions contain quadrivalent tin as SnO_3^{2-} so that, theoretically, the same quantity of electricity yields only half as much tin from alkaline as from acid or halide baths. The inferior power economy of alkaline baths is compounded by a lower cathode efficiency, i.e., about 75%, compared with $>95\%$ for acid baths and by lower conductivity of the electrolyte. The tin deposits all have a matte finish. It is technically possible but not

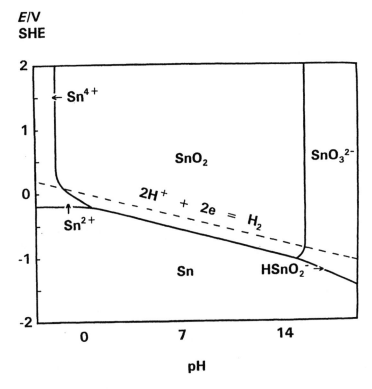

Figure 6.4
Pourbaix Diagram for the tin/water system at 298 K (25°C). $a_{Sn^{2+}} = a_{Sn^+} = a_{Sn^{2+}} = 1$.

commercially viable to deposit bright tin. Matte coatings are brightened by flash heating the substrate to above 232°C to melt the tin.

6.2.5.2 Acid Tin Baths

Acid Sulfate Baths

Acid sulfate solutions are often used for high speed deposition in tinplate manufacture. As expected, deposits from the simple Sn^{2+} ions:

$$Sn^{2+} + 2e^- = Sn \tag{6.21}$$

are coarse-grained but the structure is refined by de-activating easy nuclei with phenol or cresol, solubilized as the corresponding sulfonic acid derivatives:

$$C_6H_5OH(phenol) + H_2SO_4 = C_6H_5(OH)SO_3H(phenol\ sulfonic\ acid) + H_2O \tag{6.22}$$

supplemented by other additives such as β-naphthol and gelatine. A typical formulation is given in Table 6.1. For high speed deposition as in tinplate lines, the bath is agitated to disperse concentration polarization. For slower batch plating, a similar bath is operated at a lower temperature, 25°C, and lower current density. Pure tin consumable anodes are used to replenish the depleted tin content:

$$Sn = Sn^{2+} + 2e^- \qquad (6.23)$$

The anode current efficiency exceeds that of the cathode and to prevent drift of the tin content, the consumable tin anodes are supplemented by inert anodes.

Fluoborate Baths

Fluoborate baths also use tin(II) sulfate as the source of tin but the low pH is maintained by fluoboric acid instead of sulfuric acid as in the formulation given in Table 6.1. Proprietary additives are used instead of phenol or cresol.

6.2.5.3 Alkaline Stannate Baths

The expense of alkaline stannate baths is justified when high throwing power is essential for complex shapes. A typical formulation is given in Table 6.1. The throwing power is due to high cathodic overpotential for deposition from the complex ion SnO_3^{2-}, in which the tin is quadrivalent, by the cathodic reaction:

$$SnO_3^{2-} + 6H^+ + 4e^- = Sn + 3H_2O \qquad (6.24)$$

The depletion of quadrivalent tin ions is replenished by dissolution of consumable tin anodes:

$$Sn + 3H_2O = SnO_3^{2-} + 6H^+ + 4e^- \qquad (6.25)$$

but it is accompanied by a minority reaction producing some divalent SnO_2^{2-} ions:

$$Sn + 2H_2O = SnO_2^{2-} + 4H^+ + 2e^- \qquad (6.26)$$

that are subsequently oxidized to the quadrivalent ions by dissolved oxygen:

$$2SnO_2^{2-} + O_2 = 2SnO_3^{2-} \qquad (6.27)$$

unbalancing the anodic and cathodic reactions causing the tin content to drift upwards. This is dealt with using supplementary inert anodes and controlling the anode current density and the solution chemistry.

6.2.5.4 Halide Baths

Halide baths are used for tinplate manufacture. They are solutions of tin(II) chloride, $SnCl_2$, with alkali metal fluorides and chlorides, with β-naphthol as an additive as illustrated in the formulation given in Table 6.1. The components interact and ionize yielding the anions $SnCl_3^-$ and SnF_3^-:

$$SnCl_2 + NaCl = Na^+ + SnCl_3^- \qquad (6.28)$$

$$3SnCl_2 + 3NaF = 3Na^+ + 2SnCl_3^- + SnF_3^- \qquad (6.29)$$

and a refined structure is produced, again due to high polarization associated with deposition from complex ions:

$$SnCl_3^- + 2e^- = Sn + 3Cl^- \qquad (6.30)$$

$$SnF_3^- + 2e^- = Sn + 3F^- \qquad (6.31)$$

6.2.6 Electrodeposition of Zinc

Zinc coatings are used extensively to protect steel. They are well suited for the purpose because zinc is inexpensive, coating processes are easy and the protection is galvanic and very effective. Electrodeposition is used as an alternative to hot-dip galvanizing where the higher cost is justified by close dimensional tolerances, as in screw threads or by improved appearance. Bright deposits are technically feasible but the cost is not often justified for the utilitarian applications for which zinc coatings are used.

As with tin, zinc can be electrodeposited from acid and from alkaline baths, using consumable anodes. The former is simple to operate but the latter has higher throwing power for complex shapes.

Acid Sulfate Baths

The standard electrode potential for zinc is -0.76 V SHE that is very negative with respect to potentials for hydrogen evolution, but the hydrogen overvoltage on pure zinc is high and hydrogen evolution can be suppressed. This is favored by a formulation with a high $a_{Zn^{2+}}/a_{H^+}$ ratio, implying a high concentration of zinc sulfate, $ZnSO_4$, with the least acidity needed to keep it in solution. For this reason the pH is buffered in the range pH 3 to pH 4. Organic additives from among β-naphthol, glucose, dextrin and glycerol are added to refine the deposit. The anodes and solutes must be free from impurities like nickel and cobalt that co-deposit with the zinc, inviting hydrogen evolution by lowering the hydrogen overvoltage. A typical formulation, using sodium acetate as the pH buffer is given in Table 6.1.

Alkaline Cyanide Baths

Alkaline baths are solutions of zinc cyanide, sodium cyanide and sufficient sodium hydroxide to raise the pH > 13 to maintain a high zinc content in solution. A typical solution used without brighteners or other additives is given in Table 6.1. The improved throwing power over acid baths is probably due to high activation polarization associated with the deposition of zinc from anions. The ions present include, Na^+, ZnO_2^{2-}, $HZnO_2^-$, $Zn(CN)_4^{2-}$, CN^- and OH^-. Pourbaix diagrams are clearly of little value for solutions with so many ion species.

6.3 Hot-Dip Coatings

In hot dipping, a solid metal is immersed in a liquid bath of another metal and withdrawn with an adherent film of the liquid that solidifies on cooling. It is the simplest method of applying protective metal coatings to base metals but there are limitations. The melting point of the protecting metal must be substantially lower than that of the substrate metal and preferably low enough to allow the substrate to retain the benefit of work-hardening. There must be sufficient interdiffusion between the metals to enable a bond to form but not so much as to consume the thin layer of protecting metal that is required. In practice, the process is most useful for applying coatings of zinc, tin, or aluminum to steel. The processes for the three metals differ in detail but follow the same essential sequence of operations:

1. The steel is cleaned by pickling as described in Section 6.1.2.1.
2. It is fluxed with chlorides to promote wetting by the liquid metal.
3. It is immersed for a very short time in the liquid metal.
4. Excess liquid is drained off or wiped away by appropriate means, e.g., exit rolls for sheet or dies for wire.
5. Post-dipping treatments are applied as required.

6.3.1 Zinc Coatings (Galvanizing)

Galvanizing to give sacrificial protection to steel is the most common application of hot dipping and provides a major market for zinc. It is used to coat familiar fabricated articles, e.g., water tanks and buckets, but its main use is for semi-finished fabricated steel products including sheet and pipes, where it is an alternative to electrodeposition.

Zinc melts at 419.5°C and dipping is carried out in the temperature range 430 to 470°C. The flux is a mixture of zinc and ammonium chlorides that is applied to the steel before dipping, either by immersion in the liquid salt mixture or by pre-fluxing with an aqueous solution of the salts dried on the surface.

If pure zinc is used, an *alloy layer* is interposed between the zinc coating and the steel substrate. It comprises a sequence of brittle iron/zinc intermetallic compounds consuming some of the zinc and its thickness is manipulated to suit the application. If the zinc coating is intended as the sole protection and the material is to be fabricated, the brittle alloy layer can be virtually eliminated by adding 0.15 to 0.25% of aluminum to the liquid zinc, provided that the immersion time is short and temperature is not too high. For some purposes, e.g., for some automobile body panels, a zinc coating is used to supplement protection by paint, in which case the formation of the alloy coating is promoted to produce so-called iron/zinc (IZ) sheet that is better suited to phosphating. This is done by reheating the coated steel to allow all of the zinc to diffuse further into the iron surface.

A characteristic feature often observed on galvanized steel is the *spangle* a macroscopic grain clearly visible to the unaided eye; it has no disadvantage if appearance is unimportant.

6.3.2 Tin Coatings

Hot-dip tinning was the original process for the production of tinplate but it has been superseded by electrodeposition to apply the very thin coatings now required. It is still useful for smaller scale work especially to protect pieces of awkward shape, e.g., for equipment used in the food industry. Tin melts at 232°C and dipping is carried out in the temperature range 240 to 300°C using zinc chloride as the flux. The steel is inserted into the tin through a bath of the liquid flux confined in a frame on the tin surface, and withdrawn through a bath of palm oil also floating on the surface. The solidified tin coating has only a thin alloy layer of the compound $FeSn_2$ interposed between it and the steel.

6.3.3 Aluminum Coatings

Hot dip alumininizing is the most difficult of the three processes because the melting point of aluminum, 660.1°C, is high and the oxide film on aluminum is tenacious, so that it is not easy to produce a smooth finish. The steel is fluxed in a fused salt bath at a temperature approaching the temperature of the liquid aluminum to avoid chilling it. A thick, brittle alloy layer, based on the intermetallic compound, $FeAl_3$, is always present between the aluminum and the steel. A disadvantage of the alloy layer is that it severely limits the degree of deformation possible in any subsequent fabrication but for applications such as automobile exhaust systems, advantage can be taken of the high melting point of $FeAl_3$, 1160°C, to produce heat-resisting coatings by allowing all of the aluminum to be absorbed into the alloy layer by subsequent heating.

6.4 Conversion Coatings

Conversion coatings bear the name because the metal substrate participates in forming its own coating. Three processes are especially valuable: phosphate coatings on steels and zinc, anodic coatings on aluminum alloys, and chromate coatings on aluminum and zinc.

Phosphate coatings are an essential part of protective paint systems applied to manufactured steel products such as automobiles, bicycles, domestic, and office equipment. Without pretreatment of the metal substrate, paints adhere only by weak Van der Waal's forces and by interference keying to surface roughness. Phosphate coatings are chemically bound to the metal and their physical nature offers a firm mechanical key for paints. Because of this vital, but hidden role, they are among the most indispensable coatings in use.

Anodic coatings can greatly enhance the corrosion resistance and aesthetic appearance of aluminum and some of its alloys and they also have other useful attributes that extend the commercial application of the metal and its alloys in several directions. Moreover, in practice, the treatment is simple and, as a result, it is extensively applied.

Chromate coatings have miscellaneous applications; they are best known for enhancing the corrosion resistance of protective tin or zinc coatings applied to steel and for providing a key for paint on aluminum alloys but they can also serve as corrosion resistant coatings in their own right.

6.4.1 Phosphating

Phosphating is a commercial process, producing coatings of stable, insoluble phosphates of iron(II), zinc(II), manganese(II), nickel(II) cations or more often mixtures of them, bonded to a metal substrate, usually steels or zinc-coated steels but sometimes aluminum alloys.

6.4.1.1 Mechanism of Phosphating

Phosphoric acid is tribasic and forms a series of three iron(II) salts, primary, secondary and tertiary phosphates, $Fe(H_2PO_4)_2$, $FeHPO_4$ and $Fe_3(PO_4)_2$. The primary phosphate is appreciably soluble in water but the secondary phosphate is sparingly soluble and the tertiary phosphate is insoluble. Using the symbol, $\downarrow$, to indicate the insoluble and sparingly soluble phosphates, equilibria between the various phosphates and free phosphoric acid are:

$$Fe(H_2PO_4)_2 = \downarrow FeHPO_4 + H_3PO_4 \qquad (6.32)$$

for which

$$K_1 = \frac{a_{H_3PO_4}}{a_{Fe(H_2PO_4)_2}} \qquad (6.33)$$

and

$$3Fe(H_2PO_4)_2 = \downarrow Fe_3(PO_4)_2 + 4H_3PO_4 \qquad (6.34)$$

for which

$$K_2 = \frac{\left(a_{H_3PO_4}\right)^4}{\left(a_{Fe(H_2PO_4)_2}\right)^3} \qquad (6.35)$$

The activity quotients in Equations 6.33 and 6.35 give critical ratios of free phosphoric acid to primary phosphate needed to precipitate insoluble phosphates.

A phosphate coating is produced by treating the metal with an aqueous solution containing the soluble primary phosphate and free phosphoric acid, either by immersion or by spray. The solution is pregnant, with just sufficient free phosphoric acid to prevent spontaneous precipitation. An iron or steel article immersed in or sprayed with the solution destabilizes it by consuming free phosphoric acid:

$$Fe + 2H_3PO_4 = Fe(H_2PO_4)_2 + H_2 \qquad (6.36)$$

and presents a surface on which the insoluble phosphates nucleate. The ratio of free phosphoric acid to the total phosphate content is critical and a basic phosphating solution is most efficient if the ratio is 0.12–0.15 and it is close to boiling. Too much free acid impairs the growth of the coating and too little initiates precipitation within the solution. Zinc and manganese form equivalent series of phosphates and there are advantages in formulating phosphating solutions with combinations of phosphates of the three metals. Simple phosphate processes are very slow, even when hot, and treatments of 30 to 60 minutes are required; longer treatments are needed for iron or manganese solutions than for zinc solutions.

6.4.1.2 Accelerated Phosphating Processes

The criterion for the speed of a phosphating process is the time to cover the metal surface completely, not the time to precipitate a given coating thickness. Written in ionic form, the dissolution of iron, that drives the process is:

Anodic reaction: $$Fe \rightarrow Fe^{2+} + 2e^- \qquad (6.37)$$

Cathodic reaction: $$6H^+ + 2PO_4^{3-} + 2e^- \rightarrow H_2 + 2H_2PO_4^- \qquad (6.38)$$

The reaction creates a mosaic of anodic and cathodic sites over the metal surface. Since it is the cathodic reaction that consumes phosphoric acid, deposition of secondary and tertiary phosphates is at the local cathodic sites, where one of the reaction products is hydrogen gas that forms a surface film inhibiting the spread of the coating over the metal surface. The reaction is *polarized* by the hydrogen and it can be depolarized by adding oxidizing agents, e.g., using hydrogen peroxide, the depolarizing reaction is:

$$H_2O_2 + H_2 = 2H_2O \qquad (6.39a)$$

For sodium nitrite in acid solution the reaction is more complex and depends ultimately on the weakness of nitrous acid relative to phosphoric acid, producing the overall result:

$$2NaNO_2 + 2H_3PO_4 + H_2(adsorbed) \rightarrow 2NaH_2PO_4 + 2H_2O + 2NO(gas)$$
$$(6.39b)$$

Depolarizing the cathodic reaction also stimulates reaction at the anodic areas and the cathodic areas spread at their expense.

Most phosphate processes are accelerated, reducing treatment times to 1 to 5 minutes at 40 to 70°C, that is suitable for spray application on production lines. Coatings produced in accelerated processes are thinner, smoother and yield better paint finishes. The coating thickness is expressed by custom in mass per unit area and is typically 7.5 g m^{-2}. The condition of the metal surface influences the quality of the coating. Coatings on heavily strained metal can be uneven. Selective nuclei introduced by surface detritus or deep etching yield coarse-grained deposits but multiple nuclei derived e.g., from precipitation *within* zinc-bearing solutions, refines the grain.

There are options for producing phosphate coatings in situations where regular phosphating is impracticable, such as in on-site preparation or *ad hoc* use but the coatings are inferior. Phosphoric acid can be brushed on steel but residual free acid can cause blistering under paint. Proprietary alkali metal phosphate solutions are available that give very thin coatings suitable as a key for paint provided that subsequent exposure is benign.

6.4.2 Anodizing

Anodizing is the formation of thick oxide films on metal substrates, driven by an anodic potential applied to the metal in a suitable electrolyte. The process is important commercially for protective coatings on aluminum products but it also has niche applications for tantalum and titanium, notably for high integrity electrolytic condensers.

Films on aluminum and its alloys have the following general characteristics that extend the applications of the metal:

1. A thin layer of the film immediately adjacent to the metal surface is a dense adherent *barrier layer*.

2. The barrier layer is overlaid by a thicker layer of micro-porous material that is sealed by a post-anodizing treatment yielding an impermeable hard film.

3. The electrical resistance and dielectric constant of the film material are $4 \times 10^{15} \ \Omega \ cm^{-1}$ and 5 to 8 respectively at 20°C, that are both very high.

4. With suitable metal and treatment, the films can be transparent.

6.4.2.1   Mechanism of Anodizing

The electrolyte in which the anodic potential is applied to the metal determines which of two kinds of film is formed:

1. Non-solvent electrolytes, e.g., boric acid or ammonium phosphate that are indifferent to the anodic film.

2. Solvent electrolytes, e.g., sulfuric, oxalic and chromic acids that attack the film in the manner described below.

Non-solvent electrolytes produce barrier films with thicknesses corresponding to 1.4 nm/V of applied potential. They are used as dielectrics in electrolytic condensers but they have no applications as protective coatings because very high potentials, 400 V or more, are needed to produce films that are thick enough.

Solvent electrolytes provide a solution to this difficulty. They attack the barrier layer at discrete points. The barrier layer reforms continuously under the points of attack to restore the thickness corresponding to the applied potential, leaving a porous overlay and developing the film structure represented schematically in Figure 6.5. The pore structure is regular, imposing a cellular pattern on the outer part of the film that, in plan, appears as an array of closely packed hexagons with cell walls of about the same thickness as the barrier layer. Thus the film thickness is not potential-dependent as for non-solvent electrolytes but time-dependent and films thick enough for use as protective coatings can be grown in less than an hour at potentials in the range 12–60 V.

Diffuse electron diffraction patterns indicate that the structure of the freshly-formed oxide lacks long-range order but it spontaneously transforms slowly to γ-Al_2O_3. After anodizing, the porous film can be sealed in boiling water or steam, converting the γ-Al_2O_3 to a monohydrate, $Al_2O_3 \cdot H_2O$, that occupies sufficient extra volume to fill the pores.

6.4.2.2   Practice

General anodizing is carried out in sulfuric acid and less often in chromic acid. The procedures are not difficult and they are adaptable to continuous

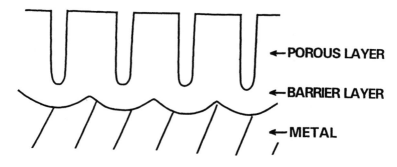

Figure 6.5
Schematic section through anodic film on aluminum, showing barrier layer and overlying porous layer.

operation for simple shapes. The acid is contained in rubber or lead-lined tanks and the workpieces are supported in jigs, with care to ensure good electrical contact. Since an anodic potential is applied to them, the jigs must be made from strongly passivating metal; aluminum and titanium are used in sulfuric acid but titanium is unsuitable for chromic acid. The anodic potential is applied between the workpieces and aluminum or lead cathodes. Since the coating is insulating, the film thickness is self-regulating and concepts of primary current density and throwing power do not apply.

Some qualities of the film make conflicting demands on the anodizing conditions and the practice is adapted to suit the application. Higher current densities, less concentrated acid and lower temperatures are needed for hard, abrasion resistant films than for transparent films. Hence close control of operating parameters is essential not only to maintain consistent results but also to determine the character of the product. The anodic current density is controlled using a dummy sample as a sensor and temperature is controlled by heating and cooling coils in the tank. Typical operating conditions are given in Table 6.2.

The metal surface is prepared for anodizing by etching, chemical brightening, or brushing as required and it must be clean, free from grease

TABLE 6.2
Typical Operating Conditions for Producing Anodic Coatings

Electrolyte	Temperature °C	Potential V	Current density A dm^{-2}	Coat thickness μm	Appearance
Sulfuric acid 10–15 wt %	15–24	10–22	1–3	3–50	Transparent or white
Chromic acid 3–10 wt %CrO$_3$	30–40	30–50	0.3–0.4	2–8	Opaque gray
Oxalic acid 3–8 wt %	20–40	30–60	1–3	10–60	Transparent yellow

Typical treatment times are 30 to 60 minutes

detritus, and crevices. After anodizing, the work is rinsed in cold water to remove electrolyte. At this stage, the porous film is absorptive and must be handled with care to preserve its appearance. It is sealed by boiling in water or exposure to steam for about an hour.

6.4.2.3 Applications

Coatings for Corrosion Protection:
Anodic films markedly improve the corrosion resistance of aluminum and its alloys. If a yellow appearance is acceptable it is enhanced by sealing in a solution of an inhibitor, e.g., sodium dichromate. Recommended film thicknesses for corrosion protection are:

Permanent outdoor exposure with no maintenance:	25 μm
Permanent outdoor exposure with good maintenance:	15 μm
Indoor exposure with good maintenance:	5 μm

A possible hazard in relying on anodic films for corrosion protection is intense attack by active/passive cells established at imperfections in the film overlying porosity and inclusions. This should not be a problem with material from reputable suppliers but it is prudent to specify that the material must be of adequate quality for anodizing. Nevertheless, for some applications, such as for thin-walled vessels holding liquids, it may be preferable to accept general corrosion of untreated metal rather than to risk perforation of anodized material.

Decorative Protective Coatings
Decorative applications depend on the transparency of the film that is produced in sulfuric acid and the facility to seal dyes in the outer porous layer of the film. Transparent films preserve the brilliance of aluminum and certain alloys brightened by mechanical and chemical polishing. Acceptance standards are high, imposing a requirement for special alloys based on high purity aluminum. Alloy components and impurities present as intermetallic compounds in the metal structure resist anodizing and are incorporated in the film, rendering it dull. Elements in solid solution are less bothersome. Iron is a particularly undesirable impurity and for the highest quality anodized finish it is limited to 0.006%. This implies the use of 99.99% purity aluminum, that is too soft even for decorative use but it may be hardened by a small magnesium content in solid solution without detriment to the brightening treatments or the transparency of the anodic film; one such alloy has the composition, 1% magnesium, 0.25% copper, balance 99.99% pure aluminum. For applications in which higher strength is required, or where the metal is machined, an age-hardening alloy is available with the composition 0.7% magnesium, 0.3% silicon, 0.25% copper, balance 99.99% pure aluminum. For less demanding mass

markets, a less expensive range of alloys is available, based on lower purity, 99.8% aluminum; they are given special heat-treatment during manufacture to keep iron in solid solution.

Miscellaneous Applications

Some specialized applications of anodic films are worth noting. Films from non-solvent electrolytes are used as dielectrics in electrolytic condensers; they are formed *in situ* in the condenser on > 99.8% pure aluminum foil, etched to increase the true surface area. Aluminum wire can be electrically insulated with regular anodic films for motor and transformer windings to exploit the heat-resistance of the film. Films 100 μm thick can be used as hard facings for moving aluminum parts and surfaces subject to hard wear, e.g., fuel pumps, camera parts, bobbins, and pulleys. Polished alloys based on high purity aluminum protected by thin anodic films are used in maintenance-free reflectors for heat, lasers and UV radiation.

6.4.3 Chromating

Chromating is a term covering several different treatments that produce firmly adherent coatings with a small but effective reserve of chromium in a high oxidation state immediately available to metal surfaces. The coatings can serve one or both of two purposes:

1. To enhance the corrosion resistance of passive metals.
2. To act as a key for paint systems.

The coatings are applied by immersing the metal or article to be coated in an appropriate solution containing a soluble chromate, usually, sodium dichromate, $Na_2Cr_2O_7$, for a few minutes. Simple neutral chromate solutions do not produce substantial deposits because they passivate the metal, inhibiting further reaction. Effective solutions therefore contain depassivating agents that may include sulfates, chlorides, fluorides, and nitrates with or without pH control to establish conditions in which passivity is broken down to allow reaction to proceed. The pH and the nature and concentrations of depassivating agents depend on the metal to be treated. This provides the scope for the development of proprietary treatments, developed for particular metals. The coatings can be produced on aluminum alloys, zinc, cadmium, and magnesium but not on iron and steel.

The composition of the coating can be quite complex and varies both from metal to metal and according to the way it is produced. The predominant active species is usually a slightly soluble hydrated chromium chromate, $Cr_2^{III}(Cr^{VI}O_4)_3$, that is the source of the oxidizing ion, $Cr^{VI}O_4^{2-}$. Other components can include the oxide of the metal treated, and other species introduced from the particular solutions used to produce the coatings.

6.4.3.1   Coatings on Aluminum

Chromate coatings on aluminum alloys have important functions as keys for paint systems, as corrosion inhibitors and as aesthetically pleasing color finishes. The term *chromating* is usually applied to chromate/phosphate as well as plain chromate coatings. They fulfil these roles economically and well, except for the most critical corrosion resistant and decorative applications, for which anodizing is preferred where its extra cost is justified. They are applied by treating the work with suitable aqueous solutions of chromate, CrO_3^{2-} and phosphate, PO_4^{3-} ions together with an activator such as fluoride ions to prevent passivation while the coatings are forming. The simplicity of processing gives chromating an advantage over anodizing for coating large and awkward work pieces. Chromating competes to some extent with alternative less effective simpler metal finishing processes, such as hydrated oxide (Boehmite) coatings produced by controlled chemical oxidation of the metal in water, reviewed by Werner, Pinner, and Sheasby cited at the end of the chapter.

Chromate and chromate/phosphate coatings have broadly similar applications but chromate/phosphate coatings are the more versatile and more widely applied. Proprietary treatments introduced by the American Chemical Paint Company are known under the name *Alodine* in the U.S.A. and *Alochrome* in Europe.

The structures of chromate/phosphate coatings are indeterminate and the compositions vary through the thickness with the highest chromium and phosphate contents towards the outside. Surface analysis indicates that they are based on hydrated chromium phosphate, $CrPO_4 \cdot 4H_2O$ and Boehmite, $AlO(OH)$ containing hydrated chromium oxide, Cr_2O_3 and chromium chromate, $Cr_2(CrO_4)_3$. Consequently, it is difficult to determine the reaction mechanisms but they must comprise balanced anodic and cathodic reactions. In principle, the dissolution of aluminum by the anodic reaction:

$$Al \rightarrow Al^{3+} + 3e^- \qquad (6.40)$$

can support the formation of chromic oxide, chromium chromate and chromium phosphate by notional overall cathodic reactions such as:

$$2Cr^{VI}O_4^{2-} + 10H^+ + 6e^- = Cr_2^{III}O_3 + 5H_2O \qquad (6.41)$$

$$5Cr^{VI}O_4^{2-} + 16H^+ + 6e^- = Cr_2^{III}(Cr^{VI}O_4)_3 + 8H_2O \qquad (6.42)$$

$$Cr^{VI}O_4^{2-} + 8H^+ + PO_4^{3-} + 3e^- = Cr^{III}PO_4 + 4H_2O \qquad (6.43)$$

Equations such as these show that the composition of the film deposited is critically dependent on the activities of CrO_4^{2-} and PO_4^{3-} ions in the solution, that are related to their concentrations, and on the activity of hydrogen ions, i.e., on the pH. The ratio of the activator to chromate activity is also critical. Satisfactory coatings are obtained only within optimum composition ranges. Outside of these ranges, the coating either does not form or has an unsatisfactory physical form. Effective solutions are obtained either by mixing chromic acid anhydride, CrO_3, phosphoric acid, H_3PO_4 and an activator source such as sodium fluoride, NaF, or by mixing sodium secondary phosphate, Na_2HPO_4, potassium dichromate, $K_2Cr_2O_7$, and sulfuric or hydrochloric acid. Two examples from among several patented compositions are:

1. H_3PO_4 (75%) 64 g dm^{-3} + CrO_3 10 g dm^{-3} + NaF 5 g dm^{-3}
2. $NaH_2PO_4 \cdot H_2O$ 67 g dm^{-3} + $K_2Cr_2O_7$ 15 g dm^{-3} + $NaHF_2$ 4 g dm^{-3} + H_2SO_4 5 g dm^{-3}

The treatment is given by immersing work pieces in the solution contained in stainless steel or aluminum vessels at a controlled temperature below 50°C. The immersion time is in the range 2 to 15 minutes, depending on the quality of coating required. After removal from the chromating solution, the work is rinsed in clean water, briefly immersed in dilute chromic or phosphoric acid and dried. The solution is maintained by removing sludge and replenishing the solutes consumed. For continuous operation, concentrated solutions can be applied hot, e.g., at 50°C by spray.

The coatings are hard and the mass deposited is typically 2×10^{-4} g dm^{-2} but it is adjusted to suit particular applications by varying the time of contact between metal and solution. They exhibit attractive pale-straw yellow to bluish green colors that can be exploited for decorative effect. The particular color depends on the alloy, the condition of the metal, the solution composition and the time of immersion, so that maintaining color match is not easy. Wrought alloys in the AA 3000, AA 5000, and AA 3000 series yield coatings of good appearance. Castings are more difficult because the coloration responds to segregation of alloy components, grain size effects and defects such as porosity and included oxides occasionally present. Alloys containing silicon do not respond well because they contain virtually elemental silicon as a eutectic component. Good corrosion resistance is associated with strong color probably because it indicates a high concentration of CrO_4^{2-} ions.

Plain chromate coatings are produced from proprietary solutions containing chromic acid, sodium dichromate and sodium fluoride but without a phosphate component and often with potassium ferricyanide described as an accelerator. These coatings are also indeterminate in composition and structure but seem to be based on mixtures of hydrated aluminum and chromium oxides. They are thinner and less aesthetically pleasing as colored finishes than chromium/phosphate coatings but they are reputed to provide a better base for paint; the process is also more difficult to control.

Arrangements can be made to apply thin chromate coatings to the cleaned surfaces of aluminum alloy strip as a preliminary treatment in continuous painting or lacquering lines. In one example of such a system, a concentrated solution of chromic acid is partially reduced, yielding a mixture of Cr^{III} and Cr^{VI} ions, and treated with colloidal silica to adjust the consistency for application to the metal strip by rubber or plastic rolls.

6.4.3.2 Coatings on Zinc

Zinc is used extensively as a cathodic protective coating for steel. Although the natural corrosion resistance of zinc in neutral environments is good it can be enhanced, especially during its initial exposure by pre-passivation with a chromate coating. The coating also prevents the formation of *white rust*, a disfiguring white bloom of zinc carbonate formed by the action of atmospheric carbon dioxide on the surface oxide film. The treatments are based on the New Jersey Zinc Company's Cronak process using an aqueous solution of 182 g dm^{-3} of sodium dichromate and 11 g dm^{-3} of sulfuric acid. Various equivalent proprietary solutions are also available. The metal is immersed in the solution for 10 to 20 seconds at ambient temperature, yielding a coating mass of about 1.5 g m^{-2}.

6.4.3.3 Coatings on Magnesium Alloys

The application of chromate coatings to magnesium alloys is important as a preparation for protective paint systems not only to provide a key for the paint but also to clear away any natural corrosion product formed on the metal during storage and prevent its reformation because it is alkaline and can degrade paint applied over it. Successful chromate coatings can be applied by proprietary treatments of such diversity that it would be tedious and unproductive to enumerate them. The source of chromate is sodium or potassium dichromate and the solutions differ in pH, in the activator used and whether the activator is applied to the metal surface before or during treatment. In a representative example, the metal is first treated in an aqueous solution of hydrofluoric acid to activate the surface with a film of magnesium fluoride, MgF_2, and then the chromate coating is formed by immersing the metal in a 10 to 15% aqueous solution of sodium or potassium dichromate with pH maintained in the range 4.0 to 5.5. All of the treatments are slow and immersion times of as much as 30 minutes are required at ambient temperatures.

6.5 Paint Coatings For Metals

Paint is the generic term for filled organic media applied to surfaces as fluids that subsequently harden into protective coatings. When applied to a metal, the hardened coating has the following functions:

1. To provide an environment at the metal surface incompatible with the mechanisms that sustain corrosion.

2. To maintain its own integrity against chemical degradation, mechanical forces and natural ultraviolet radiation.

3. To give an aesthetically pleasing, glossy, colored appearance. This is especially important for manufactured domestic consumer durables and automobiles.

The material must be amenable to the method of application and be available at the lowest cost consistent with the service envisioned. These are conflicting requirements that cannot all be satisfied by a single formulation and paint coatings are multi-layered systems comprising primers, intermediate coats and finish coats.

6.5.1 Paint Components

Paints have three principal kinds of components:

1. Organic binding media, that determine the drying characteristics of the paint and control the chemical and physical properties of the hardened coating.

2. Pigments that are insoluble fillers finely ground and mixed with the binding media.

3. Solvents to impart physical characteristics needed for application to the metal surface.

Other components are added to modify the rheology of the liquid, to assist dispersion of pigments, to accelerate drying, to impart plasticity in the hardened paint and to reduce skinning of air drying paints.

6.5.1.1 Binding Media

The product of a binding medium for a paint is a coherent continuous network structure. The precursors of this network are discrete structural units large enough to determine the form of the final structure yet small enough to be dispersed in a solvent for application to the substrate as a liquid. An essential requirement is that the structural units have *functional groups*, i.e., chemically active groups of atoms that enable them to *cross-polymerize*, i.e., link together to create a three-dimensional structure. The means by which the linking is activated is the first distinction between the various binding media available:

1. Air drying oils that *polymerize* spontaneously by absorbing oxygen from the air.

2. Synthetic resins that do not polymerize spontaneously, but do so when heated.

Air-drying oils are the natural products, linseed, soya bean, and tung oils. The rate of drying can be increased if the oils are pre-treated by heating them in the presence of catalysts to produce *boiled oils* or *stand oils*.

Synthetic resins are of several kinds, distinguished both by the organization of the basic structural units and by the functional groups responsible for cross-polymerization. There is a wide variety of functional groups that can be employed and they can be attached to various kinds of basic structural units. Detailed descriptions of the production and polymerization of these resins are available in specialized texts. For now, it is sufficient to identify them by name and to associate them with particular chemical and physical attributes they can confer on paints. They are classified according to their chemical constitutions. The following list gives the most widely applied materials but is by no means exhaustive:

Alkyd resins

Alkyd resins are *polyesters*, produced as basic structural units by the reaction of polyhydric alcohols with monobasic and dibasic organic acids. Their formation is exemplified in Figure 6.6 for the production and subsequent cross-linking of the polyhydric alcohol, glycerol, and the anhydride of dibasic phthalic acid. The links are formed between the hydroxy, –OH, groups on the alcohols and the characteristic carboxylate groups on the acid anhydride. These are both common precursors of alkyd resins. Various alcohols and acids can be used and mixed to produce a range of different resins to suit particular applications.

Amino Resins

Amino resins are characterized by the use of nitrogen atoms in the structural unit. Two of them, urea formaldehyde and melamine formaldehyde are common constituents of some of the more durable binding media. The structural unit is made by reaction of urea or of melamine with formaldehyde to add functional groups needed for cross-polymerization.

Epoxy Resins

Epoxy resins are complex molecules that have two kinds of groups available for cross-linking — hydroxyl groups and epoxide groups — one at either end. They can be cross-linked with other resins, including melamine formaldehyde, phenols, urea, and others.

Blends for Air-Drying Paints

Drying oils are essential components of binding media for air-drying paints used on site, where the option to harden them by thermal treatment is impracticable. However, paints based solely on natural drying oils lack abrasion resistance and gloss and do not weather well. The oils are therefore blended with sufficient quantities of synthetic resins to improve their

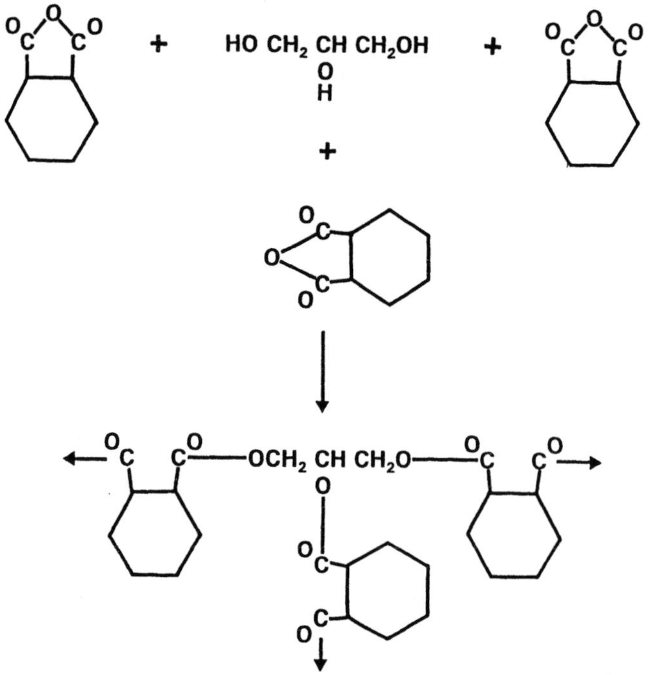

Figure 6.6
Cross-linking of phthalic anhydride with glycerol. The C–O–C ring in phthalic anhydride breaks to give two links, forming chains but glycerol has three OH linking groups to cross link the chains yielding a three-dimensional structure.

properties without losing the ability to dry spontaneously in air after application.

Air drying paints are not used for manufactured durable items like automobiles, domestic equipment and office equipment because air drying is incompatible with production line assembly, it is difficult to maintain control of the paints, the finish is not good enough and the properties of the hardened paint are usually unsuitable.

Blends for Stoving Paints

Binding media in stoving paints have a greater proportion of synthetic resins and are hardened by a thermal treatment, *stoving* or *baking*, typically for 30 minutes at 130 to 180°C, to suit the composition. Blends of resins are selected to suit applications.

6.5.1.2   Pigments and Extenders

Pigments and extenders are fillers with the following functions:

1. To give the paint *body*, the ability to form a thick coherent film.

2. To impart color either by self-color or by accepting *stainers*.

3. To protect binding media by absorbing ultraviolet light.

4. To act as corrosion inhibitors in primer paint coatings.

Pigments are classed as corrosion inhibiting pigments and coloring pigments. Extenders are used to replace part of the pigment content to reduce costs. Some common examples are:

Inhibiting pigments: Red lead, Pb_3O_4, calcium plumbate, $CaPbO_3$, zinc chromate, $ZnCrO_4$, and metallic zinc dust.

White pigments: Rutile and anastase, both forms of titanium dioxide, TiO_2.

Black pigment: Carbon black.

Red pigment: Red oxide, a naturally occurring Fe(III) oxide.

Yellow pigment: Yellow oxide, a natural hydrated Fe(III) oxide.

Organic pigments: Stable insoluble organic substances with various colors.

Extenders: Barytes, $BaSO_4$, Paris white, $CaCO_3$, dolomite, $CaCO_3.MgCO_3$, Woolastonite, $CaSiO_4$, talc and mica.

Pigments and extenders are ground to particle sizes in the range 0.1 to 50 μm, as appropriate, for intimate mixing with the binding media.

6.5.1.3   Solvents

Solvents disperse the binding media and pigments, reduce viscosity and impart characteristics needed for applying the paint to metal surfaces.

Organic Solvents

Traditional solvents for paints are volatile organic liquids that reduce the viscosity of the unpolymerized binding media, facilitating application; they subsequently evaporate, depositing the binding media/pigment mixture. They are selected from among hydrocarbons, e.g., white spirit, xylene and toluene, alcohols, e.g., methanol and ethanol, ketones, and esters.

Water

Binding media can be modified by processes called *solubilization* to disperse them in water as ionic species that can be applied to metal surfaces by electrodeposition. There are two options, the binding media can be given either anionic or cationic character, that can be deposited at metal anodes or cathodes respectively. Cationic character is the better option because deposition at a metal cathode does not damage phosphate coatings and gives paint coatings that have superior performance in service. Paints formulated with these modified resins are known as *waterborne* paints. Cationic character is imparted by treating suitable insoluble resins with an organic acid yielding the resin anion and a counter ion:

$$
\begin{array}{cc}
R & R \\
| & | \\
R-N: + R'COOH \rightarrow R-N-H^+ + R'COO^- \\
| & | \\
R & R
\end{array}
\qquad (6.44)
$$

(insoluble resin) (acid)   (soluble resin) (counter ion)

where, R represents groups of atoms based on carbon that complete the structure of the unpolymerized binding medium unit. R' is the structure characterizing the acid.

Resins that are not amenable to solubilization can be linked to those that are, so that a wide range of binding media is available for electropaints. The paint is prepared by dispersing solubilized binding media in water with finely ground pigments in suspension and other components as required.

6.5.2   Application

6.5.2.1   Traditional Paints

Paints dispersed in traditional solvents are applied by all of the obvious methods. Brush application is for on-site work but it is labor intensive and the quality of the result depends on the painter's skill. Manual spraying also depends on skill and wastes paint in overspray. A modern development of spraying, more economical in the use of paint and independent of operator skill is to deliver the paint through spinning nozzles into a strong electrostatic field to charge the paint particles so that they are attracted to earthed workpieces. Immersion and flow coating both require control of large volumes of paint to maintain its quality, allowing for changes due to solvent evaporation, selective drain off from workpieces and contamination.

6.5.2.2   Waterborne Paints

Electropainted waterborne paints are extensively applied on production lines as primer coats for automobiles and domestic equipment and for lacquers on food cans. Using paints with cationic solubilized resins, a cathodic, i.e., negative, potential of 100 to 200 V is applied to a metal workpiece against an inert counter electrode. The resin is discharged and deposited on the metal by a complex cathodic process with the overall result:

$$\begin{array}{ccc} R & & R \\ | & & | \\ R\!-\!N\!: H^+ + 3H^+ + 4e^- & \rightarrow & R\!-\!N\!: + 2H_2 \\ | & & | \\ R & & R \end{array} \qquad (6.45)$$

solubilized   deposited
resin   resin

complemented by an overall reaction at the anode of the form:

$$R'COO^- + 3OH^- \rightarrow R'COOH + O_2 + H_2O + 4e^- \qquad (6.46)$$

counter   organic
ion   acid

Pigments are entrained in the coating in the same proportions as in the paint, yielding a coating about 12 μm thick. The paint is circulated through filters and heat exchangers to keep pigments in suspension, remove detritus and remove heat generated by the flow of electric current. Drag out losses are recovered by filtering rinse water from the coated workpieces.

6.5.3 Paint Formulation

Endless combinations of binding media blends, pigments and solvents are available to devise paint formulations to match expected service conditions, methods of application and acceptable material costs. Paint formulation is a highly specialized activity based on scientific principles but also relying on experience and empiricism. Detailed formulations are difficult for the uninitiated to interpret. It is a professional skill best left to experts.

Binding media blends are customized to meet the requirements of the applications within cost constraints. If required for waterborne paints, they must possess chemical functional groups needed for solubilization. Alkyd resins are relatively inexpensive and are among the most widely used general purpose resins. Melamine formaldehyde is an expensive amino resin but it confers outdoor durability and impact resistance and is a component of stoving automobile paints. Urea formaldehyde is less costly but less durable. Epoxy and polyurethane resins confer chemical and solvent resistance needed for refrigerators and washing machines but are also expensive. Binding media for lacquers are selected for durability and clarity to add gloss to underlying paint coats and, of course, they are not pigmented.

Pigments are selected according to requirements for color and corrosion resistance and are often mixed with extenders for economy. Concern over toxicity now severely restricts the use of lead-bearing pigments and they are not used for domestic products or automobiles.

Factors that determine the choice of organic solvents for traditional paints include cost, volatility, low-toxicity, and acceptable fire hazard.

6.5.4 Protection of Metals by Paint Systems

Paint films are readily permeable to water and oxygen and when exposed to the weather they are often saturated with both. They protect the metal by blocking electrochemical activity. Outer shell electrons in binding media are mostly localized in covalent bonds so that the paint is a poor conductor of electrons and ions. A significant corrosion circuit cannot be established with the open environment but only through water and oxygen absorbed in the paint structure, where both the anodic and cathodic reactions are inhibited by polarization.

Protection can be supplemented by inhibiting pigments incorporated in the priming paint coats. Chromate pigments are probably slow-release sources of the soluble anodic inhibitor, $Cr^{VI}O_4^{2-}$ at the metal surface, along the lines explained in Section 3.2.6.2. Red lead and calcium plumbate are proven inhibitors but it is not clear why; the $Pb^{IV}O_2^{2-}$ ion is virtually insoluble in water containing dissolved oxygen so that its potential as an anodic inhibitor is not realized. One view attributes the inhibiting properties to lead soaps formed by interaction with the binding media. Zinc dust pigment protects steel sacrificially if sufficient is present to make electrical contact between the particles.

Further Reading

Rowenheim, F. A., *Modern Electroplating*, John Wiley, New York, 1974.

Wernick, S., Pinner, R., and Sheasby, P. G., *The Surface Treatment of Aluminum and Its Alloys*, ASM International, Ohio, 1990.

Waldie, J. M., *Paints and Their Applications*, Routledge Chapman and Hall, New York, 1992.

Lambourne, R. (Ed), *Paint and Surface Coatings*, John Wiley, New York, 1987.

7

Corrosion of Iron and Steels

Iron and steels are the most versatile, least expensive and most widely applied of the engineering metals. They are unequaled in the range of mechanical and physical properties with which they can be endowed by alloying and heat-treatment. Their main disadvantage is that iron and most alloys based on it have poor resistance to corrosion in even relatively mild service environments and usually need the protection of coatings or environment conditioning. This generalization excludes stainless irons and steels that are formulated with high chromium contents to change their surface chemistry as explained in Chapter 8.

An essential preliminary is to characterize the structures of steels and irons that are vulnerable to attack, by briefly reviewing some essential phase relationships and introducing the terminology used to describe them. Table 7.1 gives examples of the materials considered.

7.1   Iron and Steel Structures

7.1.1   Solid Solutions in Iron

Pure iron exists as a body-centered cubic (BCC) structure at temperatures below 910°C and above 1400°C but, in the intermediate temperature range, it exists as a face-centered cubic (FCC) structure. Both structures can dissolve elements alloyed with the iron. Solutions in BCC iron are called *ferrite*, denoted by phase symbols, α and δ, identifying solutions formed in the lower and higher temperature ranges respectively. Solutions in FCC iron are called *austenite*, denoted by the symbol, γ. The solutes stabilize the phases in which they are most soluble. Chromium, molybdenum, silicon, titanium, vanadium, and niobium are *ferrite stabilizers*. Nickel, copper, manganese, carbon, and nitrogen are *austenite stabilizers*.

In their binary systems with iron, ferrite stabilizers progressively expand the δ and α ferrite fields, usually until they merge into a single field enclosing an austenite field, the *γ-loop*, exemplified by the iron-chromium system illustrated in Figure 8.1 and described in Chapter 8.

TABLE 7.1
Some Typical Compositions of Carbon Steels and Irons

Material	Composition, Weight %				Typical Applications
	C	Mn	Si	Others	
Ultra low carbon steel	<0.005	0.20	<0.10	S < 0.003 P < 0.015	Two-piece can manufacture
Low carbon steel	0.05	0.30	<0.10	S < 0.003 P < 0.015	Automobile body panels
Low carbon steel	0.08	0.60	<0.10	S < 0.05 P < 0.025	Three-piece can manufacture
Hypoeutectoid steel	0.20	0.70	<0.15	S < 0.05 P < 0.025	General construction steel, forgings
Hypoeutectoid steel	0.30	0.80	<0.15	S < 0.05 P < 0.025	General engineering, forgings
Hypoeutectoid steel	0.40	0.80	<0.15	S < 0.05 P < 0.025	Nuts and bolts, forgings
Hypoeutectoid steel	0.50	0.80	<0.15	S < 0.05 P < 0.025	Gears
Eutectoid steel	0.80	0.80	<0.15	S < 0.05 P < 0.025	Dies, chisels, vice jaws, springs
Hypereutectoid steel	1.20	<0.35	<0.30	S < 0.05 P < 0.025	Twist drills, files, taps, woodworking tools
Gray cast iron	3.5	0.50	2.00	—	General purpose castings
Nodular cast iron	3.7	0.40	2.40	1.0 Ni, 0.05 Mg	Castings with some ductility
Malleable cast iron	3.0	0.30	0.80	—	Castings with some ductility

Austenite stabilizers expand the austenite field, introducing a peritectic reaction by which austenite forms from δ ferrite and liquid, and a eutectoid reaction in which it decomposes to α ferrite and a second phase, as in the iron-carbon system described in the next section.

7.1.2 The Iron-Carbon System

The iron-carbon system is the basis for the most prolific ferrous materials, plain carbon steels and cast irons. The standard phase diagram is given in Figure 7.1.

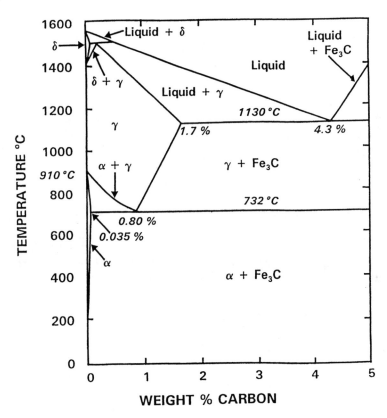

Figure 7.1
Equilibrium phase diagram for the iron-carbon system: α, α-ferrite; δ, δ-ferrite; γ, austenite; Fe₃C, cementite.

There are three invariant points:

1. A peritectic at 1492°C, 0.18 weight % carbon, in which austenite is in equilibrium with δ ferrite deposited during solidification and the residual liquid:

$$\delta + \text{liquid} = \gamma \qquad\qquad (7.1)$$

2. A eutectoid at 732°C, 0.80 weight % carbon, in which austenite is in equilibrium with ferrite and Fe_3C, *cementite*:

$$\gamma = \alpha + Fe_3C \qquad\qquad (7.2)$$

3. A eutectic at 1130°C, 4.3 weight % carbon at 1130°C in which the liquid is in equilibrium with austenite and cementite:

$$\text{liquid} = \gamma + Fe_3C \qquad\qquad (7.3)$$

The boundaries of the austenite phase field extend to the peritectic and eutectic points. The maximum solubilities of carbon are 1.7% in austenite at the eutectic temperature, 1130°C and 0.035% in ferrite at the eutectoid temperature, 732°C.

The diagram represents pseudo-equilibrium because cementite is unstable at temperatures below 800°C. It appears as a metastable phase because the true stable phase, graphite, is difficult to nucleate. A technically important exception is that graphite can be induced to form in cast iron eutectics by appropriate alloying or heat-treatments.

7.1.3 Plain Carbon Steels

The structures of steels have been the subject of an immense amount of observation and experience, commensurate with their economic importance. They are the subjects of standard texts that should be consulted for serious study. The following summary does no more than indicate the origins of the structures encountered.

The compositions of most plain carbon steels lie in the range 0.1 to 1.0 weight % carbon. Their structures are determined primarily by the effects of their carbon contents although the compositions include other components, inherited from steelmaking processes. Most steels contain 0.5 to 1.7 weight % of manganese and all contain sulfur and phosphorous, limited to < 0.05 weight % each, because higher concentrations embrittle the steel. In some special quality steels such as deep-drawing qualities and qualities to resist stress-corrosion cracking in sour gas, sulfur contents are further limited to 0.002%.

7.1.3.1 *Normalized Steels*

The production routes for wrought steels include hot deformation and heat-treatments at temperatures in the austenite phase field. If the steels cool in air through the eutectoid transformation, *normalizing*, they adopt structures that depend on the carbon content in relation to the eutectoid composition.

Eutectoid Composition

As its temperature falls, a steel of the eutectic composition, 0.80 % carbon, passes through the eutectoid point and transforms from austenite to colonies of very fine alternating laminae of ferrite and cementite, called *pearlite* from its pearlescent appearance under the microscope.

Hypo-Eutectoid Compositions

Steels with lower carbon contents, *hypo-eutectoid steels* cross the boundary of the austenite field and deposit *pro-eutectoid* ferrite as they cool through the two-phase ferrite/austenite field. Carbon rejected by the ferrite concentrates in the remaining austenite until it reaches the eutectoid composition at the eutectoid temperature and transforms to pearlite. The final structure comprises crystals of ferrite among colonies of pearlite. The fraction of pearlite in the structure rises approximately linearly with carbon content from zero for a steel with 0.035 % carbon to a completely pearlitic structure at the eutectoid composition. In general, steels become stronger, harder and less ductile as the quantity of pearlite increases.

Hyper-Eutectoid Compositions

Steels with higher carbon contents, *hyper-eutectoid steels* cross the boundary of the austenite field and deposit *pro-eutectoid* cementite as they cool through the two-phase austenite/cementite field. Carbon consumed in forming the cementite depletes the carbon content of the remaining austenite until it reaches the eutectoid composition at the eutectoid temperature and transforms to pearlite. The final structure comprises grains of cementite among colonies of pearlite. The pro-eutectoid cementite still further hardens the steel.

7.1.3.2 Quenched and Tempered Steels

If steels are reheated to a temperature in the austenite field and solutionized, i.e., allowed to transform completely to austenite and then cooled rapidly by quenching in water, the eutectoid reaction is suppressed and replaced by an alternative transformation that yields the tetragonal phase, *martensite*, given the phase symbol, α'. The martensite forms as plates in the austenite by re-alignment of the atoms without rejecting carbon and is thus a diffusionless transformation product. This transformation forms the basis of hardening steels by heat treatment. The hardness of the martensite increases with rising carbon content and steels designed for hardening usually have > 0.36 weight % of carbon. In the as-quenched condition, steels have a simple structure of martensite plates set in retained austenite and although they are hard, they are unusable because they are unstable and brittle. To stabilize the structure, alleviate the brittleness and develop useful mechanical properties, quenching is always followed by *tempering*, i.e., reheating the steel for a short time at a temperature below the eutectoid temperature. Tempering sets in train a complicated sequence of events including the precipitation of $Fe_{2.4}C$, *epsilon carbide*, within the martensite plates and decomposition of the retained austenite to *bainite*, an intimate two-phase structure of ferrite with epsilon carbide. If tempering is prolonged, the structure further transforms into a dispersion of fine spherical cementite, Fe_3C in ferrite and the steel becomes softer. Thus tempering is adaptable to produce customized structures to suit various applications.

Plain carbon steels can be hardened only in thin sections, typically < 2 cm thick because in thicker sections the structure in the interior does not cool rapidly enough to transform to martensite. This difficulty is surmounted by introducing small quantities of other metals into the composition, producing *low-alloy steels*, with martensitic transformations that are more tolerant of slower cooling; they are said to have greater *hardenability*. The metals that promote hardenability most effectively are manganese, chromium and vanadium. With judicious alloy formulation, steels with as little as 3% of alloy additions can be hardened through 15 cm thick sections.

In practice, steels are solutionized at temperatures about 50°C above the lower boundary of the austenite phase field for a period of 0.4 hour per cm of the metal thickness and quenched in cold water or in oil. Subsequent tempering is in the temperature range 400 to 550°C for periods between a few minutes and an hour or two, as required to produce the prescribed structure.

7.1.4   Cast Irons

Although cast irons are based on the iron-carbon system with compositions near the eutectic composition, they are really multicomponent alloys, usually containing a substantial quantity of silicon, a significant quantity of manganese and sometimes also phosphorous. As the name implies, they have good casting characteristics including fluidity and the ability to take sharp imprints from molds. They are inexpensive because the primary raw materials that supply the iron and most of the alloy content are blast furnace iron, cast iron scrap, and steel scrap, together with trimming additions to adjust composition.

According to the pseudo-equilibrium phase diagram in Figure 7.1, a plain iron-carbon alloy of near eutectic composition is expected to solidify as a cementite/austenite eutectic in which the austenite component subsequently transforms to pearlite on cooling through the eutectoid temperature; such a *white iron* is produced if the carbon and silicon contents are low but it is brittle and used only for special purposes, such as wear resistance. General purpose irons are formulated to produce graphite in the eutectic transformation and this is accomplished by raising the carbon and silicon contents, and sometimes adding phosphorous. Most of the carbon content is present as graphite, except for 0.8 % needed to form the cementite component of the pearlite into which the austenite transforms at the eutectoid temperature. Graphitic irons are called *gray irons* from the color of their fractures.

There are three kinds of gray iron, differentiated by the morphology of the graphite:

1. *Flake irons* are standard gray irons in which the graphite exists as flake-like structures, illustrated in Figure 7.2. They are strong in compression but weak in tension, they cannot be deformed and they are not easy to machine.

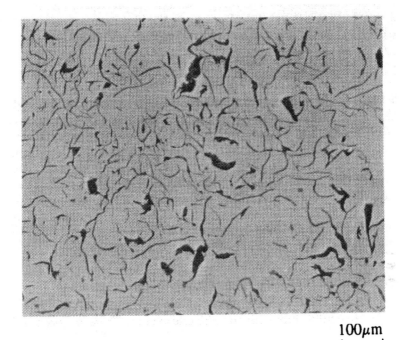

100μm

Figure 7.2
Graphite flakes in a gray cast iron.

2. *Nodular irons* are formulated to alter the morphology of the graphite component of the eutectic so that it is deposited as nodules instead of flakes, improving the mechanical properties generally. The nodular morphology is produced by adding magnesium to the metal immediately before casting. Magnesium reacts with any sulfur present and the nodular structure does not form in iron with > 0.02% sulfur.

3. *Malleable irons* yield a favorable graphite morphology by post-casting treatment. They solidify with cementite/pearlite structures and are reheated to temperatures in the range 850 to 900°C to convert the cementite to graphite; graphite produced in this way is called *temper carbon*. The thermal cycle can be adjusted to yield either a pearlite or ferrite matrix.

7.2   Rusting

Rust is not a determinate chemical substance but a complex material that changes continuously as it develops through the precipitation, evolution, and transformation of chemical species in the iron-oxygen-water system. The course of these events is conditioned by the prevailing environment.

7.2.1   Species In The Iron-Oxygen-Water System

Iron can exist in oxidation states Fe(II), Fe(III), and Fe(VI). The highest state, Fe(VI), in the ferrate ion, FeO_4^{2-} is stable only in highly oxidizing alkaline media and is seldom significant in corrosion processes or in chemistry generally. The common oxidation states are iron(II) and iron(III) formerly called *ferrous and ferric iron*, respectively. Both form oxides, hydrous oxide phases, soluble cations and corresponding series of salts. The redox potential of the Fe(III)/Fe(II) couple in dilute aqueous solution, $E^{\ominus}$, is + 0.771 V (SHE) but it is sensitive to the chemical environment of the metal atoms and in other contexts the facility with which Fe^{2+} and Fe^{3+} change their oxidation states depends on phase relationships.

7.2.1.1   *Iron(III) Oxides, Hydrous Oxides, and Derivatives*

Iron(III) is the stable oxidation state of simple compounds and aqueous solutions in equilibrium with atmospheric oxygen. There are two iron(III) oxides, the common red-brown oxide, α-Fe_2O_3 (*hematite*) and γ-Fe_2O_3. α-Fe_2O_3 has the corundum (Al_2O_3) structure with Fe^{3+} ions in the octahedral spaces of a hexagonal close-packed lattice of oxygen ions. γ-Fe_2O_3 has a structure, sometimes incorrectly called 'spinel like', with Fe^{3+} ions distributed randomly in the octahedral and tetrahedral interstices of an FCC lattice of oxygen ions; it is produced by dehydrating lepidocrocite, γ-$FeO(OH)$, or oxidizing the lower oxide, Fe_3O_4.

Iron(III) is precipitated from neutral or alkaline solution as a red-brown gelatinous material loosely described as "ferric hydroxide" but there is no clear evidence that it contains discrete hydroxyl ions and its constitution is uncertain. Some chemical equations use the notation "$Fe(OH)_3$" but it is probably a hydrous iron(III) oxide $Fe_2O_3.nH_2O$, where n approaches 3. It loses water on standing, becoming an amorphous powdery material.

Partially hydrated iron(III) oxide phases are difficult to characterize unequivocally. There are at least two brown or brown-black hydroxyoxides, α-FeO(OH) (*goethite*) and γ-FeO(OH) (*lepidocrocite*) as well as other phases that can be regarded as oxides containing variable quantities of water, $Fe_2O_3 \cdot nH_2O$. All of these compounds are only very slightly soluble in water unless it is strongly acidic or alkaline. For this reason the reaction of iron(III) with anions of weak acids often yields hydrous iron(III) oxide, $Fe_2O_3 \cdot nH_2O$. One such reaction is with the carbonate ion, usually present in natural waters:

$$2Fe^{3+} + 6CO_3^{2-} + (3 + n)H_2O = Fe_2O_3 \cdot nH_2O + 6HCO_3^- \quad (7.4)$$

7.2.1.2 Iron(II) Oxides, Hydroxides, and Derivatives

The iron(II) ion has a lower charge density than the iron(III) ion and because of this its salts are much less susceptible to hydrolysis. It forms simple salts with most common anions except when they are oxidizing enough to raise the oxidation state from iron(II) to iron(III).

Iron(II) forms an oxide, FeO (*wüstite*), with a simple cubic structure but it is unstable at temperatures below 570°C. Iron(II) does however participate in Fe_3O_4, (*magnetite*), an anhydrous mixed iron(II)/iron(III) oxide, stable at ambient temperature with the inverse spinel structure, described in Section 2.3.3.4. It can form by oxidation of iron at temperatures below 570°C or by decomposition of FeO formed at higher temperatures:

$$4FeO = Fe + Fe_3O_4 \quad (7.5)$$

Unlike iron(III), iron(II) forms a distinct hydroxide, $Fe(OH)_2$. It has the $Mg(OH)_2$ structure and is white or pale green when pure but discolors to yellow in contact with air by superficial oxidation to a hydrous iron(III) oxide:

$$2Fe^{II}(OH)_2(s) + \tfrac{1}{2}O_2(g) = Fe^{III}_2O_3 \cdot H_2O(s) + H_2O(l) \quad (7.6)$$

In contact with water, a corresponding oxidation reaction occurs more readily:

$$2Fe^{II}(OH)_2 + 2OH^- = Fe^{III}_2O_3 \cdot H_2O + 2H_2O + 2e^- \quad (7.7)$$

driven by the reduction of dissolved oxygen:

$$\tfrac{1}{2}O_2(solution) + H_2O + 2e^- = 2OH^- \quad (7.8)$$

The reaction rate is strongly pH dependent as indicated by the presence of OH⁻ in Equation 7.7 and occurs most rapidly in alkaline media. Iron(II) hydroxide is amphoteric and can dissolve in both strong acids and alkali, although it has some solubility in neutral solution yielding Fe^{2+} and OH^- ions:

$$Fe(OH)_2 = Fe^{2+} + 2OH^- \tag{7.9}$$

for which the solubility product is:

$$K_s = (a_{Fe^{2+}})(a_{OH^-})^2 = 1.64 \times 10^{-14} \text{ at } 18°C \tag{7.10}$$

In hard water containing carbon dioxide at significant pressures, iron(II) can also form iron(II) carbonate, $FeCO_3$, *siderite*, that is oxidized by dissolved oxygen to hydrous iron(III) oxide:

$$4FeCO_3 + 2nH_2O + O_2 = 2[Fe_2O_3 \cdot nH_2O] + 4CO_2 \tag{7.11}$$

7.2.1.3 *Protective Value of Solid Species*

When formed as coherent, adherent layers on the metal, the close structures of any of the anhydrous iron(III) oxides, α-Fe_2O_3, γ-Fe_2O_3 and Fe_3O_4, can passivate the metal by inhibiting the transport of reacting species between the metal and its environment. In contrast, the amorphous character of many of the hydrated forms is unsuited to the formation of protective layers.

7.2.2 Rusting in Aerated Water

The conventional Pourbaix diagram for the iron-water system, given in Figure 3.2, and corrosion velocity diagrams as given in Figure 3.13 are useful indicators of corrosion possibilities but they cannot disclose how the transformations they depict occur.

The diagram in Figure 3.2 shows that, at 25°C for $a_{Fe^{2+}} = 10^{-6}$, there is no common zone of stability for metallic iron and water across the entire pH range, so that iron and steels are unstable in the presence of water. This implies a thermodynamic *tendency* for the metal to transform but whether or not it actually does so at a significant rate depends on the routes that the transformations must follow and the natures of intermediate products. These in turn depend on pH and on the absence or presence of contaminants, notably chlorides.

7.2.2.1 *Fresh Waters*

Natural waters including rain, rivers, and supplies derived from them are near neutral aqueous media approaching equilibrium with the atmosphere, in which the absorption of oxygen is the dominant cathodic reaction, as determined by the calculation given in Section 3.2.3.1. The oxygen potentials in these environments are close to the equilibrium line for oxygen absorption* superimposed on the iron-water Pourbaix diagram given in Figure 3.2. This line is wholly within the domain of stability of Fe_2O_3 for $a_{Fe^{2+}} = a_{Fe^{3+}} = 1$ in the pH range 2 to 16, so that given time and opportunity, a phase of this composition must ultimately appear.

The transformation is, however, approached by electrochemical partial reactions at the metal surface that place the corrosion potential in the domain of stability for Fe^{2+}, so that the primary products are produced by the anodic dissolution of iron:

$$Fe = Fe^{2+} + 2e^- \qquad (7.12)$$

complemented by the cathodic absorption of oxygen:

$$\tfrac{1}{2}O_2 + H_2O + 2e^- = 2OH^- \qquad (7.13)$$

When the concentrations of Fe^{2+} and OH^- ions close to the metal surface exceed the low solubility product given by Equation 7.10, the first solid product, $Fe(OH)_2$, is produced:

$$2Fe^{2+} + 6OH^- = 2Fe(OH)_2 \qquad (7.14)$$

Precipitation of a solid from a liquid phase requires energy to create the solid/liquid interface and this is usually less on any pre-existing surface (*heterogeneous nucleation*) than from the liquid (*homogeneous nucleation*). Thus although the $Fe(OH)_2$ is produced by the interaction of species in solution, it is usually deposited on the iron surface.

The initial deposit of $Fe(OH)_2$ is only the provisional corrosion product and it is slowly oxidized to compositions approaching that of the stable iron(III) hydrous phase, by reactions based on that given in Equation 7.7. As the rust ages, its constitution changes continuously, with associated changes in color from yellow through red to brown and it may contain various species from among those described in Sections 7.2.1.1 and 7.2.1.2. Such an unpromising material, derived from the initial $Fe(OH)_2$ deposit obstructs diffusion of oxygen to the metal surface but lacks the coherence needed to protect the iron. If the metal is totally submerged and corrosion is uniform, iron is lost from the surface at a steady rate of 0.15 mm per year, after an initial period of a few days, during which the rust is first established. In general, the corrosion rate does not vary significantly with the composition or condition of clean iron and steel products, excluding stainless steels and irons.

---

* The chemical potential of oxygen is a logarithmic function of its activity, so that the oxygen potential line for water in equilibrium with air, for which $p_{O_2} = 0.21$, is only a little lower than that for $p_{O_2} = 1$ plotted in Figure 3.2.

7.2.2.2 Sea Waters

Ocean waters approximate to a 3.5 weight % solution of sodium chloride but also contain significant quantities of magnesium, calcium, potassium, sulfate, and bicarbonate ions together with lesser quantities of many other solutes. In estuaries and land-locked seas, such as the Baltic, these solutes can be diluted considerably by fresh waters delivered by rivers but polluted by man-made effluents. Oxygen contents in ocean surface water are close to equilibrium with the atmosphere and the pH is in the range 8.0 to 8.3. The surface water temperatures are in the range –2 to 35°C.

The high concentrations of ionic solutes might be expected to influence corrosion rates of iron and steels but long-term field tests have shown that the corrosion rates for small unshielded panels are remarkably consistent in all ocean waters irrespective of location and temperature and they are similar to the rates in fresh waters. Despite these similarities, sea waters can cause more problems than fresh waters in some respects:

1. The higher electrical conductivity of sea water increases the range over which differential aeration or bimetallic effects can operate.

2. Tidal action increases the corrosion rate of partially immersed structural steel. The splash zone above high tide and a zone just under low tide levels are vulnerable. Explanations lack detail but are based on enhanced differential aeration effects.

7.2.2.3 Alkaline Waters

Waters are sometimes deliberately made alkaline for corrosion control or acquire alkalinity from environments such as cement. Confining attention to ambient temperatures, information from the iron-water Pourbaix diagram for 25°C in Figure 3.2 shows that if the pH is above about 9, the triangular domain of stability for Fe^{2+} (defined in Section 3.1.5.4 as $a_{Fe^{2+}} > 10^{-6}$) is eliminated and there is a common boundary between the domains of stability for metallic iron and magnetite, Fe_3O_4. Thus corrosion potentials must lie within the domain of stability of magnetite, Fe_3O_4, or hematite, Fe_2O_3, so that one or other of these oxides is the primary anodic product, e.g.:

$$3Fe + 4H_2O \rightarrow Fe_3O_4 + 8H^+ + 8e^- \qquad (7.15)$$

It nucleates directly on the iron surface and grows into an adherent coherent layer that passivates the metal.

Successful passivation depends on forming a layer that is complete and free from attack by aggressive ions in the environment. Thus if the pH is only just high enough to passivate the metal there is a risk of local attack on areas of metal remaining active that become anodes in active/passive cells. Aggressive ions, especially chlorides can attack passive surfaces producing small local anodes that develop into pits, a common form of attack on metals depending on passivated surfaces for protection; this effect is considered in detail in Chapter 8, within the context of the more extensive evidence available for stainless steels.

7.2.2.4 Suppression of Corrosion by Applied Potentials

The information in the iron-water Pourbaix diagram suggests that in near neutral aqueous media it should be possible to suppress corrosion either by impressing a cathodic current on the metal to depress the potential of the metal into the domain of stability for metallic iron or by applying an anodic potential to raise it into the domain of stability for hematite, Fe_2O_3 to passivate the metal. Both techniques work but there are important practical constraints. Impressing a cathodic current is a standard method of cathodic protection, described in Chapter 4: it has the merit that it is fail-safe; if the current does not succeed in depressing the potential into the Fe^{2+} domain, corrosion is reduced even if not completely suppressed. An applied anodic potential is not fail-safe; if it does not succeed in raising the potential of the whole metal surface to the domain of stability for magnetite, Fe_3O_4, or hematite, Fe_2O_3, corrosion is increased. For this reason, it is dangerous to rely on an externally applied anodic potential. However, anodic protection is applied chemically by the redox potential of powerful oxidizing agents such as chromates or nitrites, when used as anodic inhibitors, as described in Section 3.2.6.2.

7.2.3 Rusting in Air

Rusting of bare iron and steel surfaces is generally slower in outside air than in water but is much more variable, ranging from near zero to over 0.1 mm per year and are less predictable. The factors that influence it differ from location to location and vary with time. They include climate season and weather, atmospheric pollutants, temperature cycles, and the initial conditions and orientations of the iron surfaces. There are so many independent variables that empirical comparisons between the results of limited field tests can be unreliable and the effort to acquire statistically significant information is protracted because rates of rusting can take several years to settle to steady values. More progress can be made by applying scientific principles and logic from accumulated local experience.

There are three sources of water to sustain the electrochemical processes, atmospheric humidity, precipitation, and wind or wave driven spray.

7.2.3.1 Rusting Due to Atmospheric Humidity

Rust can form on iron even when it is not wet to the touch; it is less in dry than in humid air and more in industrial or marine locations than in rural areas. These observations are rationalized on the basis of Evans* and his colleagues' classic experiments to clarify the effects of relative humidity and pollutants on rusting. Iron samples do not rust appreciably in pure clean air even when nearly saturated with water vapor. If as little as 0.01% by volume of sulfur dioxide is present as a pollutant, the metal rusts rapidly if the relative humidity exceeds a threshold value of about 70%. Dust contaminated with certain ionic salts has a similar effect in initiating rusting. Evans proposed a simple but elegant explanation as follows. An aqueous electrolyte is needed to produce rust and although water cannot condense spontaneously from *unsaturated* air, it can be induced to do so by hygroscopic salts. The role of pollutants is to provide particles of such salts on the metal surface; if the pollutant is sulfur dioxide, as it is in industrial locations, the salt is iron(II) sulfate, $FeSO_4$; in marine locations, the salt is sodium chloride, $NaCl$ produced from sea spray; in cities there are other possibilities including the cocktail of pollutants released into the atmosphere by automobiles. As expected, the water vapor pressures in equilibrium with saturated solutions of these various salts correspond closely with the threshold values of relative humidity needed to initiate rusting. This suggests the following sequence of events that Evans observed in experiments exposing iron to humid air polluted by sulfur dioxide:

1. Deposition of iron(III) sulfate.
2. Formation of a barely visible mist of water.
3. Initiation of rust spots.
4. Extension of the rust spots to cover the surface.

Once rusting has started it is not arrested by removing the pollutant from the atmosphere, so that contamination that initiates rusting continues to sustain it. A corollary of this observation is that the rate of rusting is influenced by the degree of pollution in the initial conditions of exposure and this is borne out by experience that rusting on a clean iron surface first exposed in winter, when pollution is at a maximum, continues at a faster rate than it does if first exposed in summer when pollution is at a minimum.

In atmospheric rusting, the electrolyte is not present in bulk and ideas of the prevailing electrochemical reactions need revision from those that apply in aqueous corrosion. Evans suggests that the orthodox cathodic reaction:

$$\tfrac{1}{2}O_2 + H_2O + 2e^- = 2OH^- \tag{7.16}$$

is replaced by the reduction of an Fe(III) species in the rust to a species containing Fe(II):

---

* Cited in the Further Reading list.

$$Fe(III) + e^- \rightarrow Fe(II) \qquad (7.17)$$

followed by its re-oxidation back to iron(III) by atmospheric oxygen:

$$2Fe(II) + \tfrac{1}{2}O_2 + H_2O = 2Fe(III) + 2OH^- \qquad (7.18)$$

Various schemes are possible, depending on what species are present in the rust as discussed in Sections 7.2.1 and 7.2.2 provided that there is a conducting path to supply electrons from the metal. Evans' own suggestion is that the reaction is between goethite, $Fe^{III}O(OH)$, and magnetite, $Fe^{II} \cdot Fe_2^{III}O_4$.

7.2.3.2 *Rusting from Intermittent Wetting*

Wetting from liquid water as rain, condensation or spray introduces aspects of aqueous corrosion, described in Section 7.2.2, with the difference that wetting is discontinuous. This contribution to rusting depends on the relative periods for which the metal is wet or dry.

If wetting is by rain or snow, the metal remains wet for the sum of the periods of precipitation and subsequent drying. The drying period depends on humidity, temperature, wind velocity, and sunlight and hence on the general climate in regions of interest and on microclimates due to local geography. Melting snow remains for a considerable period and can leave hygroscopic contaminants such as highway de-icing salts that promote subsequent rusting after the metal has dried; a familiar example is rusting due to mud poultices accumulated in traps in automobile underbodies. Further factors are the orientation of metal surfaces that allow or prevent drainage and the sponge effect of pre-existing rust.

Wetting from condensation is induced by temperature cycles above and below the dew point. It is most prevalent when there is a wide difference between day and night temperatures in temperate and tropical regions. Heavy metal sections are most vulnerable because their temperature cycles are out of phase with air temperatures.

7.2.4 Rusting of Cast Irons

Despite their vulnerability, cast irons have given remarkably good service when used unprotected. Historically, they were used extensively for buried pipes and structures, tanker fittings and other services, starting when they were one of the few inexpensive materials available for such duties. Some such systems can still survive after a century. Part of the reason for longevity is that cast iron sections are thicker than steel counterparts and in former times they were overdesigned with greater structural reserves to absorb wastage. More recent systems benefit from experience in identifying the most suitable applications and applying protective coatings or cathodic protection.

7.2.4.1   Gray Flake Irons

The general purpose gray flake irons, differ from steels in their response to wet environments in several respects.Graphite is an electrical conductor and flakes outcropping at the surface can act as cathodic collectors on a micro-scale, as explained in Section 4.1.3.3, stimulating corrosion of the adjacent metal structure. Corrosion of this kind, developing along the flakes is called *graphitic corrosion*.

The metal is leached out from among the network of graphite flakes, illustrated in Figure 7.2, leaving the surface contour of the metal intact. If the matrix is pearlite, as it usually is, carbonaceous material from dissolution of its cementite component becomes trapped in the mesh of residual graphite together with corrosion products and the composite mass acts as a resistance to diffusion delaying reaction. The value of this residue in protecting the iron depends on its consolidation and that in turn depends on the graphite morphology in the unattacked iron. Fine flakes produce stronger residues and protect the metal better than coarse flakes and residues from irons containing phosphorous are reinforced by iron phosphide.

In near-neutral aerated waters, corrosion is primarily by oxygen absorption but corrosion by hydrogen evolution is serious when stimulated by sulfate-reducing bacteria in anaerobic waters containing sulfates that can prevail in locations such as polluted estuary waters and sewers. This form of attack can afflict other metals but it is particularly associated with gray cast irons because it is easy to identify. The bacteria produce an enzyme, *hydrogenase* that enables them to use hydrogen, accelerating corrosion by depolarizing the cathodic reaction:

$$SO_4^{2-} + 4H_2 \rightarrow S^{2-} + 4H_2O \tag{7.19}$$

Accumulation of sulfides trapped in the mesh of graphite residue provides evidence for bacteriological stimulation of corrosion.

7.2.4.2   Other Irons

Nodular and malleable and white irons lack the graphite network that is responsible for the effects described in Section 7.2.4.1, and therefore corrode in more orthodox fashion.

Normal gray irons have poor resistance to acids and compositions have been developed to meet these conditions. Raising the silicon content substantially produces irons for which initial corrosion leaves a residue of silica, SiO_2 on the surface that confers subsequent resistance to most common acids except hydrofluoric acid, hydrochloric acid and also for some obscure reason, sulfurous acid, H_2SO_3. For acid resistance the silicon content must be > 11 weight %, preferably 14%. Their most useful attribute is resistance to sulfuric acid in all concentrations and all temperatures. The irons are hard brittle and lack tensile strength and their use is justified only when their superior acid resistance is needed.

Cast irons containing high nickel or nickel/copper contents have been developed for improved corrosion and heat resistance. Their main disadvantage is that the nickel content must be at least 18% or 13.5% with some substitution of copper for nickel, so that they are much too expensive for general application. The structures are essentially graphite in an austenite matrix. Their acid resistance is better than that of regular gray flake irons but not so good as that of the irons with high silicon contents. They do, however, perform well in aggressive waters and find applications in special situations.

7.3 The Oxidation of Iron and Steels

The oxidation of iron at oxygen pressures near atmospheric pressure is worthy of study both because it is technically important and because it is complex and exemplifies many of the principles which underlie the oxidation of metals generally. The oxidation behavior can be described within four contexts, that are mutually consistent:

1. Crystallographic and defect structures of the oxides.
2. Phase equilibria within the iron-oxygen system.
3. Oxygen potential diagrams, i.e., $\Delta G^{\ominus} - T$ diagrams
4. Wagner's theory of oxidation mechanisms.

7.3.1 Oxide Types and Structures

There are three stable oxides of iron:

1. FeO (wüstite), stable at temperatures $> 570°C$, is a simple cubic cation vacant p-type oxide. The only mobile specie is Fe^{2+} but its diffusivity is high.
2. Fe_3O_4 (magnetite) is an inverse spinel, in which iron fulfills both divalent and trivalent roles. The mobile species are Fe^{2+}, Fe^{3+} and O^{2-}.
3. Fe_2O_3 (hematite) is a rhombohedral n-type oxide with both cation interstitials and anion vacancies. The defect populations are much smaller than those in FeO and Fe_3O_4. The mobile species are Fe^{3+} and O^{2-}.

7.3.2 Phase Equilibria in the Iron-Oxygen System

Figure 7.3 gives a part of the phase diagram for the iron-oxygen system relevant to the oxidation of the solid metal. There are four solid phases, pure iron and the oxides FeO, Fe_3O_4 and Fe_2O_3. The solubility of oxygen in iron is so small that the solid solution phase field is vanishingly narrow and is not indicated in the figure. The FeO phase field is relatively wide, corresponding to the ability of the oxide to sustain an exceptionally high population of cation vacancies; the maximum composition range is 0.51 to 0.54 mole fraction of oxygen at 1370°C, that does not include the stoichiometric mole fraction, 0.50, indicating that the oxide always contains at least 0.02 fraction of vacant cation sites. The composition ranges for Fe_3O_4 and Fe_2O_3 are much narrower but detectable at high temperatures.

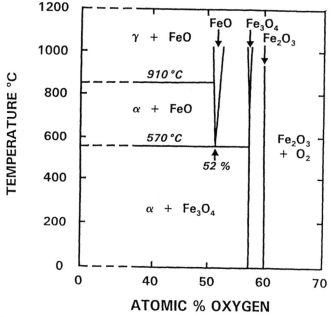

Figure 7.3

Equilibrium phase diagram for the iron-oxygen system. α, ferrite; γ, austenite; FeO, wüstite; Fe_3O_4, magnetite; Fe_2O_3, hematite.

FeO decomposes to iron and Fe_3O_4 by a eutectoid reaction at 570°C:

$$4FeO = Fe + Fe_3O_4 \qquad (7.20)$$

The reaction is reflected in the plots of standard Gibbs free energy of formation versus temperature, *oxygen potential diagrams*, illustrated in Figure 7.4. The lines for FeO → Fe_3O_4 and for Fe → FeO intersect at 570°C, corresponding to the eutectoid temperature. At higher temperatures, FeO is more stable than Fe_3O_4 because the free energy of formation of FeO is more negative than that of Fe_3O_4 and so it is the more stable oxide but at lower temperatures it becomes unstable with respect to a mixture of iron and Fe_3O_4.

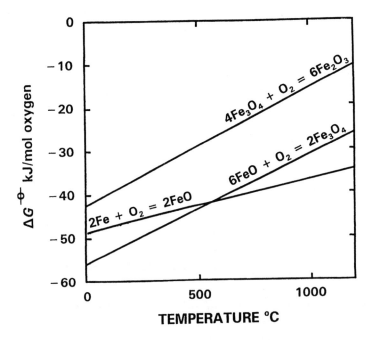

Figure 7.4
Gibbs free energy of formation for iron oxides as functions of temperature. Lines in descending order indicate increasing stability. Note the reversal of the relative stabilities of FeO and Fe$_3$O$_4$ at 570°C.

The solubility of oxygen in solid iron is so small that it has an negligible effect on the α-Fe to γ-Fe transformation which is represented in Figure 7.3 by the horizontal line at 910°C across the iron/FeO phase field. FeO is in equilibrium with α-Fe and γ-Fe at temperatures below and above 910°C, respectively.

7.3.3 Oxidation Characteristics

7.3.3.1 Nature of Scale

Scales formed in ambient air comprise layers of the stable oxides in a sequence of increasing oxygen activity from the metal surface to the atmosphere, illustrated schematically in Figure 7.5. Scales formed at temperatures > 570°C have *three* layers but scales formed at temperatures < 570°C, where FeO is unstable have only *two* layers.

The scale layers are of different thicknesses, reflecting their different rates of formation, that depend on the relative diffusion rates of the reacting species, due *inter alia* to the very different defect populations referred to earlier. The Fe$_3$O$_4$ is thicker than the overlying outer layer of Fe$_2$O$_3$ and for temperatures > 570°C, the innermost layer of FeO is the thickest of all. As oxidation proceeds, the total thickness of the scale increases but the ratios

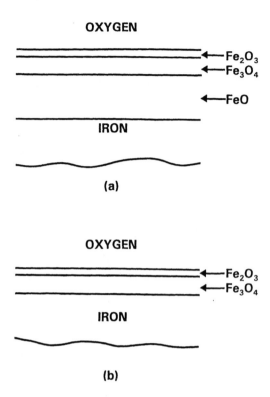

Figure 7.5
Sequence of oxide layers formed on iron. Temperature (a) > 570°C; (b) < 570°C.

of the thicknesses of the different layers remain approximately the same. Quantitative measurements of the actual scale thickness are inconsistent. Values given for the $Fe_2O_3/Fe_3O_4/FeO$ ratios at temperatures > 570°C range from 1/10/100 to 1/10/50 and the value for the Fe_2O_3/Fe_3O_4 ratio at temperatures < 570°C is approximately 1/10.

7.3.3.2 Oxidation Rates

At all temperatures, the oxidation kinetics are parabolic, indicating that the oxidation is diffusion-controlled. The temperature-dependence of the parabolic rate constant for the total scale thickening, K_P, obeys an Arrhenius-type relation that is also consistent with diffusion control, as described in Chapter 3. For oxidation in still dry air, the expression is:

$$K_P = 6.1 \exp(-17000/RT) \ cm^2 sec^{-1}. \qquad (7.21)$$

Diffusion through the outer layer of Fe_2O_3 that is the least defective oxide is rate-controlling. Consequently, there is no abrupt discontinuity in the temperature-dependence of K_P at the eutectoid temperature, below which the FeO layer disappears.

7.3.3.3 *Diffusing Species and Growth Interfaces*

Sustained oxidation depends on the mechanisms by which oxygen from the atmosphere and iron from the metal can reach growth interfaces, where new oxide is formed.

The inner oxide layer, cation-vacant FeO, is permeable to iron in the form of Fe^{2+} ions but impermeable to oxygen because there is no diffusion path for O^{2-} ions; FeO can therefore grow at the FeO/Fe_3O_4 interface but not at the Fe/FeO interface.

The middle layer, the spinel, Fe_3O_4, is permeable to both iron and oxygen in the forms of the ions, Fe^{2+}, Fe^{3+} and O^{2-}. Thus it can grow at either interface and also transmit species needed for growth of the FeO and Fe_2O_3 layers on either side of it.

The outer layer, the *n*-type oxide, Fe_2O_3 with both cation interstitials and anion vacancies can transmit both iron as Fe^{3+} ions and oxygen as O^{2-} ions but the latter species dominates.

These considerations lead to the following view on the locations of the growth interfaces:

1. There is virtually no scale growth at the Fe/FeO interface.
2. FeO grows at the FeO/Fe_3O_4 interface
3. Fe_3O_4 grows at both the FeO/Fe_3O_4 and Fe_3O_4/Fe_2O_3 interfaces.
4. Fe_2O_3 grows predominantly at the Fe_3O_4/Fe_2O_3 interface, supplemented by some growth at the Fe_2O_3/atmosphere interface.

7.3.3.4 *Application of the Wagner Theory*

The Wagner theory of oxidation described in Chapter 3 for a single oxide layer bounded by the metal/oxide and oxide/atmosphere interfaces can be extended to a multilayered oxide system, by assuming that local equilibrium is also maintained at internal interfaces. For the scale forming on iron, the sequence comprises the Fe/FeO, FeO/Fe_3O_4, Fe_3O_4/Fe_2O_3 and Fe_2O_3/atmosphere interfaces. This implies that there is a composition gradient across every oxide layer corresponding to the width of its phase field shown in Figure 7.3. For example, at 900°C, the mole fractions of oxygen in equilibrium with iron at the Fe/FeO interface and with Fe_3O_4 at the FeO/Fe_3O_4 interface are about 0.051 and 0.053, respectively, giving a difference of 0.02 mole fraction of oxygen across the FeO layer. This corresponds with the difference in the population of cation vacancies across the layer that provides the driving force for the diffusion of Fe^{2+} ions. Analogous considerations apply to the Fe_3O_4 and Fe_2O_3 layers.

7.3.4 Oxidation of Steels

It is usually the oxidation of steels rather than pure iron that is of most interest. Plain carbon steels contain small concentrations of carbon, manganese, silicon, and excess aluminum from deoxidation; alloy steels may contain increased concentrations of these elements and nickel, chromium, molybdenum and vanadium as alloying components. The interplay between these elements can be complex but the following brief summary of the effects of individual elements can be used as an empirical guide.

7.3.4.1 Effects of Alloying Elements and Impurities

Carbon oxidizes to carbon monoxide and hence may be expected to have a disruptive effect on the oxide. For steels with high carbon contents at temperatures > 700°C, this prediction is correct but at lower temperatures, carbon is found to be beneficial rather than deleterious. Silicon promotes the formation of glassy silicates in oxide layers formed at high temperatures, e.g., $2FeO \cdot SiO_2$ (fayalite), that tend to stifle oxidation and provide some resistance to sulfur-bearing gases that stimulate oxidation. If silicon is present in high concentrations, e.g., 3% or 5% silicon steels for electrical applications, it promotes some internal oxidation, i.e., some oxide is nucleated in the outer layers of the metal, under the true scale. Manganese and nickel resemble iron in oxidation behavior and in moderate concentrations have little effect on the oxidation rate, as would be anticipated from Hauffe's valency rules. Although it does not have a significant influence on the oxidation rate, nickel can have a curious effect on the scale structure formed in air. In the early stages of oxidation, FeO and NiO both form because they have oxygen potentials that are close and much lower than that of air. As the scale thickens and the oxygen supply recedes, the FeO controls the oxygen potential at a value for which NiO is reduced back to metallic nickel that is found in the scale. Aluminum can improve the oxidation resistance because it can form a protective layer of $FeAl_2O_4$ under the main scale but the effect of concentrations normally present is small. Molybdenum and vanadium yield oxides that form low melting-point eutectics with other scale components. Thus they can flux away the scale, destroying its protective function and thereby promoting catastrophic oxidation as explained in Chapter 3; the effects can be confusing because molybdenum is sometimes inexplicably beneficial. Sulfur and phosphorus in the small residual concentrations normally present have little effect.

Chromium is of such importance and value in formulating iron-base alloys for oxidation resistance, e.g., stainless steels and chromium irons, that it requires the detailed discussion, given in Chapter 8.

7.3.4.2 Influence on Metal Quality of Scales Formed During Manufacture

Scales formed on plain carbon steels are adherent but are easily detached when the steel surface is increased in the early stages of hot-rolling at temperatures > 1000°C. Thus scaling of ingots during pre-heating for rolling can *remove* pre-existing surface blemishes. In contrast, scales formed on some alloy steels are only partially detached and become fragmented and imprinted on the steel surface as hot-rolling continues. This effect *creates* surface blemishes. Hence experience of the oxidation behavior of particular steels is essential in producing steels for applications where surface finish is important.

7.3.4.3 Oxidation in Industrial Conditions

Irons and steels are used at temperatures up to about 600°C in industrial conditions, but the idealized conditions for which the value of K_P given in Equation 7.21 applies do not necessarily or even usually prevail. The complicating factors include the composition and flow of the gaseous environment and the thermal history.

Atmospheres

In applications such as flues, furnace structures, steam pipes, oil stills and fume treatment facilities, steels may be exposed to various gases or mixtures of them from among air, carbon monoxide, carbon dioxide, sulfur dioxide, and live steam that can modify the composition of the scale, introducing for example, sulfates, sulfides, or hydroxides; there may also be abrasion or chemical attack from ash particulates. Further complications are introduced if the gas is flowing or if its composition is variable. These variables cover such a wide range that every case must be considered on its merits. Experience and comparative experimental tests simulating service environments have produced a large body of comparative information which is apparently reliable but difficult to correlate. For an index of specific information, the reader is referred to specialized texts such as the monographs by Kubaschewski and Hopkins and by Shreir, cited at the end of the chapter.

Thermal History

So far, it has been implicitly assumed that the scale remains intact but if the temperature varies randomly or cyclically, differential expansion and contraction between the metal and the scale can cause it to buckle or flake, reducing the protection it affords. Thick scales produced by prolonged oxidation can also buckle or flake under the shear forces built up within them; for this reason, it is unsafe to extrapolate the results of short-term tests to provide information for long-term service.

7.3.5 Oxidation and Growth of Cast Irons

Cast irons exposed to air at high temperatures form surface scales as do steels but by exploiting the metal structure, oxidation has additional significance in contributing to a form of degradation known as *growth*.

7.3.5.1 Growth of Cast Irons

At temperatures above about 500°C in clean air, cast irons, especially gray flake irons are susceptible to dimensional changes that distort the metal, due to two effects:

1. Internal oxidation.
2. Graphitization of cementite, Fe_3C.

Since both effects are cumulative and operate simultaneously, they are best considered together as different aspects of the same problem.

Internal Oxidation

In a gray iron, scale forms not only on the metal surface but also as intrusions penetrating into the metal interior along the interfaces between graphite flakes and iron, where oxygen can percolate through fissures propagated by the intruding scale. Shallow penetration improves oxidation resistance if the iron surface is stressed in compression, because it keys the external scale to the surface, excluding direct access of air to the metal. Deeper penetration, illustrated in the photomicrograph given in Figure 7.6, is however, a source of weakness if the surface is under tension because the metal is expanded and weakened by a wedging action of the scale that occupies a greater volume than the volume of metal consumed.

Raising the carbon content increases internal oxidation because there is more graphite in the structure. Refining the graphite flakes has the same effect because it increases the area of the vulnerable graphite/iron interface. Nodular and malleable irons are less susceptible because the modified shapes of the graphite give lower surface to volume ratios and reduce coalescence of the scale intrusions evident in Figure 7.6.

Sulfur dioxide and other atmospheric pollutants that accelerate surface oxidation also accelerate scale penetration into cast irons. This increases the risk of premature failure of components exposed to heavily polluted hot gases such as the products of combustion from coals with high-sulfur contents and off-gases from pyritic smelting operations. Examples are fire grates, furnace doors and supports for furnace roofs.

Graphitization

Cementite tends to decompose to iron and graphite because of its instability at temperatures below about 800°C:

$$Fe_3C = 3Fe + C \tag{7.22}$$

In a gray iron as cast, carbon is distributed between graphite in the flakes and cementite as a constituent of the pearlite matrix. The silicon content of the iron that is added to promote the deposition of carbon as graphite in

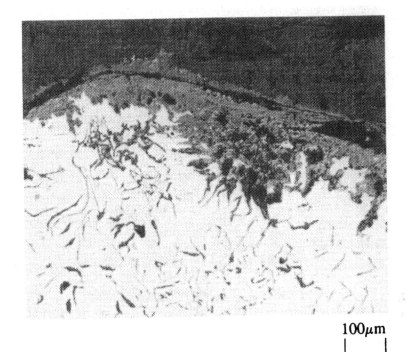

100μm

Figure 7.6
Oxide penetration along graphite flakes in gray cast iron.

the eutectic transformation during casting also accelerates the conversion of the cementite fraction in the pearlite to graphite when the iron is held at temperatures in the range, 500 to 800°C. The conversion expands the iron because the graphite occupies a greater volume than its cementite precursor. Thermal cycling through a temperature range around 732°C, accelerates the growth by mechanisms associated with the austenite/ferrite transformation.

Further Reading

Honeycombe, R. W. K., *Steels: Microstructures and Properties*, Edward Arnold, London, 1981.
Evans, U. R., *Br. Corr. J.*, 7, 10, 1972.
Nicholls, D., *Complexes and First Row Transition Elements*, Macmillan, London, 1981.
Kubaschewski, O. and Hopkins, B. E., *Oxidation of Metals and Alloys*, Butterworths, London, 1962.
Shreir, L. L., *Corrosion and Corrosion Control*, John Wiley, New York, 1985.

8

Stainless Steels

Stainless steels are alloys of iron and chromium with or without other added components. They are the most widely applied and versatile of the corrosion-resistant alloys formulated from the anodic passivating metals described in Chapter 3. Other alloys from within the nickel-chromium-iron-molybdenum system are specialist materials, designed to meet specific difficult environments, and are excluded from general application by their high costs. This constraint does not apply to alloys in which iron and chromium predominate because chromium is available as an inexpensive iron-chromium master alloy, *ferro-chrome*, suitable for alloying with iron. As a consequence, a wide range of corrosion-resistant alloys is available at moderate cost; some of them are mass products, manufactured in integrated steel plants.

For stainless steels, more than for any other group of alloys, structures, properties, corrosion resistance, and costs are so closely related that none of these aspects can be properly appreciated without detailed reference to the others. The essential requirement to confer the expected minimum passivating characteristics on a stainless steel is a chromium content of not less than 13 wt%. To modify passivating characteristics or to manipulate phase equilibria to produce metallurgical structures matched to various specific applications, other components are required. The need to maintain or enhance the passivating characteristics restricts the most useful additional alloy components to nickel, molybdenum, and carbon. Certain other elements may also be added: titanium or niobium to control the effects of carbon, manganese or nitrogen to economize on the use of nickel, and copper to induce precipitation-hardening. There is a further constraint because of the relative costs of alloy components, summarized in Table 8.1.

Components added to steel can influence the phase equilibria in two respects:

1. They alter the relative stabilities of austenite and ferrite. Carbon, nickel, manganese, nitrogen, and copper stabilize austenite; chromium, molybdenum and silicon stabilize ferrite.

2. They can precipitate any carbon present if their own carbides are more stable than the iron carbide, Fe_3C (cementite). Metals that form carbides selectively in steels include chromium, molybdenum, titanium, and niobium but not nickel, copper, or manganese. Nitrogen, if present, can partly replace carbon, forming carbo-nitrides.

TABLE 8.1
Raw Material Costs for Stainless Steels

Material	Price $/tonne
Iron (as scrap)	149[a]
Chromium (as high carbon ferro-chromium)	924–1073[a]
Nickel (as the pure metal)	6755[b]
Molybdenum (as ferro-molybdenum)	10000[a]
Manganese (as ferro-manganese)	396[a]
Copper (as the pure metal)	2323[b]
Titanium (as ferro-titanium)	3465[a]
Niobium (as ferro-niobium)	15000[a]

Note: Costs are for alloy content, e.g., the cost given is per tonne of chromium not per tonne of ferro-chromium.

[a] Private communication from suppliers, December, 1996.
[b] London Metal Exchange, December, 1996.

8.1   Phase Equilibria

In the following brief review, equilibria in multi-component stainless steels is approached by considering how successive additions of other components modify equilibria in the iron-chromium binary system. Pickering, cited in the further reading list, gives a more comprehensive treatment.

8.1.1   The Iron-Chromium System

The equilibrium phase diagram for the iron-chromium system is reproduced in Figure 8.1. Chromium and iron both crystalize in the BCC structure and their atom sizes do not differ much, so that they have extensive mutual solubility and a ferrite, α, phase field extends over most of the diagram.

There are two other significant features:

1. A phase field, the γ-*loop*, in which austenite, γ, is the stable phase, spanning the temperature range 840 to 1400°C, with a maximum chromium content of 10.7 wt % at 1075°C. The γ-loop is surrounded by a narrow two-phase field in which austenite and ferrite coexist, with a composition range of 0.8 wt % chromium at its widest.

2. A phase field with a maximum at 821°C, in which a tetragonal, non-stoichiometric FeCr compound, the sigma phase, σ, is stable. The sigma phase field is flanked by wide two-phase,(α + σ), fields so that sigma phase can be precipitated in alloys containing 25 to 70% chromium by prolonged heat-treatment at temperatures in the range 500 to 800°C. Permissible chromium contents in stainless steels based on the binary system are limited to the range 12 to 25% to secure adequate passivation without risk of precipitating the sigma phase, which embrittles the steel.

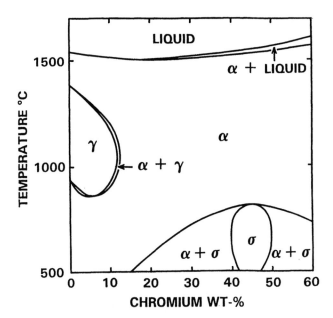

Figure 8.1
Equilibrium phase diagram for the iron-chromium system.

8.1.2 Effects of Other Elements on the Iron-Chromium System

8.1.2.1 Carbon

Carbon is introduced naturally during steelmaking and its content in the finished steel can be adjusted to any required value. It modifies the iron-chromium system in two respects:

1. On a mass for mass basis, it is the most efficient austenite stabilizer, expanding both the γ-loop and the $(\alpha + \gamma)$ phase field, due to the much higher solubility of carbon in austenite than in ferrite. The effect is limited by the maximum carbon content that austenite can hold in solution, about 0.6% C at 1300°C, that extends the γ-loop and $(\alpha + \gamma)$ phase field to 18% and 27% Cr, respectively, as illustrated in Figure 8.2.

2. At temperatures below the boundary of the γ-loop, carbon in excess of its very small solubility in ferrite tends to precipitate as the complex carbides, Cr_3C, $Cr_{23}C_6$ and Cr_7C_3. Since the carbides are rich in chromium, the precipitation can severely deplete the chromium content of the adjacent metal.

8.1.2.2 Nickel

Nickel forms a complete series of solid solutions with γ-iron, with which it shares a common FCC structure but it has only limited solubility in the

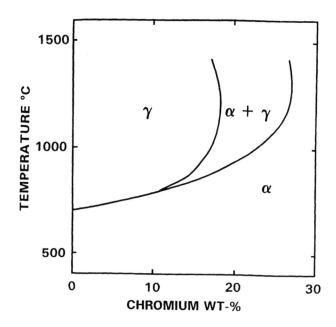

Figure 8.2
Section through the iron-chromium-carbon ternary system at 0.05% carbon showing extended γ and (α + γ) phase fields.

structurally incompatible BCC α-iron. As a consequence it is a powerful austenizing element and its addition to the iron-chromium system progressively extends the γ-loop, especially towards higher chromium contents and lower temperatures. Comparison of Figure 8.3 with Figure 8.1 shows the extent of the expansion for an addition of 8% nickel. Although the γ-loop does not extend to ambient temperatures, the transformation is sluggish at low temperatures and in practice austenite can be retained as the metal cools. The nickel content for retention of a fully austenitic structure depends on the chromium content and has a minimum value of 10 to 11% at an optimum chromium content of 18%. This balance of chromium and nickel contents is economically important in formulating austenitic steels, as described later in Section 8.2.2.2, because the cost of nickel is much higher than the cost of chromium.

A disadvantage of added nickel that must be accepted along with its advantages is that it enlarges the sigma phase field and its associated two- and three-phase fields, as evident in Figure 8.3.

8.1.2.3 Nickel and Carbon Present Together

The austenizing powers of carbon and of nickel are additive, but the assistance given by carbon is limited by its low solubility in iron-chromium-nickel austenites, as indicated by Table 8.2. Even then, to exploit its considerable austenizing power, the carbon content must be retained in solution by

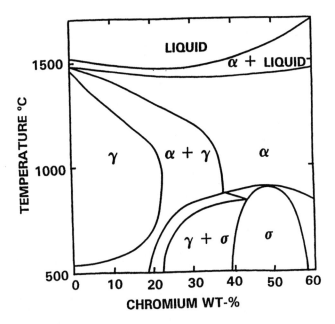

Figure 8.3
Section through the equilibrium phase diagram for the ternary iron-chromium-nickel system at 8% nickel.

quenching the steel from a high temperature, *circa* 1050°C. If the steel is not quenched, carbon contents in excess of 0.03% tend to precipitate as carbides.

Nickel and carbon introduce pseudo-peritectic reactions into the iron-chromium-carbon-nickel system that control the structure of metal solidified from the liquid. The first solid deposited is δ-ferrite that subsequently reacts with residual liquid to form austenite, γ:

$$\text{Liquid} + \delta = \gamma \qquad (8.1)$$

The reaction is incomplete in industrial casting and welding practices, leaving up to 10% residual δ-ferrite. This remains as a structural feature of near net-shape castings and in welds but it is converted to austenite by the mechanical and thermal treatments used in producing wrought materials, so that there is less than 1% of δ-ferrite in plate or sheet products.

TABLE 8.2
Solubility of Carbon in Austenite with 18% Chromium + 8% Nickel

Temperature °C	700	800	900	1000	1050
Solubility wt-%	0.03	0.03	0.04	0.06	0.09

8.1.2.4 Molybdenum

Molybdenum stabilizes ferrite and when added to improve corrosion resistance to a steel formulated to be fully austenitic, its effect must be compensated by reducing the chromium content and increasing the nickel content. Molybdenum also replaces some of the chromium in the complex carbides, $Cr_{23}C_6$ and Cr_7C_3 described in Section 8.1.2.1. Unfortunately, high molybdenum contents promote sigma-phase formation in austenitic as well as in ferritic steels and therefore molybdenum contents in stainless steels are normally restricted to <3.5%.

8.1.2.5   Other Elements

Titanium and niobium can be added as scavengers for carbon in steels designed for welding.

Manganese can be used as a substitute to reduce nickel contents in austenitic steels.

Copper is added to certain steels is to participate in precipitation hardening.

Nitrogen can enhance the mechanical properties and assist austenite stabilization.

8.1.3   Schaeffler Diagrams

Schaeffler diagrams named for their originator, provide means of estimating the combined effect of all of the components on the relative proportions of austenite, ferrite, and martensite present in a multicomponent stainless steel. The empirical coefficients given in Equations 8.2 and 8.3 are used to condense the composition variables into two parameters representing the austenite or ferrite stabilizing effects of the components in terms of the equivalent effects of chromium or nickel:

$$\text{Ni equivalent} = \text{wt-\% Ni} + 30 \times \text{wt-\% C} + 25 \times \text{wt-\% N} + 0.5 \times \text{wt-\% Mn} \qquad (8.2)$$

$$\text{Cr equivalent} = \text{wt-\% Cr} + \text{wt-\% Mo} + 1.5 \times \text{wt-\% Si} + 0.5 \times \text{wt-\%Nb} + 1.5 \times \text{wt-\%Ti} \qquad (8.3)$$

This reduces the system to a section of the iron-rich corner of a pseudo-ternary system, as in Figure 8.4. The phase fields identify the phases observed in steels quenched to ambient temperature from 1050°C. Fields labelled for martensite represent compositions which pass through a γ-loop on cooling and transform to the metastable phase martensite, α' instead of the stable phase, ferrite. Schaeffler-type diagrams were devised to assess the structures of welds. Although useful for more general application, they are only approximations and alternative equations for nickel and chromium equivalents are sometimes advocated.

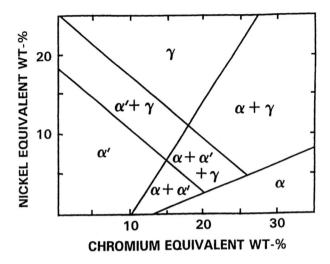

Figure 8.4
Schaeffler diagram for estimating the constitutions of stainless steels. Ni equivalent = wt-% Ni + 30 wt-% C + 25 wt-% N + 0.5 wt-% Mn. Cr equivalent = wt-% Cr + wt-% Mo + 1.5 wt-% Si + 0.5 wt-% Nb + 1.5 wt-% Ti. α = Ferrite; α' = martensite; and γ = austenite. (Reproduced with permission from Argus Business Media, Queensway House, Redhill, Surrey, UK.)

Example 1

Using the Schaeffler diagram given in Figure 8.4, suggest the phases probably present in the following steels:

Steel 1 — Analysis: 17 % Cr, 12 % Ni, 2.5 % Mo, 0.06 % C, 2 % Mn, 0.6 % Si.

Steel 2 — Analysis: 12 % Cr, 0.5 % Ni, 0.15 % C, 0.4 % Mn, 0.4 % Si.

Steel 3 — Analysis: 25 % Cr, 6 % Ni, 3 % Mo, 0.06 % C, 0.25 % N.

SOLUTION:

The probable structures are assessed from the Schaeffler equivalents:

Ni equivalent = wt-% Ni + 30 × wt-% C + 25 × wt-% N + 0.5 × wt-% Mn.

Cr equivalent = wt-% Cr + wt-% Mo + 1.5 × wt-% Si.

Steel 1

Ni equivalent = 12 + (30 × 0.06) + (0.5 × 2) = 14.8 %
Cr equivalent = 17 + 2.5 + (1.5 × 0.6) = 20.4 %

These coordinates are in the γ phase field, so the steel is probably wholly austenitic.

Steel 2

Ni equivalent = 0.5 + (30 × 0.15) + (0.5 × 0.4) = 5.2 %
Cr equivalent = 12 + (1.5 × 0.4) = 12.6 %

These coordinates are in the α' phase field, so the steel is wholly or mainly martensitic.

Steel 3

Ni equivalent = 6.0 + (30 × 0.06) + (25 × 0.25) = 14.05 %

Cr equivalent = 25 + 3 = 28 %

These coordinates are in the middle of the $\alpha + \gamma$ phase field, so the steel is duplex, probably with equal quantities of austenite and ferrite.

8.2   Commercial Stainless Steels

8.2.1   Classification

The American Iron and Steel Institute (AISI) specifications use a three-digit code to identify wrought stainless steels by structure. The first digit defines the following classes:

AISI 300 series: Austenitic steels with nickel as the primary austenite stabilizer.

AISI 200 series: Austenitic steels with manganese and nitrogen as nickel substitutes.

AISI 400 series: Ferritic and martensitic steels with little or no nickel.

The other digits identify the steels and letter suffixes indicate other features; in particular L, denotes low carbon content. Duplex and precipitation-hardening steels are specified in different formats. Table 8.3 gives specifications for representative standard steels.

8.2.2   Structures

8.2.2.1   Ferritic steels

Standard ferritic stainless steels with 11% to 30% chromium, <1% carbon and little or no nickel have single-phase BCC structures and are less ductile than the austenitic steels that have FCC structures. They are susceptible to grain coarsening on heating to high temperatures and can be embrittled if heated in the temperature range 340 to 500°C, because the homogeneous ferrite can precipitate a chromium-rich α phase from the main α phase: the effect known as *"475 embrittlement"*. There is, of course, the further risk that some sigma phase can form in ferritic steels with more than 25% chromium on prolonged heating in the temperature range 500 to 800°C.

TABLE 8.3
Selection of Specifications for Stainless Steels

Designation	Type	Composition weight-%						
		Cr	Ni	Mo	C	Mn	Si	Others
AISI 300 Series:								
AISI 304	Austenitic	18–20	8–10.5	—	<0.08	2	1	—
AISI 304L	Austenitic	18–20	8–12	—	<0.03	2	1	—
AISI 321	Austenitic	18–20	8–10.5	—	<0.08	2	1	Ti = 5 × wt-% C
AISI 347	Austenitic	18–20	8–10.5	—	<0.08	2	1	Nb = 10 × wt-% C
AISI 316	Austenitic	16–18	10–14	2.0–3.0	<0.08	2	1	—
AISI 316L	Austenitic	16–18	10–14	2.0–3.0	<0.02	2	1	—
AISI 317	Austenitic	18–20	11–15	3.0–4.0	<0.08	2	1	—
AISI 317L	Austenitic	18–20	11–15	3.0–4.0	<0.03	2	1	—
AISI 310	Austenitic	24–26	19–22	—	<0.25	2	1.5	—
AISI 330	Austenitic	17–20	34–37	—	<0.08	2	1.5	—
AISI 200 Series:								
AISI 201	Austenitic	16–18	3.5–5.5	—	0.15	5.5–7.5	1	+ 0.25 N
AISI 202	Austenitic	17–19	4–6	—	0.15	7.5–10	1	+ 0.25 N
AISI 400 Series:								
AISI 409	Ferritic	10.5–11.7	<1	—	<0.08	1	1	—
AISI 430	Ferritic	16–18	<1	—	<0.08	1	1	—
AISI 434	Ferritic	16–18	<1	0.8–1.2	<0.08	1	1	—
AISI 410	Martensitic	11.5–13.5	<1	—	0.15	1	1	—
AISI 431	Martensitic	15–17	1.2–2.5	—	0.20	1	1	—
Duplex Steels:								
Steel 1	Duplex	25	6.0	3.0	<0.08	—	—	1.5 Cu + 0.25 N
Steel 2	Duplex	25	5.5	3.0	<0.08	—	—	—
Precipitation-Hardening Steels:								
Steel 1	Precipitation	16	4.2	—	0.04	0.5	0.5	3.5 Cu
Steel 2	Precipitation	15	4.5	—	0.04	0.3	0.4	3.5 Cu + Nb

Note: All steels: %P < 0.04%, %S < 0.03%.

Ferritic steels with ultra-low carbon and nitrogen contents, >0.01% have improved ductility. Versions with 26 to 29% chromium and 1 to 4% molybdenum have been produced but they are expensive.

8.2.2.2 Austenitic Steels

Commercial austenitic steels are formulated to realize almost completely austenitic structures at minimum cost, consistent with satisfactory performance. In practice, this implies minimizing the nickel content because nickel is one of the most expensive alloying elements. Carbon is available free of cost, because of the nature of the steelmaking process, and if it is retained in solid solution by quenching from a high temperature it can replace a small but economically significant fraction of the nickel content. The maximum carbon content that it is convenient to retain in solution is 0.06%, and the composition for a fully austenitic steel with the minimum nickel content is 18 wt-% chromium, 9 wt-% nickel, and 0.06% carbon.

Versions of austenitic steels are formulated at extra cost for welding, to counter carbide precipitation that can cause local corrosion, as explained later. There are two approaches: (1) steels such as AISI 304L, AISI 316L, and AISI 317L with low carbon contents and compensating higher nickel contents to stabilize the austenite, and (2) steels such as AISI 321 and AISI 347 with titanium or niobium as carbon scavengers.

Austenitic steels containing molybdenum to enhance corrosion resistance, AISI 316, AISI 316L, and AISI 317 have lower chromium contents to avoid the sigma phase and higher nickel contents to offset the effect of molybdenum in stabilizing ferrite.

8.2.2.3 Martensitic Steels

The chromium content of a martensitic stainless steel must be above the minimum needed for passivation but below the maximum of the γ-loop, so that the steel can be converted to austenite for transformation to martensite on cooling. For a nickel-free steel such as AISI 410, a high carbon content is needed both to expand the γ-loop to accommodate enough chromium for passivation and to harden the martensite. The chromium content can be higher if a small nickel content is added to expand the γ-loop further, as for AISI 431. It is usually sufficient for the steel to cool in air to effect the martensitic transformation. After the martensite transformation, the steels must be *tempered* to alleviate brittleness, i. e. given controlled thermal treatment in the range 200 to 450°C that allows the martensite to transform into hard but less brittle decomposition products. Care is needed not to exceed 450°C, because carbides precipitate above this temperature, reducing both corrosion resistance and toughness.

8.2.2.4 Duplex Steels

Duplex steels contain about 28% chromium equivalent and 6% nickel equivalent, as defined for the Schaeffler diagram, to produce structures with about equal proportions of austenite and ferrite. The alloy components are distributed unequally between austenite and ferrite: the austenite has a higher proportion of the nickel, carbon, and copper contents and the ferrite has a higher proportion of the chromium and molybdenum contents.

8.2.2.5 Precipitation-hardening steels

Precipitation-hardening steels have compositions with a latent potential for the martensitic and other transformations that can be exploited for precipitation-hardening by thermal and mechanical treatments.

8.3 Resistance to Aqueous Corrosion

8.3.1 Evaluation from Polarization Characteristics

8.3.1.1 Relevance to Corrosion Resistance

Stainless steels resist or succumb to corrosion according to whether or not they succeed in maintaining a passive surface. Information on conditions favoring passivation of particular steels is given by anodic polarization characteristics, i.e., the current flowing at the surface of a steel as a function of the potential applied to it. These are not inherent properties of the steels but system characteristics, because they are environment-dependent.

In a natural system, the potential of the steel is determined by mutual polarization of the prevailing anodic and cathodic reactions. This potential, the *mixed potential* and the corresponding current exchanged between anodic and cathodic reactions, can be assessed by displaying the net polarization characteristics of the steel and of cathodic reactions in the same format as for corrosion velocity diagrams described in Section 3.2.3 of Chapter 3.

The essential feature of the polarization characteristics for an anodically passivating metal, such as a stainless steel, is a *passive range* over a potential interval, as in the examples illustrated in Figures 8.5 through 8.12, discussed later. The objective in matching a steel to an environment is to ensure that balance struck between anodic currents from the steel and the currents due to the prevailing cathodic reactions yields a mixed potential that is stable and lies in the passive range.

8.3.1.2 Determination

Polarization characteristics can be determined empirically using a potentiostat, that is a customized instrument designed to maintain a constant potential on a small sample of the steel in any selected aqueous medium. It can simulate any mixed potential that the steel may experience in a natural system. The instrument monitors deviations from an adjustable preset potential and corrects them by applying an external potential to a counter electrode remote from the sample. The information required is the current density at the sample surface as a function of potential. The preset potential is scanned through the potential range of interest and the potential/current density relationship is recorded. The procedure is standardized.*

Fortunately, the active ranges of most stainless steels lie almost wholly above the equilibrium potential for hydrogen evolution, even for acids with low pH, so that the measurement does not suffer significant interference from it. Interference by reduction of dissolved oxygen is circumvented by continuously purging oxygen from the test solution with a stream of very pure nitrogen. Under these conditions, an experimentally determined polarization plot for the steel merges smoothly with plots for hydrogen evolution at potentials below the range of interest and for oxygen evolution at potentials above it.

8.3.1.3 Presentation

The current coordinate is usually plotted on a semi-logarithmic coordinate, not for any fundamental reason but to circumvent the difficulty in presenting both the passive current and the maximum current at the peak of an active range that can differ by three orders of magnitude. The following terms and symbols describing anodic polarization characteristics are marked on the schematic example for a stainless steel in an acid, given in Figure 8.5:

Active range	A potential range where the metal dissolves as ions in a low oxidation state, e.g., Fe^{2+} and Cr^{3+}.
Passive range	A potential range where passivity prevails.
Transpassive range	A potential range where the metal dissolves as ions in a higher oxidation state, e.g., CrO_4^{2-}.
Rest potential, E_R	The potential for zero current.
Passivating potential, E_P	The lower limit of the passive range.
Breakdown potential, E_B	The upper limit of the passive range.
Critical current density, i_{CRIT}	The active current density at the passivating potential.
Passive current density, $i_{PASSIVE}$	The current density in the passive range.

---

* Designation G5, *Annual Book of ASTM Standards*, American Society for Testing and Materials, Philadelphia, 1996.

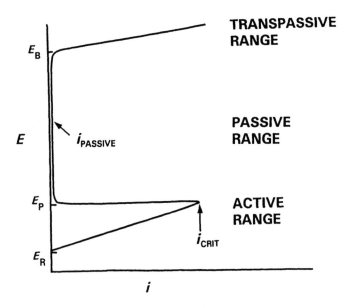

Figure 8.5
Schematic anodic polarization characteristics for a stainless steel in an acidic aqueous medium. E_R, Rest potential. E_P, Passivating potential. E_B, Breakdown potential. i_{CRIT}, Critical current density. $i_{PASSIVE}$, Passive current density.

8.3.1.4   Influence of the Environment

Two factors in particular influence the polarization characteristics for a stainless steel, pH and the presence or absence of Cl^- or Br^- ions. Figures 8.6 through 8.9 show schematic polarization characteristics in neutral, acidic, neutral chloride, and acid chloride aqueous media, presented on schematic corrosion velocity diagrams with cathodic polarization characteristics for oxygen absorption and for hydrogen evolution for later reference.

Acidic Media

For acidic media, there are active, passive and transpassive ranges in sequence with rising potential, as in Figure 8.6. Typical critical values for an AISI 304 austenitic steel in 0.1M sulfuric acid are given in Table 8.4.

TABLE 8.4
Polarization Characteristics of AISI 304 Steel in 0.05 M Sulfuric Acid (pH 1.2) Critical Potentials and Current Densities

Rest potential, E_R	= -0.2 V (SHE)
Passivating potential, E_P	= -0.0 V (SHE)
Transpassive breakdown potential, E_B	= $+1.3$ V (SHE)
Critical current density, i_{CRIT}	= 10^{-3} A cm^{-2}
Passive current density, $i_{PASSIVE}$	= 2×10^{-6} A cm^{-2}

Benign Neutral Media

There is no active range for neutral media and the steel is either immune or passive at all potentials below the breakdown potential, as in Figure 8.7. Corrosion in neutral media at the high potentials in the transpassive range is rarely encountered.

Acid Halide Media

Chloride, Cl^-, or bromide, Br^-, ions introduced into an acid modify the anodic polarization characteristics as follows, illustrated for an arbitrary Cl^- ion concentration of 0.1 M in Figure 8.8:

1. The active peak is extended to a higher critical current density, i_{CRIT}.
2. The passive range is prematurely terminated by the intervention of a breakaway current at a potential E_{PP}, below the normal breakdown potential, E_B. Steels polarized to potentials above this potential are found to be pitted by intense local attack and on this account, it is called the *pitting potential*, E_{PP} and the line tracing the current densities after the breakaway is called the *pitting branch*.

Neutral Halide media

Cl^- or Br^- ions introduced into neutral media also produce pitting potentials as illustrated in Figure 8.9 but there is, of course, no active peak for them to influence.

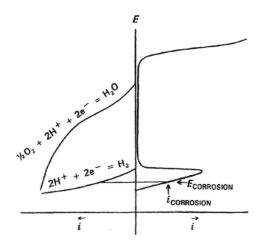

Figure 8.6
Corrosion velocity diagram for a stainless steel in dilute acidic media. In oxygen-free acid, the steel corrodes at an active potential, $E_{CORROSION}$, established by hydrogen evolution. In air-saturated acid, it is passivated at a potential in the passive range imposed by reduction of dissolved oxygen.

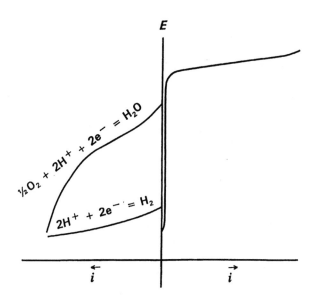

Figure 8.7
Corrosion velocity diagram for a stainless steel in neutral aqueous media. The steel is immune or passive at all potentials imposed by hydrogen evolution or the reduction of oxygen dissolved in air-saturated media.

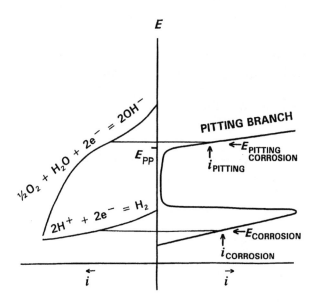

Figure 8.8
Corrosion velocity diagram for a stainless steel in dilute acid chloride media. In oxygen-free acid, the steel corrodes at an active potential, $E_{CORROSION}$, established by hydrogen evolution. In air-saturated acid, it corrodes by pitting at a potential $> E_{PP}$ imposed by reduction of dissolved oxygen.

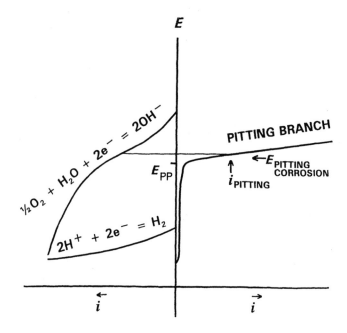

Figure 8.9
Corrosion velocity diagram for a stainless steel in neutral chloride media. In oxygen-free media, the steel is passive. In air-saturated media, it corrodes by pitting at a potential $> E_{PP}$ imposed by reduction of dissolved oxygen.

A pitting potential with a value significantly less positive than the breakdown potential is observed if the Cl^- ion concentration exceeds about 0.01M and it moves to progressively lower values as the concentration is raised. In an acid, the pitting branch intercepts the active range as it approaches 1M, eliminating the passive range, as illustrated in Figure 8.10.

8.3.1.5   Influence of Steel Composition and Condition
Raising the chromium content confers the following benefits:

1. Lower passivating potentials.
2. Reduced passive current densities.
3. Higher breakdown and pitting potentials.
4. Lower critical current densities.

The addition of nickel decreases the critical current density without affecting the passivation potential but the most beneficial additional alloying element is molybdenum because of its strong effects in lowering critical current densities and raising pitting potentials, as illustrated by the polarization characteristics for representative alloys given in Figure 8.11.

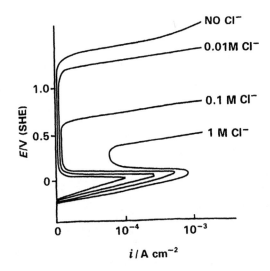

Figure 8.10
Typical anodic polarization characteristics for AISI 304 steel in 0.05 M sulfuric acid solutions containing sodium chloride, showing progressive lowering of the pitting potential and expansion of the active peak as the chloride ion concentration is raised. *Note* the current density is plotted on the semi-logarithmic scale generated by a standard potentiostat. This scale is linear for $i < 10^{-4}$ A cm^{-2} and logarithmic for $i > 10^{-4}$ A cm^{-2}.

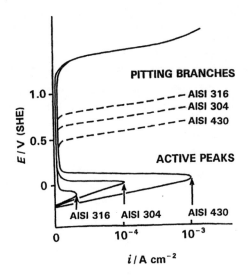

Figure 8.11
Typical anodic polarization characteristics in acid and acid chloride media at 20°C for some standard stainless steels quenched from 1050°C. Full lines: polarization characteristics in 0.05 M sulfuric acid. Dotted lines: Pitting branches for 0.05 M sulfuric acid with 0.01 M sodium chloride. Expanded active peaks omitted for clarity. AISI 430 — 17% Cr, 0.05% C (ferritic steel). AISI 304 — 18 % Cr, 9.5% Ni, 0.05% C (austenitic steel). AISI 316 — 17% Cr, 11% Ni, 2.5% Mo, 0.05% C (austenitic steel).

Polarization characteristics are inseparable from phase equilibria and microstructures. Austenitic steels have higher nickel and chromium contents than martensitic and ferritic steels and yield lower passive currents and critical current densities. Precipitated carbides deplete adjacent metal of chromium and molybdenum, curtailing the passive range and reducing pitting potentials, as described in Section 8.3.2.4, in the context of an effect called *sensitization*.

8.3.2 Corrosion Characteristics

The corrosion resistance of stainless steels improves in the ranking order: martensitic, precipitation-hardening, ferritic, duplex and austenitic steels. Martensitic and precipitation hardening steels are exploited for their strength or hardness but their use is restricted by their limited corrosion resistance. Ferritic steels are better but austenitic and duplex steels are required for more aggressive environments. All stainless steels are susceptible to stress-corrosion cracking but ferritic and duplex steels are less susceptible than austenitic steels. These matters were considered earlier in Chapter 5.

8.3.2.1 Corrosion Resistance in Acids

Non-oxidizing Acids

To resist corrosion in a non-oxidizing acid, the potential of a stainless steel must be raised from the active into the passive range, either by oxygen or some alternative oxidizing agent, such as Fe^{3+} or NO_3^- ions. Oxygen dissolved from the air is generally sufficient but if it is the only oxidizing agent, access to air must be continuous and unrestricted because even passivated steel slowly reacts with dissolved oxygen and, unless replenished, the oxygen concentration can be depleted to an extent that passivity breaks down and the steel corrodes. The role of oxygen can be explained using the critical values for AISI 304 steel given in Table 8.4. In an acid of pH 2 that is *completely free from dissolved oxygen*, the only possible cathodic reaction is hydrogen evolution:

$$2H^+ + 2e^- = H_2 \tag{8.4}$$

for which the equilibrium potential is:

$$E' = -0.0591 \ pH = -0.0591 \times 2 = -0.118 \ V \ (SHE) \tag{8.5}$$

This potential lies between the passivating potential, $E_P = 0.0$ V (SHE), and the rest potential, $E_R = -0.2$ V (SHE), so that the balance struck between the anodic and cathodic currents yields a mixed potential in the active range and the steel corrodes. If, now, the acid is allowed to equilibrate with oxygen in the air, hydrogen evolution is replaced by oxygen reduction as the dominant cathodic reaction:

$$\tfrac{1}{2}O_2 + 2H^+ + 2e^- = H_2O \tag{8.6}$$

for which the equilibrium potential is:

$$E' = E^{\ominus} - \frac{0.0591}{2} \log \frac{1}{\left(a_{H^+}\right)^2 \left(a_{O_2}\right)^{1/2}} \tag{8.7}$$

inserting values, $E^{\ominus} = 1.228\,V$ (SHE), pH $= 2$ and $a_{O_2} = 0.21$ for equilibrium with air yields the value $E' = 1.1\,V$ (SHE).

This potential is so much higher than the passivation potential that even though the oxygen absorption reaction is heavily polarized, the current density available at the passivation potential is greater than the critical current density for the steel so that a balance cannot be struck between anodic and cathodic currents in the active range. Consequently, the mixed potential rises into the passive range and the steel passivates. The essence of these calculations is illustrated graphically in Figure 8.6.

Provided that these conditions are satisfied, austenitic stainless steels without molybdenum, such as AISI 304, can resist most aerated dilute mineral and organic acids in concentrations up to about 0.5 M at ambient temperatures. Corresponding steels with molybdenum contents, such as AISI 316, can resist acids in concentrations up to 1 or 2 M, at ambient temperatures and dilute acids at moderately elevated temperatures.

Oxidizing Acids

Oxidizing acids raise the potential on a stainless steel into the passive range without assistance and therefore austenitic stainless steels have excellent resistance to nitric acid and mixtures of nitric and sulfuric acids in almost all concentrations and at temperatures up to about 80°C. Attack can occur at carbide sites if titanium is used as a carbon scavenger as in AISI 321 steel formulated for welding.

Phosphoric Acid

Austenitic stainless steels resist phosphoric acid in all concentrations even at temperatures up to about 80°C. Passivation is probably assisted by the sparingly soluble iron secondary and tertiary phosphates, referred to in Section 6.4.1.

Hydrochloric Acid

As explained in Section 8.3.1.4, the progressive introduction of chloride ions into an acidic medium curtails and ultimately eliminates the passive range. As a result, the corrosion resistance of stainless steels in hydrochloric acid at ambient temperature is unreliable and limited to concentrations of less than about 0.05 M.

8.3.2.2 Pitting Corrosion

Pitting corrosion is intense local attack at isolated sites, where the passivity has been breached. These sites are very small and they are anodic to the surrounding areas because of the electric field across the passive layer. The resultant active/passive cells stimulate intense attack on the small anodes, driving the growth of the pits and inhibiting the nucleation of new pits in the immediate vicinities of existing ones. This establishes a pattern of attack at discrete sites widely distributed over the metal surface. The pits nucleate after a finite induction time and grow progressively deeper into the metal. They can perforate thin gauge metal, rendering stainless steel containers unserviceable even if general attack is insignificant.

The pitting is induced by halide ions, most commonly the Cl^- ion or its precursor, the hypochlorite ion, ClO^- but also sometimes the bromide, Br^-, ion. The steels are susceptible to pitting in most aqueous environments containing these ions, including near-neutral media in which corrosion is otherwise not usually very serious. Hence, stainless steels can be attacked by pitting in contact with seawater, marine atmospheres, acid/chloride mixtures in chemical processes, residual bleach-based disinfectants, and photographic materials. The most effective pitting ion is chloride. The bromide ion is less effective but fluoride, F^- and iodide, I^- are not normally pitting ions.

For pits to form, the pitting ion must be present at sufficient concentration in the environment to yield a pitting potential, $E_{PP,}$ and the potential on the metal must exceed it. The potential is imposed by prevailing cathodic reactions and in natural situations, the most likely effective reaction is the reduction of oxygen dissolved in an acid or in water from the air. For samples of AISI 304 steel with the basic polarization characteristics given in Table 8.4, potentiostatic determinations yield typical values for pitting potentials as follows:

0.1 M solution of sodium chloride (Cl^- ions): $E_{PP} = +0.4$ V (SHE)

0.1 M solution of sodium bromide (Br^- ions): $E_{PP} = +0.7$ V (SHE)

The equilibrium potential for reduction of oxygen dissolved from the air, by Equation 8.7 is well above the pitting potential for the 0.1 M chloride solution and the consequent mixed potential places the current on the pitting branch, as illustrated in Figures 8.8 and 8.9. It is also above the pitting potential for the 0.1 M bromide solution. Hence pitting is expected in both solutions but inflicting more extensive damage in the chloride solution than in the bromide solution.

Once they have nucleated, some pits grow but others can re-passivate. This is sometimes attributed to changes in the micro-environment within the pits that influence a balance between dissolution and passivation. To arrest pit growth once it has started, the potential on the metal must be reduced to a value significantly below the pitting potential, the so-called *protection potential*. There are sundry other influences on pit growth; for example a gravitational effect favors pit growth more on horizontal than on vertical surfaces. This also seems to be associated with the micro-environment in pits.

Pits often grow at the sites of microscopic inhomogeneities outcropping at the steel surface, such as carbide and sulfide particles and grain boundaries. Although these sites are preferred, they are not essential, because pits can nucleate elsewhere. These features probably cause weaknesses in the passive surface, due to associated local chromium depletion. Mechanical factors such as heavy cold-work and rough surface finish also encourage pit nucleation.

Detailed theories of pit induction on stainless steels are inevitably inconsistent because of the conflicting views on the nature of anodic passivity. Recalling the discussion in Section 3.2.4.3, the film and absorption theories of passivity are not necessarily incompatible, but the implications for pitting are quite different. The film theory envisions that pits are initiated by breaching an established film of tangible thickness, whereas the absorption theory assumes that pitting ions are already present in the passivating monolayer and passivity breakdown is due to their selective removal at the pitting potential.

The tendency of the different halogen ions to induce pitting does not correspond with their increasing electronegativity and their sequence in the Periodic Table, i.e., iodine, bromine, chlorine, fluorine. From the values of pitting potentials for chlorine and bromine ions given above, a pitting potential for the fluoride ion might be expected at some value less positive than that for the chloride ion, + 0.4 V (SHE). However, whereas chloride and bromide ions are *structure breaking*, the fluoride ion is *structure forming* and therefore its ability to interact with a passivated surface is hindered by a solvation shell. Iodide ions are also ineffective as pitting ions for stainless steels but for a different reason; they are easily oxidized to iodine and cannot exist in significant concentrations at high positive potentials, as illustrated by the following example.

Example 2

A stainless steel considered for use in plant processing 0.1 M solutions of iodide salts for photographic materials exhibits the following characteristics at the prevailing pH:

$$\text{Rest potential, } E_R = -0.2 \text{ V (SHE)}$$
$$\text{Passivating potential, } E_P = 0.0 \text{ V (SHE)}$$
$$\text{Breakdown potential, } E_B = +1.3 \text{ V (SHE)}$$

Information is available that for this particular steel, 0.1 M solutions of Cl^- or Br^- ions introduce pitting potentials $E_{PP}(Cl^-) = +0.4$ V SHE and $E_{PP}(Br^-) = +0.7$ V SHE, respectively. Does this suggest that the steel is likely to suffer pitting corrosion in the application contemplated?

SOLUTION:

Pitting potentials in the ascending order Cl^-, Br^-, I^- would be logical, in view of the decreasing reactivities of the halogens in descending order in the Periodic Table, so that a hypothetical pitting potential for a 0.1 M solution of iodide ions would be $> +0.7$ V SHE. However, the low standard electrode potential for the reaction:

$$I_2 + 2e = 2I^-$$

given in Table 3.3:

$$E^{\ominus} = +0.536 \text{ V (SHE)}$$

suggests that iodine ions may not survive at such a high potential. Applying the Nernst equation to the reaction for a 0.1 M solution of iodide:

$$E' = E^{\ominus} - \frac{0.0591}{2} \log \frac{\left(a_{I^-}\right)^2}{a_{I_2}}$$

$$= +0.536 - \frac{0.0591}{2} \log(0.1)^2 = +0.595 \text{ V (SHE)}$$

assuming unit activity for the precipitated iodine.

Hence the I^- ions are progressively removed as the potential is raised $> + 0.595$ V SHE, i.e., well below any hypothetical pitting potential, so that pitting corrosion is not expected.

8.3.2.3 Crevice Corrosion

Crevice corrosion is due to enhanced anodic activity in oxygen-starved crevices in a metal surface as it is for active metals. For stainless steels, the root cause is easier to envision, because local oxygen starvation leads to passivity breakdown and the establishment of active/passive cells in which the large open passive surface stimulates corrosion on the oxygen-starved active area in a crevice. Chloride contamination can assist depassivation where the degree of oxygen starvation would otherwise be marginal.

8.3.2.4 Sensitization and Intergranular Corrosion

Standard austenitic steels, such as AISI 304 and AISI 316 are supplied by the manufacturer with the carbon retained in supersaturated solution by rapid cooling from a final heat-treatment at 1100°C, and they are intended to be used in this condition. If they are subsequently reheated and cooled slowly through a critical temperature range, they become *sensitized*, a condition in which they are susceptible to *intercrystalline corrosion*, i.e., corrosion along the grain boundaries. Intercrystalline corrosion due to sensitization in the heat-affected zones close to welds is known as *weld-decay*. The effect and suitable countermeasures are well-known and it poses no problem to those who are aware of them.

The condition is due to precipitation of chromium or chromium/molybdenum carbides at grain boundaries that produces a thin continuous network of metal so severely depleted in chromium that it cannot passivate. Active/passive cells formed where this network intercepts the passive surface stimulate intense attack penetrating down the grain boundaries. Sensitized steel can be identified from anodic polarization characteristics, as illustrated in Figure 8.12.

Three factors explain why the chromium depletion is so severe and why it occurs selectively at the grain boundaries:

1. Carbon can trap chromium equivalent to 16.6 times its own mass as $Cr_{23}C_6$ or 10.1 times its own mass as Cr_7C_3.
2. Carbon diffuses in austenite orders of magnitude faster than chromium does and is therefore available from a large catchment volume.
3. Carbides precipitate preferentially at grain boundaries, where nucleation is favored.

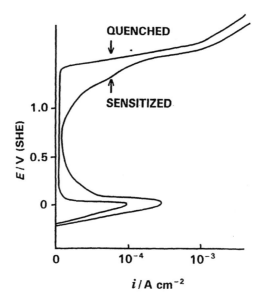

Figure 8.12
Effect of sensitizing wrought AISI 304 steel containing 0.06 % carbon on its anodic polarization characteristics in 0.05 M sulfuric acid. (1) Steel as supplied by the manufacturer, quenched in water from 1050°C to retain carbon in solution. (2) Steel sensitized by reheating for 20 minutes at 650°C.

Steels with carbon contents above 0.03% are at risk and the temperature range to avoid is 500–800°C. At lower temperatures, carbon diffuses too slowly and at higher temperatures, chromium diffuses fast enough to replenish some of the local depletion.

Molybdenum can partially replace chromium in the carbides as $(Cr,Mo)_{23}C_6$ and $(Cr,Mo)_7C_3$, so that molybdenum-bearing grades of stainless steel are less susceptible to sensitization than the corresponding molybdenum-free steels but not immune to it.

Sensitization can be avoided by using the steels customized for welding listed in Table 8.3. There is a choice between steels with low carbon contents, i.e., AISI 304L, AISI 316L, and AISI 317 L, accepting the cost penalty of higher nickel contents and *stabilized* steels containing titanium or niobium as carbon scavengers, such as AISI 321 and AISI 347, in which precipitation of chromium carbides is preempted by a preliminary heat-treatment at 900 to 920°C to trap carbon as TiC or NbC; niobium is a better scavenger for steels exposed to highly oxidizing acids because TiC can be attacked, inducing *knife-line* corrosion.

In theory, sensitization can be eliminated by prolonged post-welding heat-treatment to replenish the depleted zones by chromium diffusion from the bulk of the metal but it is expensive, impracticable for large welded structures and it eliminates strength developed by work-hardening that is one of the benefits of using austenitic steels.

8.3.2.5 *Corrosion in Aggressive Chemical Environments*

In some processes in the chemical industry, particularly aggressive liquors must be handled, including mixtures of undiluted mineral acids and concentrated solutions of aggressive ions, often at high temperatures even approaching boiling. Such media cannot be considered in the same way as aqueous solutions and are difficult to treat theoretically. Alloys based on various combinations of metals selected from among iron, nickel, chromium, molybdenum and cobalt have been customized empirically to meet various application-specific requirements. They are mostly austenitic alloys that include some stainless steels but more often iron is a minor component or is absent. Information on these alloys and their applications is given in suppliers' brochures.

8.4 Resistance to Dry Oxidation

Stainless steels are among the best standard commercial alloys for resistance to oxidation, an attribute conferred by protective films formed by preferential oxidation of chromium.

There are five oxides in the iron-chromium-oxygen system, FeO (wüstite), Fe_3O_4 (magnetite), Fe_2O_3 (hematite), Cr_2O_3 and $FeCr_2O_4$ (chromite). Chromium can be accommodated in Fe_2O_3 as the mixed oxide, $(Fe,Cr)_2O_3$ and in Fe_3O_4 as the mixed spinel, $Fe^{II}(Fe,Cr)^{III}_2O_4$ but it is only sparingly soluble in FeO.

The relative stabilities of these oxides are functions of alloy composition, as described in Section 3.3.4, but evolution of the scale depends on the intervention of other factors, i.e.:

1. Transport of Fe^{2+}, Fe^{3+}, Cr^{3+} and O^{2-}, ions within the various oxides.
2. The structural incompatibility of FeO with $FeCr_2O_4$ and Cr_2O_3.
3. The structural compatibility of Fe_3O_4 with $FeCr_2O_4$ and of Fe_2O_3 with Cr_2O_3.
4. Progressive reduction in chromium activity at the metal surface by selective oxidation.
5. Stresses developed within and between the layers as the scale thickens.

Consider the isothermal oxidation in clean air, at say 800°C, of iron-chromium alloys, starting from pure iron and progressively raising the chromium content. The scale structure characteristic of iron, $FeO/Fe_3O_4/Fe_2O_3$, persists until the chromium content reaches 2 or 3% and there is little effect on the scaling rate. As the chromium content is raised further, the spinel, $Fe^{II}Cr^{III}_2O_4$, appears as discrete islands within the

FeO layer, as illustrated in Figure 8.13. At the same time, the total thickness of the scale diminishes and so does the relative thickness of the FeO layer. This indicates that the transport of Fe^{2+} ions in FeO is reduced that would be consistent with a reduction in the composition range over which it is stable. The islands of mixed spinel are incorporated in the Fe_3O_4 layer when the FeO/Fe_3O_4 interface sweeps over them as it advances into the FeO layer. The chromium is passed on to the outermost $(Fe,Cr)_2O_3$ layer as the Fe_3O_4/Fe_2O_3 interface advances in its turn.

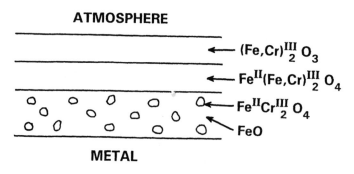

Figure 8.13
Morphology of oxide layers formed on iron-chromium alloys in air.

As the chromium content rises, the protection afforded by the chromium-enriched scale improves progressively until at about 13% chromium it is adequate for isothermal exposure to air at temperatures up to about 700°C. For a higher degree of protection, the chromium content must be raised still further. Opinions differ on whether the oxide effective in affording protection is chromium enriched $(Fe,Cr)_2O_3$ or the iron-chromium spinel.

In the long term, stainless steels oxidize by a succession of breakaways when the scale periodically cracks and reseals. The results of short-term tests can therefore be misleading.

Austenitic stainless steels are more oxidation-resistant than binary iron-chromium alloys; for example, AISI 304 with 18% chromium + 9% nickel resists oxidation as well as an binary iron-25% chromium alloy. The beneficial influence of nickel has been variously attributed to entry of nickel into the spinel as $(Fe,Ni)^{II}(Cr,Fe)^{III}O_4$ and to nickel enrichment in the metal surface by preferential oxidation of chromium and iron. In moderate amounts, molybdenum in standard stainless steels is reputed to improve the oxidation resistance by formation of an underlay of MoO_2 and manganese is considered harmful if it is high enough to contribute the spinel $MnCr_2O_4$ to the scale.

To serve at high temperatures, an alloy needs not only oxidation resistance but also other essential characteristics, including resistance to creep-rupture and immunity from embrittlement during service by sigma phase or carbide precipitation. Taking all of this into account, special steels with high chromium and nickel contents, such as AISI 310 and AISI 330 listed in Table 8.3 are manufactured for high-temperature service. They are, of course, expensive and are used only for the highest temperatures or most onerous conditions, where their special attributes are needed. For temperatures below about 900°C, steels such as AISI 304 are usually satisfactory.

8.5   Applications

The selection of stainless steels for various applications is an interesting exercise in balancing the requirements of corrosion resistance, mechanical properties, ease of fabrication and economy. The correct steel is always the one that serves its purpose *at minimum cost.*

8.5.1   Ferritic Steels

Ferritic steels have moderate to good corrosion resistance that improves with rising chromium content. Their main attribute is that they are the lowest cost stainless steels. Another merit is that they are less susceptible than austenitic steels to stress-corrosion cracking. They are ductile enough to be cold formed into simple shapes but they cannot withstand the severe deformation in forming techniques such as deep drawing. They are used for moderately demanding service and where retention of brightness or hygiene in mild environments is important. AISI 409 is applied for automobile mufflers and emission control equipment.

AISI 430 is a general purpose ferritic steel with various uses, including uncomplicated plant for handling benign liquors, hygienic surfaces, domestic equipment, and automobile bright trim.

The manufacture of domestic sink units is a homely illustration of how economies can be made by using ferritic steel. The bowl must be deep-drawn from a very ductile austenitic steel, such as AISI 304, and it is welded to the draining surface that can be made from a less costly ferritic steel, such as AISI 430.

8.5.2   Austenitic Steels

8.5.2.1   *Austenitic Steels Without Molybdenum*

The austenitic steel, AISI 304, and its derivatives, such as AISI 304L, AISI 321, etc., are so versatile and popular that the annual U.S. production of them is approaching a million tonnes. They are some of the most cost-effective mass-produced metallic materials for overall corrosion resistance. A limitation on their use is susceptibility to stress-corrosion cracking when stressed in hot aqueous chloride media, as described in Section 5.1.3.

The steels are ductile and can be easily shaped cold. In the form of softened thin rolled sheet, they can be deep drawn or pressed into hollowware for which they are especially suitable because they develop high strength from work-hardening. Representative applications include:

1. Vessels and other equipment handling acids for conditions outlined in Section 8.3.2.1.
2. Equipment for food processing, such as dairy products, brewing, and wine-making.
3. Domestic equipment including sink basins, cutlery, and cookware.
4. Cryogenic equipment.

The balance of chromium, nickel and carbon contents in AISI 304 and its derivatives yield austenite that is stable at temperatures as low as $-196°C$, even after cold work entailed in shaping them; this accounts for the cryogenic applications. Some variants of these steels, not listed in Table 8.3, are formulated to produce metastable austenite that can be hardened by a martensitic transformation induced by cold work.

8.5.2.2   Austenitic Steels With Molybdenum

These steels have similar uses to the molybdenum-free steels but their considerable extra costs are justified only when their better corrosion resistance to acids and chlorides is actually needed; it is a mistake to select them for environments where less expensive steels without molybdenum are equally satisfactory. They have many applications in chemical industrial plant handling reducing acids or chloride-contaminated liquors where the integrity of the plant is of concern. Although they are not immune to chloride-induced pitting, they are more suitable than molybdenum-free austenitic steels for marine use. A less obvious but important application is decorative external architectural and related use, where loss of visual appeal is a perfectly proper criterion for corrosion failure; molybdenum-free steels are acceptable in benign atmospheres, if regular cleaning is possible, but it is advisable to use a molybdenum-bearing steel, such as AISI 316, if maintenance is difficult or if the atmosphere is industrial, coastal or tropical, because accumulated dirt screens the metal from oxygen and also traps aggressive ions such as chlorides and sulfur compounds.

8.5.2.3   Austenitic Steels With Manganese Substitution

Manganese is used to reduce nickel contents but the economic benefit is doubtful because manganese substitution detracts from the good working characteristics and corrosion resistance that are the main recommendations of the nickel-chromium austenitic steels they replace. They do, however, provide contingency for interruption in the supply of nickel that is vulnerable to labor or political problems because it is dominated by two principal sources.

8.5.3 Hardenable Steels

Martensitic steels are hard enough for applications to resist wear and to retain cutting edges. They are the least corrosion-resistant stainless steels because of the restricted chromium and nickel contents and the high-carbon contents needed for the martensite transformation. They are used for valve seats, turbine parts, knives, and scalpels, but their corrosion resistance does not extend to aggressive environments.

The various precipitation-hardening steels were formulated to secure strength and hardness without sacrificing too much corrosion resistance. They contain 14 to 17% chromium, some carbon, nickel, manganese, silicon, and selected additions from among cobalt, titanium, copper, molybdenum, and aluminum. By appropriate thermal treatment they can yield metastable austenitic or martensitic structures at ambient temperature amenable to precipitation-hardening by subsequent thermal and mechanical treatments.

8.5.4 Duplex Steels

Duplex steels are replacements for austenitic steels in applications in which the two-phase structure has advantages. They have better resistance to stress-corrosion cracking because cracks advancing through austenite encounter the more resistant ferrite. They are less susceptible to sensitization because ferrite adjacent to grain boundaries is a good source of chromium to replenish depletion by carbide precipitation. A further advantage is good resistance to pitting in chloride media for duplex steels containing molybdenum.

Lower raw material costs for duplex steels than for austenitic steels are offset by higher hot-working costs incurred to avoid sigma phase due to high chromium/molybdenum contents.

8.5.5 Oxidation Resistant Steels

The discussion on the influence of alloy composition on isothermal oxidation in clean air, given in Section 8.4 is useful in indicating which alloys are possible and which need not be considered for use at a prescribed temperature but real service conditions can impose additional, sometimes overriding demands.

High temperature industrial environments include not only air, but also products of combustion, superheated steam, process fumes, and abrasive or reactive dust. Products of combustion can be various mixtures of carbon monoxide, carbon dioxide, water vapor, with or without excess oxygen;

they can be contaminated with sulfurous gases or loaded with ash. Process fumes can be from metal extraction and refining, glass making, oil refining, and other chemical processes. Temperatures may be neither uniform nor steady and in batch operations they may cycle between ambient and operational temperatures.

Any of these chemical physical or thermal effects can undermine an otherwise protective scale. Examples of the factors that may have to be faced are:

1. The effects of sulfur incorporated in scales formed on steels exposed to products of combustion containing sulfur dioxide; oxidation is accelerated, especially if combustion of the fuel is incomplete.

2. Catastrophic oxidation, as defined in Section 3.3.3.7 stimulated by vanadium pentoxide, V_2O_5 in fuel ash that can flux away protective scales.

3. Cracking and spalling of protective scales by differential expansion and due to temperature fluctuations.

4. Attrition of protective scales by abrasive dust.

Steels for high temperature industrial application are therefore selected by experience on an application-specific basis. These applications can include furnace parts such as roof suspension hangers, flues, stator blades in gas turbine compressors, parts for steam turbines, and nuclear installations.

Further Reading

Pickering, F.B, The physical metallurgy of stainless steels, *International Met. Rev.*, 21, 227, 1976.

Sendriks, A. J., *Corrosion of Stainless Steels,* John Wiley, New York: 1979.

Hansen, M. *Constitution of Binary Alloys*, McGraw-Hill, New York, 1958, p. 527.

Columbier, L. and Hochmann, J.,*Stainless and Heat Resistant Steels,* Edward Arnold, London, 1967.

Monypenny, J. H. G. *Stainless Iron and Steel,* Chapman & Hall, 1951, p. 61.

Schaeffler, A. L., *Met. Prog.*, 100B, 77, 1960.

Peckner, D. and Bernstein, J. M., *Handbook of Stainless Steels*, McGraw-Hill, New York, 1977.

Localized Corrosion, National Association of Corrosion Engineers, Houston, Texas, 1974.

Fontana, M. G. and Green, N. D., *Corrosion Engineering*, McGraw-Hill, New York, 1967, Chap. 3.

Problems For Chapter 8

1. A steel contains 17.0% Cr, 3.5% Mo, 0.05% C, and 1.5% Mn, together with nickel. Using the concept of Schaeffler equivalents, estimate the minimum nickel content needed to develop a fully austenitic structure. If a version of the steel with 0.02% C were required for welding, how much more nickel would be required to retain the austenitic structure and how much would this add to the cost of raw materials?

2. Taking into account the functions, costs, and probable methods of manufacture, suggest suitable steels from the list given in Table 8.3 for the following applications: (a) washing machine drums, (b) exterior architectural fittings for a hotel facing the Atlantic Ocean at Miami Beach, Florida, (c) razor blades, (d) beer barrels made by welding together hollow half barrels deep-drawn from flat circular blanks, (e) roof supports over a swimming pool disinfected by chlorinating the water, and (f) containers for catalytic converters on automobiles.

3. A particular stainless steel is being considered for use as a container for a dilute neutral aqueous solution of chloride open to the atmosphere. The pitting potential for the steel in the solution is found to be 0.95 V SHE. Is the steel likely to suffer pitting corrosion? Recall that for the reaction, $\frac{1}{2}O_2 + 2H_2O + 2e^- = 2OH^-$, $E^{\ominus} = + 0.401$ V (SHE).

Solutions To Problems For Chapter 8

Solution to Problem 1.

Applying the Schaeffler principle:

$$\text{Chromium equivalent} = \text{wt-\% chromium} + \text{wt-\% molybdenum}$$
$$= 17\ \% + 3.5\ \% = \mathbf{20.5\ \%}$$

The Schaeffler diagram, given in Figure 8.4, indicates that a minimum nickel equivalent of 15.0 % is needed for the coordinates representing the nickel and chromium equivalents to lie within the austenite phase field. Excluding the nickel content itself:

Nickel equivalent = 30 × wt-% carbon + 0.5 × wt-% manganese
$$= (30 \times 0.05) \% + (0.5 \times 1.4) \% = \textbf{2.2 \%}$$

Hence the minimum nickel content required is 15.0% − 2.2 % = **12.8%**

If the carbon content is reduced to 0.02%, its contribution to the nickel equivalent is reduced from 1.5% to 0.6%, so that 0.9% more nickel is needed to retain the austenite. This requires an extra 0.009 tonnes of nickel per tonne of steel, adding $60 to the cost of raw materials per tonne of steel at the price of nickel given in Table 8.1, i.e., $6755 per tonne.

Solutions to Problem 2.

(a) *Washing-machine drums*

The drums are lightly formed thin sheet parts assembled by welding for a competitive domestic appliance market. Exposure conditions are not severe and the main factor is customer perception that the drum retains a hygienic appearance. A ferritic steel, AISI 430, is suitable.

(b) *Architectural fittings for a hotel on Miami Beach*

The steel experiences severe conditions. It is subject to wind-driven sea spray with intermittent drying, exposing it to depassivation effects induced by chloride, including pitting. Maintenance and cleaning of fittings above ground level is difficult and yet an an attractive appearance is an essential element in generating hotel revenue, justifying the capital cost of selecting a molybdenum-bearing austenitic steel such as AISI 316.

(c) *Razor blades*

The essential feature required of razor blades is the ability to accept and retain a sharp edge. They are disposable items and conditions of exposure are mild so that an inexpensive hardenable martensitic steel, such as AA 410 is suitable.

(d) *Beer barrels*

The sheet material used for beer barrels must accept severe deformation in the deep-drawing operation by which the two halves are fabricated. It must develop strength by work hardening in the same operation so that the thinnest gauge can be used to minimize weight, facilitating handling in use. These requirements can be met with austenitic steels. Typical pH values for beers are about 4, so that the conditions are slightly acidic but essentially free from chloride ions. The barrels are assembled by welding but experience shows that sensitization is not a serious problem for austenitic steels with carbon contents < 0.06% in the relatively light gauges used. Taking all of these factors into account, AISI 304 is the most suitable choice of steel.

(e) *Roof supports for swimming pool*

The over-riding concern for steels used as load-bearing members of roof supports is susceptibility to stress-corrosion cracking. Chlorine used as disinfectant is the source of the specific agent, chloride ions. Chlorides and water condensate are transported to the roof structure as described in detail in Section 13.7. Steels with the highest molybdenum contents are least vulnerable to stress-corrosion cracking. They include the austenitic steel AISI 317 and duplex steels with 3% molybdenum that have further advantage in the more resistant ferrite they contain, such as Steels 1 and 2, cited in Table 8.3.

(f) *Containers for automobile catalytic converters*

Containers for catalytic converters are part of semi-permanent emission systems and must retain their integrity and remain clean to protect the expensive catalyst assembly. They run hot enough to avoid condensates that would stimulate aqueous corrosion but must resist dry oxidation at moderate temperatures. Costs bear down on automobile components and an inexpensive steel is required. Ferritic steels provide adequate oxidation resistance and can accept the light forming operations applied during manufacture. Restricting the chromium content reduces susceptibility to embrittlement at the moderate elevated service temperatures, as described in Section 8.2.2.1. These considerations suggest AISI 409.

Solution to Problem 3.

The steel is unlikely to suffer pitting corrosion if the potential imposed by the available cathodic reaction:

$$\tfrac{1}{2}O_2 + 2H_2O + 2e^- = 2OH^-$$

is significantly below the pitting potential, $E_{PP} = +0.95$ V (SHE). In neutral water in equilibrium with the atmosphere, $a_{OH^-} = 10^{-7}$ and $a_{O_2} = 0.21$, assuming that oxygen is an ideal gas.

Applying the Nernst equation and substituting for $E^{\ominus}$, a_{OH^-} and a_{O_2}:

$$E = E^{\ominus} - \frac{0.0591}{z} \log \frac{\left(a_{OH^-}\right)^2}{\left(a_{O_2}\right)^{1/2}}$$

$$E = +0.401 - \frac{0.0591}{2} \log \frac{\left(10^{-7}\right)^2}{(0.21)^{1/2}}$$

$$= +0.814 \text{ V (SHE)}$$

This is the maximum potential that the reaction can apply and since it is significantly lower than the pitting potential, the steel is unlikely to suffer pitting corrosion.

9

Corrosion Resistance of Aluminum and Aluminum Alloys

Aluminum is a relatively expensive metal because its extraction from the mineral, bauxite, is energy intensive. Pure alumina is extracted from bauxite in the *Bayer* process, by dissolution in sodium hydroxide from which it is precipitated, calcined and used as feedstock in the *Hall-Herault* high-temperature electrolytic reduction process to recover aluminum metal.

Aluminum has low density, high ductility, high thermal, and electrical conductivity, good corrosion resistance, attractive appearance, and it is nontoxic. Despite its cost, this remarkable combination of qualities makes it a preferred choice for many critical applications in aerospace, automobiles, food handling, building, heat exchange, and electrical transmission.

The pure metal is deficient in two respects, mechanical strength and elastic modulus and aluminum alloy development was driven by the need to improve them without sacrificing other qualities, e.g., to improve strength for aerospace, marine, and civil engineering applications without losing corrosion resistance. The success of these endeavors has secured the status of aluminum alloys as second only to steels in economic value.

Solid aluminum has an invariable face-centered cubic lattice and there is no counterpart to the structural manipulations exploited for iron alloys, so that different philosophies apply in alloy formulation.

9.1 Summary of Physical Metallurgy of Some Standard Alloys

Copper, lithium, magnesium, and zinc have fairly extensive solubilities in solid aluminum but the solubility of many other elements is very limited. Most binary systems with aluminum are characterized by limited solubility of the second metal and a eutectic in which the components are the saturated solution and an intermetallic compound, Al_xM_y. Aluminum is prolific in the number and variety of intermetallic compounds it can form because of its strong electronegativity and high valency. Critical features of some common binary systems are summarized in Table 9.1.

TABLE 9.1

Phase Relationships for Some Binary Aluminum Alloys

System	Maximum Solid Solubility Wt % Alloy Element	Eutectic Composition Wt % Alloy Element	Eutectic Temperature °C	Eutectic Components
Aluminum-copper	5.7	33.2	548	Al* + CuAl$_2$
Aluminum-iron	<0.1	1.7	655	Al* + FeAl$_3$
Aluminum-lithium	5.2	9.9	600	Al* + LiAl
Aluminum-magnesium	14	35	450	Al* + Mg$_2$Al$_3$
Aluminum-manganese	1.8	1.9	660	Al* + MnAl$_6$
Aluminum-silicon	1.65	12.5	580	Al* + Si†
Aluminum-zinc	82.8	94.9	382	Al* + Zn‡

* Containing maximum solubility of second element in solid solution
† Containing 0.5 wt % of aluminum in solid solution
‡ Containing 1.1 wt % of aluminum in solid solution

Commercial alloys are based on multicomponent systems that reflect characteristics of the component binary systems but also exhibit features due to interaction between the alloying elements, especially in introducing additional intermetallic compounds and in modifying solubilities.

In alloy formulation, strength can be imparted by:

1. Reinforcement with intermetallic compounds.
2. Solid solution strengthening.
3. Work hardening.
4. Precipitation hardening (aging).

There are more than a hundred current wrought and cast alloy compositions but they belong to a relatively few series with particular characteristics. A representative selection is given in Table 9.2, using the internationally recognized AA (Aluminum Association) designations.

9.1.1 Alloys Used Without Heat Treatment

Commercial Pure Aluminum Grades (AA 1100 Alloy Series)

Commercial grades of pure aluminum are actually dilute alloys. The iron and silicon contents of the aluminum product from the Hall-Herault process when it was first introduced raised the properties of the metal to values suitable for a wide variety of general applications, including domestic and catering utensils, packaging foil, some chemical equipment, and architectural applications. Although modern practice produces higher purity metal, grades of aluminum with similar iron and silicon contents are still offered as general purpose materials. Conventionally, these alloys are designated by specifying the aluminum content, e.g., the most

TABLE 9.2
Nominal Chemical Compositions of Representative Aluminum Alloys

Alloy	Alloy Elements, Weight %						Corrosion Resistance	Examples of Applications[a]
	Silicon	Copper	Manganese	Magnesium	Zinc	Others		
AA 1050	Minimum of 99.50% aluminum — principal impurities iron and silicon						good	architectural applications
AA 1100	Minimum of 99.00% aluminum — principal impurities iron and silicon						good	cooking utensils, foil
AA 2024	—	4.4	0.6	1.5	—	—	poor	aircraft panels
AA 3003	—	0.12	1.2	—	—	—	good	general purpose alloy, foil
AA 3004	—	—	1.2	1.0	—	—	good	beverage cans
AA 5005	—	—	—	0.8	—	—	good	architectural applications
AA 5050	—	—	—	1.4	—	—	good	general purpose alloy
AA 5182	—	—	0.35	4.5	—	—	good	beverage can ends
AA 5456	—	—	0.8	5.1	—	0.12 Cr	good	transportation, structures
AA 6061	0.6	0.28	—	1.0	—	0.20 Cr	good	extrusions, beer containers
AA 7075	—	1.6	—	2.5	5.6	0.23 Cr	poor	aircraft stringers and panels
AA 7072	—	—	—	—	1.0	—	good	cladding to protect 7075 alloy
AA 319	6.0	3.5	—	—	—	—	poor/moderate	general purpose castings
AA 380	8.5	3.5	—	—	—	—	poor/moderate	die castings
AA 356	7.0	—	—	0.35	—	—	moderate/good	age-hardenable castings
AA 390	17	4.5	—	0.55	—	—	not applicable	automobile cylinder blocks

[a] Selected applications; most alloys are multi-purpose.

common alloy, AA 1100 in Table 9.2, has a minimum aluminum content of 99.0% and typical iron and silicon contents of 0.45% and 0.25%, respectively. The metal is strengthened by dispersed intermetallic compounds, notably Fe_3SiAl_{12}, $FeAl_3$ and the metastable $FeAl_6$, raising the 0.2% offset yield and tensile strengths to the values given in Table 9.3. Additional strength can be and usually is imparted by work hardening as illustrated by the values for 75% cold reduction in thickness, given in the same table.

TABLE 9.3
Typical Strengths of Some Aluminum Alloys at 25°C

Alloy	Condition	0.2 % yield strength MPa	Tensile strength MPa
99.99 %	Annealed (softened by heating)	10	45
AA 1100	Annealed	35	90
	75% cold reduction (by rolling)	150	165
AA 3003	Annealed	40	110
	75% cold reduction	185	200
AA 3004	Annealed	70	180
	75% cold reduction	250	285
AA 5005	Annealed	40	125
	75% cold reduction	185	200
AA 5050	Annealed	55	145
	75% cold reduction	220	200
AA 5182	Annealed	130	275
	75% cold reduction	395	420
AA 5456	Annealed	160	310
	Fully strain hardened*	255	350
AA 6061	Annealed	55	125
	Solutionized at 532°C. Aged 18 h at 160°C	310	275
AA 2024	Annealed	75	185
	Solutionized at 493°C. Aged 12 h at 191°C	450	485
AA 7075	Annealed	105	230
	Solutionized at 482°C. Aged 24 h at 121°C	505	435

*   Using special temper procedure to minimize structural instability.

The Aluminum-Manganese Alloy AA 3003

The alloy, AA 3003, has applications that overlap those of AA 1100 but it has greater strength as indicated in Table 9.3. It is produced by adding 1.2% of manganese to the AA 1100 composition. The intermetallic compounds are modified to $(Fe,Mn)_3SiAl_{12}$ and $(Mn,Fe)Al_6$. Compounds containing manganese are also present as a fine dispersoid distributed throughout the aluminum matrix.

Aluminum-Magnesium Alloys (AA 5000 series)

Magnesium both imparts solid solution strengthening and enhances the ability of the metal to work harden, illustrated for alloys AA 5005, 5050, 5182, and 5456 in Table 9.3. They have practical advantages in their good formability, high rate of work hardening, weldability and corrosion resistance. A reservation on the magnesium content is set by the instability of the solid solution with respect to precipitation of Mg_2Al_3 that can be significant for magnesium contents greater than about 3%, during long storage at ambient temperatures; the precipitation is at grain boundaries and can render the metal susceptible to stress-corrosion cracking and to intergranular corrosion. Unfortunately, the instability is increased by work hardening that is one of the alloys most useful attributes but it can be ameliorated with some loss of strength by special tempering procedures.

The Aluminum-Magnesium-Manganese Alloy AA 3004

The alloy AA 3004 is strengthened by both magnesium and manganese and has an optimum combination of formability, and work hardening for application as beverage cans that are deep drawn and must have sufficient strength in very thin gauges to withstand forces imposed on them by filling and by internal gas pressure.

9.1.2 Heat Treatable (Aging) Alloys

The strongest aluminum alloys are strengthened by precipitation hardening. The principle exploits the diminishing solubility of certain solutes with falling temperature. The strength is developed by controlled decomposition of supersaturated solid solutions. An alloy is heated to and held at a high temperature to allow the solutes dissolve, an operation called *solutionizing*, and then rapidly cooled to retain them in supersaturated solution. Rapid cooling is usually accomplished by immersing the hot metal in cold water, i.e., *quenching*. The metal is subsequently reheated to a constant moderate temperature and the unstable supersaturated solution rejects excess solute in the form of finely dispersed metastable precursors of the final stable precipitate. These precursors are of suitable forms, sizes, and distributions to impede the movement of dislocations in the lattice that effect plastic flow. The process occupies a considerable time and is called *aging*. The improvement in mechanical properties is illustrated by the values given in Table 9.3 for annealed and age-hardened versions of alloys AA 6061, A 2024, and AA 7075.

Not all aluminum alloy systems are amenable to strengthening in this way. Standard commercial alloys are based on the aluminum-copper-magnesium, aluminum-zinc-magnesium and aluminum-magnesium silicon systems. In these multicomponent systems, the aging process depends on the decomposition of supersaturated solutions of more than one solute from which complex intermetallic compounds are precipitated. A typical sequence of events during aging is:

1. Assembly of solute atoms in groups, called *Guinier Preston (GP)* zones, dispersed throughout the matrix at spacings of the order of 100 nm.

2. Loss of continuity between the zones and the metal lattice in some but not all crystallographic directions, yielding transition precipitates usually designated by primed letters, sometimes with their formulae in parenthesis.

3. Complete loss of continuity with the matrix, yielding particles of stable precipitates.

The sequence is conveniently described by an equation of the general form:

$$\text{solid solution} \rightarrow \text{GP} \rightarrow \theta' \rightarrow \text{precipitate} \tag{9.1}$$

The detail varies widely from system to system and is more complicated than Equation 9.1 indicates, depending on metal composition, precipitate morphology and aging temperature.

The metal becomes stronger as dislocation movement is inhibited by matrix lattice strains building up around the GP zones and transition precipitates but as these transition species are progressively replaced by equilibrium precipitate, the lattice strains are relaxed and the metal softens. There is thus an optimum aging period for maximum strength, i.e., *peak hardness*. Metal that has been aged for less than this period is said to be *underaged* and metal that has passed the peak is *overaged*. These conditions have important implications in corrosion and stress-corrosion cracking.

The formation of transition species and their replacement by equilibrium precipitates are both accelerated at higher temperatures. The consequence is that as the aging temperature is raised, the peak hardness is reached in progressive shorter periods of time but its value diminishes. There is thus an optimum thermal cycle to secure good results economically.

Aluminum-Copper-Magnesium Alloys (e.g., AA 2024)

The aging sequence is usually represented by:

$$\text{solid solution} \rightarrow \text{GP} \rightarrow S'(Al_2CuMg) \rightarrow S(Al_2CuMg) \tag{9.2}$$

where $S'(Al_2CuMg)$ and $S(Al_2CuMg)$ are transition and stable precipitates, respectively.

The classic examples of alloys in this system are AA 2024 and its variants that are staple materials for airframe construction. The solutionizing temperature is restricted to 493°C to avoid exceeding the solidus temperature and aging is at 191°C for a period of between 8 and 16 hours to suit the form of the material.

Aluminum-Zinc-Magnesium Alloys (e.g., AA 7075)

The alloys based on this system are some of the strongest produced commercially and are also staple materials for airframe construction; the characteristic alloy is AA 7075. Particular compositions, solutionizing treatments and aging programs have been developed to secure practical benefits of strength, consistency of properties, and minimum susceptibility to stress-corrosion cracking. Detailed assessment of the aging sequences is difficult because the transition and final species encountered depend on composition, initial microstructure, speed of quenching, and aging temperature. Schemes suggested include:

$$\text{solid solution} \rightarrow \text{GP} \rightarrow \eta' \rightarrow \eta(MgZn_2) \text{ or } T(Mg_3Zn_3Al_2) \quad (9.3)$$

The heat treatment is critical. Alloy AA 7075 is typically solutionized at 482°C, just below the solidus temperature, and aged for 24 hours at 121°C.

Aluminum-Magnesium-Silicon Alloys (e.g., AA 6061)

The precipitating phase in this system is Mg_2Si and the aging scheme is:

$$\text{solid solution} \rightarrow \text{GP} \rightarrow \beta'(Mg_2Si) \rightarrow \beta(Mg_2Si) \quad (9.4)$$

The system does not produce alloys with strengths as high as those based on the aluminum-copper-magnesium and aluminum-magnesium-zinc systems and they are regarded as medium strength alloys. However, they have the following compensating advantages:

1. They are easy to hot extrude into sections.
2. In the soft condition they are ductile and accept deep drawing into hollow shapes.
3. They are the most corrosion resistant of the aging alloys.

These attributes recommend them for uses including architectural fittings such as window frames, food handling equipment, beer barrels, and transport applications. The most widely applied alloy is AA 6061 but there are other related formulations. AA 6061 is solutionized at 532°C and aged for 18 hours at 160°C.

Aluminum Alloys Containing Lithium

In recent years, a range of alloys has been developed for aerospace applications based on alloys containing lithium. One of the objectives is to match the properties of AA 2024 and AA 7075 alloys but with lower density and higher modulus, both of which are promoted by the use of lithium. The basic aging sequence:

$$\text{solid solution} \rightarrow \text{GP} \rightarrow \delta'(Al_3Li) \rightarrow \delta(AlLi) \tag{9.5}$$

is complicated by the presence of other alloy components needed to secure the required properties, notably copper and magnesium, that lead to schemes that include additional hardening phases, $T'(Al_2CuLi)$ and $S'(Al_2CuMg)$.

These alloys are expensive and justified only where mass saving is a critical economic factor, as in aerospace applications. Interest in them has temporarily diminished pending reassessment of their long term integrity but they remain as possibilities for future exploitation.

9.1.3 Casting Alloys

The most common casting alloys are based on the eutectic aluminum-silicon system and are characterized by fluidity of the liquid metal and low contraction on solidification. There are many composition variants and Table 9.2 lists alloys AA 319, AA 380, and AA 356 as examples in widespread use; the function of the copper content in AA 319 and AA 380 is to improve strength and machinability at some sacrifice of corrosion resistance. The magnesium content in alloy AA 356 offers the facility of precipitation hardening by Mg_2Si. Alloy AA 390 is a customized special alloy for automobile cylinder blocks.

9.2 Corrosion Resistance

Aluminum exists only in the oxidation state Al(III) in its solid compounds and aqueous solution and in this state it is one of the most stable chemical entities known, as illustrated by the high value of the Gibbs free energy of formation for the oxide:

$$2Al + 1\tfrac{1}{2}O_2 = Al_2^{III}O_3$$

$$\Delta G^{\ominus} = -1117993 - 10.96T\log T + 244.5T \quad J \tag{9.6}$$

and of the standard electrode potentials for formation of ions in acid aqueous solution:

$$Al = Al^{3+} + 3e^-$$

$$E^{\ominus} = -1.64 \text{ V (SHE)} \tag{9.7}$$

and in alkaline aqueous solution:

$$Al + 4OH^- = AlO_2^- + 2H_2O + 3e^-$$

$$E^{\ominus} = -2.35 \text{ V (SHE)} \tag{9.8}$$

In the context of Table 3.3 these values show that aluminum is intrinsically more reactive than any other common engineering metal except magnesium and its relative permanence is attributable to protection afforded by the oxide films that form upon it.

9.2.1 The Aluminum-Oxygen-Water System

The Pourbaix diagram for the aluminum-water system given in Figure 3.4 is constructed on the simplifying assumption that the interactions can be represented symbolically by four simple species, Al, Al^{3+}, AlO_2^- and Al_2O_3. In reality, these species are more complex and are now considered in greater detail.

9.2.1.1 Solid Species

Anhydrous Oxide

Corundum Al_2O_3 is the only anhydrous solid oxide. At temperatures below about 800°C, it is predominantly stoichiometric in the terms set out in Section 2.3.6.1. The structure is hexagonal-rhombohedral, with aluminum ions occupying two-thirds of the octahedral interstitial sites in a close packed hexagonal structure of oxygen ions. It occurs naturally in some igneous and metamorphic rocks and it can be produced by heating hydrated oxides to high temperatures but, once formed, it resists rehydration, even though it is unstable in water or normal air at temperatures below about 450°C with respect to hydrated oxides.

Trihydroxides

The stable phase in contact with water or atmospheric air at ambient temperatures is one of the trihydroxides, $Al(OH)_3$. There are three known variants of $Al(OH)_3$, gibbsite, bayerite, and nordstrandite. Gibbsite is a common naturally occurring mineral form of $Al(OH)_3$ and is also the form produced during commercial bauxite purification. The structures of all of the variants are stacks of the same basic structural unit, $[Al_2(OH)_6]_n$, comprising aluminum ions in octahedral geometry sandwiched between two layers of hydroxyl ions, similar in some respects to the brucite structure described in Section 2.3.3.5 except that aluminum ions occupy only two thirds of the available sites. Gibbsite, bayerite, and norstrandite differ only in the stacking patterns, e.g., in gibbsite, the hydroxyl groups in adjacent layers are directly opposed and alternate layers are inverted. These variants have similar energies and which of them is the true thermodynamically stable form has not been unequivocally established.

Oxyhydroxides

There are two oxyhydroxides with the empirical formula AlO(OH), boehmite and diaspore. Both occur naturally as minerals.

Boehmite has a layered structure with a basic structural unit, [OH–AlO–AlO–OH]$_n$ in cubic packing; the hydroxyl ions of every successive layer nest in the pockets between hydroxyl ions in the underlying layer. It usually contains more than the 15% by mass of water that the formula AlO(OH) would suggest and it is not clear whether the excess is due to intercrystalline water or from the presence of some trihydroxide. Diaspore has a structure of hexagonal close packed layers of hydrogen bonded oxygen ions, [–O$\cdots$H–O$\cdots$]$_n$, with aluminum ions in the octahedral interstices.

Gels

Trihydroxides are initially precipitated from aqueous solution as gelatinous or colloidal precursors with empirical formulae Al(OH)$_3 \cdot n$H$_2$O, where n is at least 3. The gels are largely 'amorphous' materials but with vestiges of crystallinity. They are unstable and on aging, they slowly transform to crystalline gelatinous boehmite and finally recrystallize to bayerite.

Dehydration Sequences

On heating them to progressively higher temperatures, trihydroxides begin to lose water at a little over 100°C. The materials pass through a series of structural changes, ultimately becoming the anhydrous oxide corundum.

The intermediate structures depend on the initial particle size, rate of heating and ambient humidity. A representative sequence is:

$$100°C \qquad 400°C \qquad\qquad\qquad\qquad 1150°C$$
$$\text{Gibbsite} \rightarrow \text{boehmite} \rightarrow (\gamma\text{-alumina} \rightarrow \delta\text{-alumina} \rightarrow \theta\text{-alumina}) \rightarrow \text{corundum}$$

The γ, δ, and θ forms are so-called transition aluminas with distorted spinel structures stabilized by a small hydroxyl ion content.

9.2.1.2 Soluble Species

Information on the true natures of the Al^{3+} and AlO$_2^-$ ions and their relationships to insoluble species can be deduced from the viscosity, osmotic properties, and electrical conductance of their solutions as functions of pH.

Aluminum Cations

In acid media with pH values below about 4, hydroxides and oxyhydroxides dissolve yielding the cationic species usually written as $Al(H_2O)_6^{3+}$. The discrete ions exist only at very low pH values and in less acidic solutions they associate as polymeric complexes. Their solubility is also influenced by agents that bind aluminum cations, e.g., some organic acids.

Aluminum Anions

In alkaline media with pH values above about 8, hydroxides and oxyhydroxides dissolve to form aluminate ions, usually described as AlO_2^-, but they also associate as various polymeric species. Discrete aluminate ions do not exist at pH values < 13 and even in such strong alkalis, the main species is probably $Al(OH)_4(H_2O)_2^-$ in which a central Al^{3+} ion is in octahedral co-ordination with four hydroxyl and two water molecules.

Relation of Soluble to Insoluble Species

Dissolution-precipitation reactions between the soluble ions and solid phases in the system are ill-defined. The tendency of the soluble ions to associate in complexes is progressively more pronounced as pH values tend towards the range within which the solid species are conventionally considered stable, e.g., pH 3.9 to 8.6 for $a_{Al^{3+}} = a_{AlO_2^-} = 10^{-6}$, and there is no sharp demarcation between the polymeric ions, colloidal material and alumina gels.

9.2.2 Corrosion Resistance of Pure Aluminum in Aqueous Media

9.2.2.1 Passivation

It follows from the invariant valency of aluminum in its ions, i.e., III, and its very negative standard electrode potential that the passivity protecting it in aqueous media depends only on pH and not on potential. Potential-dependent concepts such as anodic passivity, pitting potentials, etc., do not apply as they do to passivating alloys such as stainless steels composed of metals exercising variable valencies

In water, the thin air-formed oxide film on aluminum transforms to gelatinous trihydroxide that is then in principle subject both to further growth and to dissolution at the film/water interface. The establishment of passivity therefore depends on whether the solubility of the film is low enough for the filmed surface to be a virtually permanent condition. Typical expressions given for the solubility products for the hydrated oxides at 20°C are:

$$K_{Al^{3+}} = (a_{Al^{3+}})/(a_{H^+})^3 = 5.0 \times 10^5 \qquad (9.9)$$

$$K_{AlO_2^-} = a_{AlO_2^-} \times a_{H^+} = 2.5 \times 10^{-15} \qquad (9.10)$$

Example

To determine the minimum solubility of the trihydroxide film.

By Equations 9.9 and 9.10, $a_{Al^{3+}}$ and $a_{AlO_2^-}$ are opposite functions of a_{H^+}. Their combined minimum activities correspond to the value of a_{H^+} for which they are equal.

From Equation 9.9:
$$a_{Al^{3+}} = 5 \times 10^5 \times (a_{H^+})^3$$

From Equation 9.10:
$$a_{AlO_2^-} = (2.5 \times 10^{-15})/(a_{H^+})$$

For $a_{Al^{3+}} = a_{AlO_2^-}$:
$$5 \times 10^5 \times (a_{H^+})^3 = 2.5 \times 10^{-15}/(a_{H^+})$$

Whence:
$$a_{H^+} = 8.4 \times 10^{-6}$$

and
$$pH = -\log(8.4 \times 10^{-6}) = 5.1 \tag{9.11}$$

Inserting $a_{H^+} = 8.4 \times 10^{-6}$ into Equations 9.9 and 9.10 yields:

$$a_{Al^{3+}} = a_{AlO_2^-} = 3.0 \times 10^{-10} \tag{9.12}$$

Equations 9.9 and 9.10 have been used to construct Figure 9.1 expressing the solubility of trihydroxides as a function of pH, showing that they dissolve readily in strongly acidic and in strongly alkaline solutions but their solubilities are very low at near neutral pH values. Adopting the convention introduced for metals in Section 3.1.5.4, i.e., that a substance shall be deemed to dissolve in the case that the activity of any of its soluble ions exceed 10^{-6}, Figure 9.1 identifies a pH range of about 4 to 8.5, corresponding to the domain for stability of Al_2O_3 in the Pourbaix diagram, given in Figure 3.4.

9.2.2.2   Corrosion Resistance in Natural Waters

Fresh Waters

The values obtained in Equations 9.11 and 9.12 are instructive because the film is virtually insoluble in an aqueous medium with a pH value of 5.1 that is close to the pH value, 5.6, calculated in Section 2.2.9 for water in equilibrium with atmospheric carbon dioxide. The solubility increases as the pH rises but Figure 9.1 shows that even at pH 8.3, characteristic of hard waters, also calculated in Section 2.2.9, the activity of the soluble ion, AlO_2^-, is still only 5×10^{-7}. These simple calculations suggest that the film should resist dissolution in most fresh natural waters and supplies derived from them. The pure metal has excellent corrosion resistance under these conditions, confirming that the film is a good barrier to reacting species.

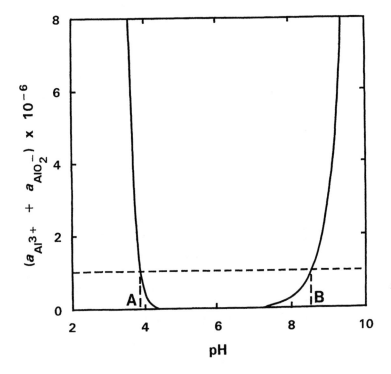

Figure 9.1
pH dependence of aluminum trihydroxide, $Al(OH)_3$, solubility expressed as the sum of Al^{3+} and AlO_2^- activities in solution. Intercepts at A and B define the conventional pH range for stability, i.e., where the activities of species in solution are less than 10^{-6}.

Sea Water

Pure aluminum is also found to have good corrosion resistance to seawater. This is perhaps unexpected in view of the strong depassivating effect of chlorides on stainless steels described in Section 8.3.2.2. The passive film resists attack by chlorides probably because of the much greater chemical affinity of aluminum for oxygen than for chlorine.

9.2.2.3 *Corrosion Resistance in Acidic and Alkaline Media*

The susceptibility of the pure metal to corrosion in acids or alkalis depends *inter alia* on how far the acid or alkali shifts the pH outside of the range 4.0 to 8.5. Thus aluminum is expected to dissolve much more readily in 0.1 M solutions of sulfuric acid (pH 0.9), hydrochloric acid (pH 1.1) and sodium hydroxide (pH 13.0) than in 0.1 M solutions of acetic acid (pH 2.9) or ammonium hydroxide (pH 11.1).

The counter ion of an acid or alkali can also influence susceptibility to dissolution of the metal. Thus oxidizing ions such as chromates can inhibit it and ions producing soluble complexes such as the AlF_4^- ion formed in hydrofluoric acid, can enhance it.

9.2.3 Corrosion Resistance of Aluminum Alloys in Aqueous Media

In use, bare aluminum and aluminum alloy surfaces often develop an unattractive dull gray patina but it is trivial and can be disregarded unless appearance is important; incidentally, it is normal and nontoxic on cooking vessels. Damaging corrosion in general purpose applications is not usually a problem and the remarks below apply to the more critical applications and to special circumstances.

9.2.3.1 *Chemical Composition of the Metal*

Table 3.3 illustrates that except for magnesium and the unreactive metalloid, silicon, all of the common alloying elements in aluminum alloys have much less negative standard electrode potentials than aluminum and are thus capable of stimulating local galvanic attack on the adjacent matrix if distributed heterogeneously. Thus, as a general rule, alloys containing these elements can be expected to be less corrosion resistant than the pure metal.

Iron and Manganese

The corrosion resistance of the various commercial grades of pure metal in the AA 1000 series improves progressively as the iron content diminishes from about 0.45% in the widely applied AA 1100 alloy through AA 1050 alloy to about 0.002% in AA 1099 (99.99% pure metal) but for most purposes the effect is seldom sufficient to justify the loss of mechanical properties. The presence of manganese in the form of $(Mn,Fe)Al_6$ is also relatively innocuous. Common experience is that all of these alloys give very good service in every day situations; examples include the integrity of kitchen foil as thin as 20 μm and the very long lives of domestic utensils and rainwater gutters and conduits fitted to residences.

Magnesium

The standard electrode potential of magnesium, $E^{\ominus} = -2.370$ V (SHE), is so much more negative than that for aluminum, $E^{\ominus} = -1.663$ V (SHE), that even if segregated as heterogeneities or compounds in an alloy, there is no prospect of local galvanic corrosion on the aluminum matrix. The stable corrosion product of magnesium is the hydroxide, $Mg(OH)_2$, and although this substance is fairly soluble in near-neutral aqueous media, the magnesium contents in wrought AA 5000 series alloys do not detract from corrosion resistance but improve it. The same is true of the medium strength age-hardening alloys based on magnesium and silicon such as AA 6061. Aluminum alloys containing magnesium resist sea water and are applied in the superstructures of large ships and the hulls of smaller ones, taking advantage of their inherent medium strengths and suitability to welded construction.

Silicon

Silicon is used as a major component of cast alloy formulations primarily to confer good casting characteristics but in essentially binary alloys, such as AA 356, it also enhances corrosion resistance because the elemental silicon in the eutectic is unreactive. The advantage of improved corrosion resistance is largely dissipated in aluminum-silicon alloys that also contain high copper contents, such as AA 319 and AA 380.

Zinc

In dilute solution, zinc contributes $Zn(OH)_2$ *pro rata* to the passive film and does not detract from the corrosion resistance of aluminum but in multicomponent alloys it participates in intermetallic compounds contributing to some of the structural effects described below.

Copper

It is unfortunate that copper seriously diminishes the corrosion resistance of aluminum alloys because it is an essential component in formulating the high strength aging alloys, characterized by AA 2024 and AA 7075. These alloys must be protected for the applications in aircraft construction with which they are particularly associated.

The most effective protection is a thin coating of pure aluminum or an appropriate alloy applied by roll-bond cladding during hot rolling. The cast ingots of the high strength alloy, the *core*, are machined flat and plates of the protective material, the *cover plates*, are placed on top and underneath. The composite is heated to the normal rolling temperature for the alloy and presented to the hot-rolling mill. During hot reduction, relative movement at the interfaces between the core and cover plates due to differential plasticity of the two materials breaches the oxide film and allows a direct bond to form between them under pressure from the rolls. The cladding thickness in the material at final gauge must be selected to provide the protection required without sacrificing too much of the strong alloy section. It ranges from typically 5% of the total thickness per side for sheet less than 1.5 mm thick to 1.5% for plate over 4.5 mm thick. The cladding is not only itself corrosion-resistant but is selected to provide cathodic protection at edges or abrasions. Pure aluminum serves this function for aluminum-copper-magnesium alloys such as AA 2024 but not for aluminum-zinc-magnesium alloys in the AA 7000 series, such as AA 7075, that are clad with the 1% zinc alloy, AA 7072, listed in Table 9.2 specially formulated for this purpose.

It is easy to produce clad flat products but impracticable for sections and forgings. They must be protected by paint systems applied over chromating or anodizing, processes whose principles and applications to aircraft are dealt with in Sections 6.4 and 10.1, respectively.

9.2.3.2   Structure Sensitivity

Premature corrosion of aluminum alloys is unlikely if they are properly selected, applied and, if necessary, protected but when it does occur, it is more structure sensitive than for most other alloy systems because the highly reactive metal depends on passivity for protection and is vulnerable to reactive pathways provided by the proliferation of intermetallic compounds more noble than the aluminum matrix. For this reason, the form the attack takes is influenced not only by alloy composition but by the influence of heat-treatments and of working operations on the microstructure.

Pitting

When commercial grades of aluminum and aluminum alloys suffer corrosion, they do so by multiple pitting rather than by general dissolution. Various explanations offered for the detailed mechanisms producing pitting are somewhat speculative but there is general agreement that pits are initiated in the vicinity of small particles of intermetallic compounds outcropping at the metal surface. There is, of course, no concept of local passivity breakdown at characteristic pitting potentials, as there is for stainless steels, because the passivity depends only on the value of pH. Initiation may be associated with either weak passivity over the intermetallic compounds or to local galvanic cells between the particles of compound and the adjacent matrix. Once pits are nucleated their growth can be autocatalytic, i.e., self-accelerating because of the well-known changes that can occur in the micro-environment within pits. Metal cations produced at an isolated local anode in the confined space of a pit:

$$Al = Al^{3+} + 3e^- \tag{9.13}$$

coordinate with the existing OH^- population, disturbing the local $a_{H^+} \times a_{OH^-}$ equilibrium, leaving a preponderance of hydrogen ions that shifts the pH in the acidic direction. If it diminishes to a value less than about 4, the base of the pit cannot passivate. The pits can grow into self-arresting shallow depressions or penetrate more deeply.

Intercrystalline and Exfoliation Corrosion

If intermetallic compounds are present as connected chains around grain boundaries they can establish paths for corrosion advancing along lines of local action electrochemical cells created by potential differences between the compounds, the matrix and grain boundary zones depleted of the solute lost from solid solution in forming the precipitates. Alloys susceptible to this form of damage include most of the age-hardening high-strength alloys, e.g., AA 2024 and AA 7075. The effect can be controlled by heat treatments designed to minimize solute depletion adjacent to grain boundaries and to induce widespread rather than localized precipitation. Medium strength alloys depending on Mg_2Si for age-hardening are less susceptible. Alloys with minimal precipitation or whose precipitates have similar electrochemical properties to the matrix, including commercial grades of pure aluminum in the AA 1100 series, the aluminum-manganese alloy AA 3003, the can stock alloy AA 3004 and alloys with low magnesium contents in the AA 5000 series are the least susceptible. Alloys in the AA 5000 series with higher magnesium contents are not immune and are susceptible both to intercrystalline corrosion and its variant, exfoliation corrosion, if Mg_2Al_3 is allowed to precipitate; for this reason, a special stabilized temper is available for alloy AA 5456 used in boat hulls.

Susceptible metals can experience an extreme form of intercrystalline attack known as *exfoliation* or *layer* corrosion, in which the metal is opened up along grain or sub-grain boundaries running parallel to the surface. In heavily worked metal such as thin plate and sheet, the grains are elongated in the direction of working and very thin in the short transverse direction, so that they assume a layered morphology with planar grain boundaries parallel to the surface. The voluminous products from intergranular corrosion along these paths separates multiple layers of grains, causing the metal to swell. The layers open away from a neutral axis, i.e., the metal exfoliates. This suggests an idea that there may be a stress-corrosion component due to residual internal stresses in crack opening mode, inherited from mechanical working. The attack can be initiated at cut edges or from the base of pits penetrating from the surface into the system of parallel grain boundaries.

9.2.3.3 Stress-Corrosion Cracking

Principles underlying stress-corrosion cracking of high strength aluminum alloys were introduced in Section 5.1.2, where it was convenient to deal with them in the context of SCC theories of environmentally sensitive cracking in alloy systems generally. Alloy manufacturers and aircraft constructors have a different perspective because they have the responsibility of eliminating risk of failure in flight. It must be assumed that the specific agent, chlorides, will inevitably be encountered and since the first failures were identified in the 1930s, immense effort has been expended to avoid the effect.

Advances usually follow experience and a large body of service history is available. Alloy specific practices have evolved that enable the high strength alloys to be used with confidence by attention to alloy compositions, heat-treatment procedures and limits on design stresses. One guiding principle in the design of heat treatments is to minimize electrochemical differences between grain boundary precipitates by rapid quenching after solutionizing and aging to an overaged condition. In all of this, the objective is to contain susceptibility to stress-corrosion cracking rather than to exploit the maximum mechanical properties that the alloys can deliver.

Cracking normally proceeds in crack-opening mode along grain boundaries and so boundaries normal to the resolved tensile component of the prevailing stress are the most vulnerable. For this reason, stress-corrosion cracking is least likely in thin sheet material stressed in longitudinal and long transverse directions and the greatest danger is to thick sections and forgings stressed in the short transverse direction. These matters are taken into account in design stress analyses.

9.2.3.4   Galvanically Stimulated Attack

The standard electrode potential of aluminum is so negative that when it or one of its alloys is coupled to another engineering metal, either directly or indirectly, it is almost invariably the member that is susceptible to galvanically stimulated attack. Whether it actually suffers attack depends on initial breakdown of its passivity. The causes of the attack are explained in Chapter 4 and some examples are illustrated in Figures 4.2 and 4.4. Accelerated corrosion of aluminum alloys from injudicious combination with other metals must be rated as a design fault because, if the hazard is appreciated, it need never be realized. Unfortunately, when such problems occur, they are sometimes through ignorance attributed to faulty manufacture of the aluminum product.

9.2.4   Corrosion Resistance of Aluminum and Its Alloys in Air

9.2.4.1   Nature of the Air-Formed Film

The very thin film that protects aluminum in normal air at ambient temperatures can be detected and its thickness measured from its electrical resistance and by electron optical techniques. It grows on a freshly exposed aluminum surface by a version of the Cabrera-Mott kinetics, mentioned in Section 3.3.2, and attains a temperature-dependent limiting thickness after an hour or so. The limiting thickness at 20°C is in the range 2 to 5 nm that is less than 25 atom layers and does not exceed 100 nm even at temperatures as high as 500°C.

Insofar as it is permissible to relate a bulk structure to such a thin film, the expected oxide species is the aluminum trioxide, bayerite, that is stable in equilibrium with aluminum and normal moist air at ambient temperatures. Electron diffraction fails to disclose the structure and it is usually given the unsatisfactory description, 'amorphous Al_2O_3'. Water can be detected in the film by the following experiment. Hydrogen is removed from a freshly machined sample of aluminum by heating it in vacuum; the degassed sample is exposed to normal air for an hour or so and reheated in vacuum. The water is detected by collecting and identifying hydrogen produced in its reaction with the metal substrate:

$$2\,Al + Al_2O_3 \cdot 3H_2O = 2Al_2O_3 + 3H_2 \qquad (9.14)$$

9.2.4.2 Weathering

The protection the air-formed film affords the pure metal and alloys without significant copper contents such as commercial grades of pure metal, AA 3003, AA 3004, AA 5050, and other aluminum-magnesium alloys in the AA 5000 series is remarkably good. Metal loss in dry indoor environments is almost undetectable. In weathering out-of-doors in locations around the world, the metal surface is roughened by shallow pitting and discolored to dull gray but the metal does not thin appreciably after as much as ten years. As expected, degradation is worst in marine and polluted industrial atmospheres as assessed by loss of tensile strength of 1 mm sheet samples caused by the stress-raising propensity of pitting but even so it is seldom more than 10% in ten years. Surface treatments such as chromating or anodizing are applied for external exposure when preservation of appearance is important, where thin gauge material is used and for long term service as in architectural applications.

The high strength aluminum-copper magnesium alloys aluminum-zinc-magnesium alloys typified by AA 2024 and AA 7075 and also casting alloys such as AA 319 and AA 380 containing substantial copper contents have considerably worse weathering characteristics and usually need surface protection such as paint systems applied over chromate coatings. Aluminum-zinc-magnesium casting alloys that contain more than 6% of zinc can suffer stress-corrosion cracking during atmospheric exposure but the influence of metal composition is not fully established.

Aluminum alloys have little hot strength and can scarcely be regarded as engineering metals at temperatures much over 200°C and since the air-formed film is very thin, high temperature oxidation during service is not an important issue. There is, however an aspect of high temperature oxidation that sometimes claims attention. Sheet products of aluminum alloys with high magnesium contents are annealed to soften them in the course of manufacture. This entails raising the metal temperature typically to 340°C in a furnace with an air atmosphere and magnesium can oxidize selectively, thickening the oxide film and contributing magnesium oxide to it. If not removed, the magnesium oxide can be converted to carbonate by atmospheric carbon dioxide when the metal is subsequently stored:

$$MgO + CO_2 = MgCO_3 \qquad (9.15)$$

The carbonate appears as a white deposit that gives an undeserved poor image of the anticipated corrosion resistance but is not otherwise deleterious.

9.2.5 Geometric Effects

Geometric considerations apply to aluminum and aluminum alloys, as they do to other metals, for example, configurations that trap water from precipitation or condensation extend the fraction of time during which the metal is wet. There is, however, a difference of emphasis between some effects for aluminum and its alloys and many other metal systems.

9.2.5.1 *Crevices*

Quiescent Crevices

Crevices in submerged metal are not so deleterious as they are for iron or stainless steels because the passivity is not wholly dependent on maintaining a sufficient oxygen concentration to film the surface or to maintain anodic passivity. Corrosion is accelerated in very fine crevices, perhaps due to depassivation by pH drift caused by local disturbance of the $a_{H^+} \times a_{OH^-}$ equilibrium.

Crevices with Relative Movement

The acceleration of corrosion in crevices between aluminum surfaces in relative motion is an entirely different proposition. It is usually inadvertent and a well-known example is fretting or faying by repeated slight movement between the mating surfaces of rivetted joints subject to condensation; it is matter of serious concern in aircraft, as described in Chapter 10. The cause of the damage is continuous mechanical disturbance of the protective passive film so that metal surfaces in the crevice can become active local anodes.

9.2.5.2 *Impingement, Cavitation, and Erosion-Corrosion*

The passive film is vulnerable to damage by energy delivered by water moving across the metal surface. The terminology that describes consequent corrosion damage at the activated surface is related to the mechanical action of the water at the metal surface:

1. Impingement attack is due to deflection of a stream of water or steam at sharp changes in section or direction of channels in pipes, and ancillary fittings.
2. Cavitation is due to energy released in the collapse of bubbles generated in turbulent flow. A characteristic example is the output of ill-designed water pumps.
3. Erosion-corrosion is due to energy imparted by water moving over the metal surface at velocities exceeding about 3 m s^{-1}, wearing grooves in the metal by conjoint mechanical/chemical action. The effect has been observed on aluminum alloy hulls of high-speed vessels.

Damage of the kind described can be serious in its own right but it can also initiate consequential damage such as fatigue failure due to the creation of stress-raising artifacts.

Further Reading

Phillips, H. W. L., *Annotated Equilibrium Diagrams of Some Aluminium Alloy Systems*, The Institute of Metals, London, 1959.

Hatch, J. E. (ed.), *Aluminum Properties and Physical Metallurgy*, American Society for Metals, Metals Park, Ohio, 1984.

Mondolfo, L. F., *Aluminum Alloys: Structure and Properties*, Butterworths, London, 1976.

Mondolfo, L. F., Structure of the aluminum magnesium zinc alloys, *Metallurgical Reviews*, 16, 95, 1971.

Polmear, I. G., *Light Alloys*, Edward Arnold, London, 1981.

Godard, H. P. (ed.), *The Corrosion of Light Metals*, John Wiley, New York, 1967.

Godard, H. P., Examining causes of aluminum corrosion, *Materials Performance*, 8, 25 1969.

10

Corrosion and Corrosion Control in Aviation

A commercial airplane is designed and operated for maximum yield, i.e., capacity to deliver (e. g., expressed as passenger km) as a function of operating and capital costs. Among attributes contributing to an aircraft's yield are its size, shape, speed, mass, and fuel efficiency. Some of these are constrained by aerodynamic considerations and external factors but mass and fuel efficiency are strongly influenced by the selection and application of the materials of construction.

The importance of minimizing mass is illustrated by a specific example. The take-off mass of a typical fully laden wide-bodied passenger aircraft is 370 tonnes, comprising 160 tonnes of structure including engines, 170 tonnes of fuel but only 40 tonnes of payload. Mass saved in the structure can be either transferred to the payload or used to secure some alternative economic benefit. Since the mass ratio of the structure to the payload is four, the effect of increments in mass on yield is magnified in proportion, e.g., a 1% change in mass theoretically produces a 4% change in yield. Therefore all contributions to mass are scrutinized. For example, metal sections and thicknesses are the reasonable minima needed to carry the maximum anticipated stresses safely and excessive structural reserves to compensate for *avoidable* corrosion are unacceptable. Even protective coatings, such as paints, contribute some mass and must be applied with discrimination. It is within this context that resistance to corrosion and other degradation agencies must be considered.

The control of corrosion has various implications:

1. Safety is paramount and corrosion damage must not initiate structural failure in flight.
2. Aircraft represent heavy capital investment that must be protected; the costs of degradation and rectification must be within allowances made in planned amortisation.
3. Airline schedules must not be disrupted through unplanned grounding.

These requirements must be met in the competitive context of airline operations and under the supervision of aviation authorities acting in the public interest such as the FAA and their equivalents in other countries.

This imposes strict disciplines on the selection, fabrication, surface treatment and service history records of the structural materials.

10.1 Airframes

10.1.1 Materials of Construction

For economy in mass, cost benefit analysis reveals that the most desirable properties of airframe materials are low density, high modulus, and damage-tolerance. Following years of experience, the materials that best meet these requirements are the age-hardening aluminum alloys in the AA 2000 and AA 7000 series, with specifications listed in Table 9.2, particularly AA 2024, and AA 7075 alloys. Typically, fuselage skins are fabricated from rolled sheet of the more damage-tolerant alloy, AA 2024, supported on frames usually made from the stronger alloy, AA 7075. A modern practice is to machine wings to shape from AA 2024 alloy plate to produce integral stiffeners and to optimize mass by varying the thickness to keep stress levels consistent. There is some replacement of aluminum alloys by carbon fiber composites where the imposed stress system is suitable. Aluminum alloys containing lithium offer density and modulus advantages for possible future use but they are not generally accepted for aircraft at present, although they are applied in space vehicles.

10.1.2 Protective Coatings

The high copper and zinc contents that confer strength on AA 2024 and AA 7075 alloys reduce the protection afforded by the natural passivation of aluminum and they must be protected by surface coatings. Rolled sheet metal is protected by roll-bonded cladding applied to both sides by the metal manufacturer; AA 2024 and AA 7075 are clad with 99.3% pure aluminum and an aluminum-1 % zinc alloy, respectively, to confer galvanic protection, using a cladding coating of 2 ½ to 10% of the sheet thickness on each side depending on the gauge. Cladding is not possible on machined surfaces and paint protection systems based on chromate-inhibited primer must be used, applied by electrostatic guided spray over anodized coatings produced in sulfuric or phosphoric acid. The same applies to panels that have been formed by shot-peening.

10.1.3 Corrosion of Aluminum Alloys in Airframes

Water causing corrosion on the inside of the fuselage structure is accumulated from vapor from human sources condensing against the cold skin and where insulating materials can obstruct drainage. The corrosion is insidious, developing out of sight and eating into structural reserves. The decisive criterion is area loss, reducing the ability of the material to transfer loads correctly within the structure. It is unusual to find substantial damage on open flat panels and corrosion is generally associated with specific features of the structure.

Corrosion around faying surfaces, rivets, joints, and crevices is particularly troublesome. The incidence and severity of corrosion in these places is sensitive to details of construction. Errors in assembly that encourage corrosion include:

1. Crevices between fuselage skins and frames
2. Dry, i.e., unsealed, joint assembly.
3. Gaps where sealant fails to protect interfaces between multiple skins applied to reduce stress levels around cut-outs such as door-frames.
4. Stringer sections disposed so as to form gravity traps for water.

Based on experience, designs and construction techniques have been modified to reduce corrosion damage. These expedients include:

1. Eliminating multiple skins at cut-outs, by providing the extra thickness needed for reinforcement from thicker sheets, profiled by chemical milling.
2. Provision of drainage holes closed by spring-loaded valves to prevent accumulation of water at critical places.
3. Assembly of joints with sealants to prevent ingress of water.

Certain zones are corrosion free. Examples are the insides of pylon box sections, that are heated by the engines.

The whole of each wing structure forms a fuel tank and it inevitably collects water through the temperature and humidity cycles experienced. If the water remained, there would be a possible biological corrosion hazard by the growth of a fungus, *Gladisporum resoni*, at the water/fuel interface that generates acids. In modern aircraft, any such problem is eliminated by *sump pumping* the fuel from low remote points in the tank where any water could collect; this extracts the water and delivers it harmlessly to the engines. Provided that this is done, there are no residual corrosion problems on the *insides* of wings.

Concorde is exceptional in that the whole airframe is clean and virtually corrosion-free inside, because it flies warm from frictional heating at its cruising speed.

10.1.4   External Corrosion

10.1.4.1   Corrosion of Aluminum Structures

At low altitudes, the airframe is exposed to the natural atmosphere; it can be made more aggressive locally by chlorides contributed from de-icing salts and marine environments. Nevertheless, the general corrosion on the outside is less than on the inside except in particular areas where it can be quite severe on some aircraft. The vulnerable areas include the external forward faces of the front and rear wing spars and undercarriage bases; this may be associated with erosion-corrosion induced by the air flow. External damage is easier to detect and rectify than internal corrosion.

10.1.4.2 Corrosion of Steel Components

Where necessary, some airframe fittings are made from high strength steels, protected by electrodeposited coatings of cadmium for bi-metallic compatibility with aluminum alloys. They include flap-track fittings and pylon pins together with their terminating attachment points that unite the engines with the wings. Corrosion of these components would be particularly serious and must be monitored carefully and rectified because of possible interference with their vital functions.

10.1.5 Systematic Assessment for Corrosion Control

Incidents in flight due to corrosion in airframes are very rare but in one incident a large section of a fuselage was lost and fortunately the airplane survived. This alerted the industry and stimulated collaboration that established current mandatory standards of prevention, monitoring and control with special reference to aging aircraft.

Following the event, the U.S. Congress considered legislation to ground aircraft as they reach design-life goals. However, design life goals, based on flight cycles, calendar time, or flight hours, are guiding concepts in aircraft design and applying them as definitive operational safety limits would have grounded a significant part of the World's fleet of aircraft without necessarily contributing to safety, because operators have found that structural integrity is determined by fail-safe rather than safe-life criteria. A meeting of industry leaders was convened to determine the issues that characterize aircraft ageing. Five issues were identified and subsequently a sixth was added:

1. Local fatigue cracking.
2. Corrosion.
3. Repairs.
4. Maintenance programs.
5. Significant structural inspection details.
6. Widespread fatigue damage.

Corrosion assessment was assigned to a separate stand-alone program formulated after a four-year study by a structures working group drawn from leading operators, manufacturers and regulators, in which operators assumed a vitally important role due to their accumulated experience derived from observations during maintenance and innumerable service bulletins on aging aircraft. Corrosion control was deemed to be too important to be conducted on an uncoordinated basis and a systematic industry-wide inspection program based on a zonal system, was imposed for all aircraft.

In this zonal system, the whole structure, irrespective of aircraft type, is regarded as an assembly of zones and sub-zones. Every zone in which there is a corrosion concern is accessed at mandatory intervals and inspected specifically for corrosion. The inspection frequency is prescribed zone by zone. The extent of corrosion and the action to be taken are recorded by the following Industry standard reporting levels:

Level 1. Corrosion within prescribed tolerable limits.

Action — Clean corroded area and resume flying service.

Level 2. Corrosion outside of tolerable limits but rectification is permissible.

Action — Repair.

Level 3. Corrosion too extensive for repair.

Action — Refer for assessment of airworthiness.

A further part of the program is the application of proprietary corrosion prevention fluids, conforming to industry standards. One kind is wax based with joint-penetrating and water-repellant properties. The other kind contains inhibitors for long-term protection. A minor disadvantage of the practice is that these fluids encourage the accumulation of dust on the surface, obscuring observations in subsequent inspections.

Compliance with the program in its entirety is mandatory and subject to verification. The structures working group meets annually to review it.

10.1.6 Environmentally Sensitive Cracking

10.1.6.1 *Fatigue*

Fatigue cracking is a matter of great concern. Although not necessarily a corrosion problem, the possibilities of enhancement by an aggressive environment, as described in Chapter 5, is recognized. The cyclic stresses are imposed in various ways, by flight maneuver loads, gust loads, landing loads, and pressurization; the maximum loads are in landing and banking. Therefore the fatigue cracks are predominantly propagated by imposition of variable high-stress, low-cycle loading, with occasional peak loads. The effects of these loads can be traced from markings on fractures. Fatigue is monitored under the auspices of the structural working group in a stand alone program.

10.1.6.2 Stress-Corrosion Cracking

If incorrectly applied, age-hardening aluminum aircraft alloys are susceptible to stress-corrosion cracking when exposed to sources of chlorides
from marine atmospheres or de-icing salts, but the problem is averted by
assiduous attention.

The characteristics of stress-corrosion cracking are well-known for the
standard alloys and the risk can and must be eliminated. In airframe
design, stress analyses must confirm that the threshold stresses for stress
corrosion cracking will not be exceeded in service. In manufacture, heat-
treatment procedures are optimized for minimum susceptibility and must
be meticulously controlled to ensure that the assumed threshold stessess
are actually realized. In assembly, care is required to avoid adding internal
stresses. Minimizing susceptibility to stress-corrosion cracking takes pri-
ority over exploiting the maximum strength of an alloy. Stress corrosion
cracking is structure-sensitive because of the intergranular crack path so
that alloys are most susceptible in the short-transverse direction of forged
components; information on the reduced stress thresholds for this direction is available and must be taken into account in design.

The high strength AA 7075 alloy is much more susceptible to stress-corrosion cracking than AA 2024 type alloys, but the newer AA 7079 alloy is
more tolerant.

10.2 Gas Turbine Engines

Whittle's pioneering work on gas turbine engines in the nineteen forties
was inevitably based on relatively uncomplicated materials that were generally available at the time. The most exacting requirement was for turbine
blade and disc materials to withstand the stresses and corrosive environments to which they were exposed and the engine efficiencies were limited
by the relatively low temperatures at which these materials could be used.
This imposed low thrust/mass ratios that restricted aircraft performance
and inefficient conversion of thermal to mechanical energy. Since then,
continuing commercial and military demands for more and more efficient
engines have stimulated the development of customized alloys to withstand ever higher operating temperatures. These endeavors illustrate how
corrosion resistance and strength cannot be separated but must be considered together as different aspects of overall high temperature performance.

10.2.1 Engine Operation

Air collected at the front of the engine is compressed by a ratio of up to 30:1
reaching a pressure of 3 MN. It is delivered into a combustion chamber,

mixed with vaporized liquid fuel and burnt, yielding a high-pressure gas stream that enters a turbine to extract energy to drive the compressor through a shaft. The balance of energy in the gas stream is discharged as a reactive thrust through a nozzle at the rear of the engine.

There are four basic engine arrangements:

1. **Turbojets** in which the turbine drives only the compressor and most of the energy is delivered as jet thrust. These are high performance engines, used e.g., on Concorde.

2. **Turbofans** in which the turbine extracts additional energy to drive a large fan at the front of the engine, generating a low pressure air-flow that partially bypasses the core of the engine and supplements the hot exhaust gases. This reduces the total jet velocity, increasing the efficiency of propulsion and reducing noise and fuel consumption.

3. **Turboshaft** and **turboprop** engines in which the turbine is designed to extract as much energy as possible and deliver it to a power output shaft. Turboprop engines have been used to drive propellers for slower fixed wing aircraft and turboshaft engines provide power for helicopters.

Figure 10.1 schematically illustrates the layout of a large three-shaft turbofan that is a standard engine type in large commercial aircraft. The turbines and compressors comprise blades of aerofoil section attached to

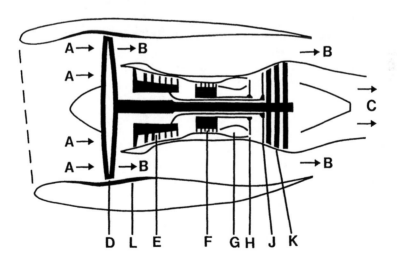

Figure 10.1
Basic layout of a three-shaft turbofan. Guide vanes omitted for clarity. A, Air intake; B, by-pass thrust; C, jet thrust; D, fan; E, first compressor section; F, second compressor section; G, combustion chambers; H, high-pressure turbine; J, intermediate pressure turbine; K, low-pressure turbine; L, fan casing.

discs mounted on the shafts; rows of static guide vanes are interposed between the successive sets of rotating blades. There are three successive sections in the turbine, the high (HP), intermediate (IP), and low (LP) pressure sections. Every section has a dedicated function in the compressor, the LP turbine section drives the fan, the IP turbine section drives the first compressor section, and the HP section drives the final compressor section. The rotating assembly of a turbine section, its counterpart in the compressor and the drive shaft carried on bearings is a *spool*; the drive shafts of the three spools are concentric. Turbine and compressor sections are multi-stage, e.g., the low-pressure turbine section driving the fan may have as many as five stages.

10.2.2 Brief Review of Nickel Superalloys

Nickel base *superalloys* were developed for arduous duties in the engine, notably turbine blades and discs and combustion chambers where the material properties required are strength at the high temperatures and corrosion resistance in the severe environment. Aspects of strength that are essential are (1) resistance to *creep* and *creep rupture*, i.e., progressive deformation of the metal under steady stresses experienced in service and associated mechanical failure by decohesion of the microstructure and (2) resistance to fatigue failure under cyclic stresses.

On inspecting a list of specifications, such as that given in Table 10.1, it is difficult to see what such diverse multicomponent alloys have in common. They were developed from an alloy of 80% nickel with 20% chromium that consist of a single FCC phase, γ, protected from oxidation at high temperatures by a film of chromium oxide, Cr_2O_3, formed selectively by the principles explained in Section 3.3.4.2.

The first modification was to add aluminum to introduce an intermetallic phase γ' based on Ni_3Al, precipitated as a fine dispersion throughout the γ matrix that confers strength by impairing dislocation movement. The characteristics of the precipitate phase and its relationship to the matrix have driven the development of the superalloys. Because both γ and γ' have FCC structures with similar lattice parameters (< 1% mismatch), the precipitate is coherent with the matrix and has low interfacial energy and good long term stability at high temperatures. If the volume fraction is low, dislocations can bypass the precipitate particles by looping around them but if it is high, the dislocations cannot loop and must pass through them, where they are impeded by the ordered structure of the phase. Titanium, tantalum and niobium can participate in the formation of γ' and by suitable alloying it is possible to obtain volume fractions of γ' in excess of 90%. The matrix can be further strengthened by adding elements such as tungsten or molybdenum that enter solid solution, stiffening the matrix and slowing the diffusion of metallic solutes that could undermine high temperature stability. By exploiting all of these effects, modern alloys such as Mar M002 can deliver high strength at temperatures approaching 1150°C.

Strength conferred on the matrix exposes microstructural weaknesses that must be addressed. One of these is relative movement between adjacent grains, allowing creep and creep rupture. This is controlled in polycrystalline materials by incorporating carbon that precipitates metal carbides at the grain boundaries to pin them. The problem is circumvented for blades operating at high temperatures by eliminating grain boundaries, using the directional solidification and single crystal casting techniques described in Section 10.2.5.1.

The extensive alloy range now available and illustrated by the examples given in Table 10.1 is based on the principles described above. It has been evolved over several decades to provide materials to suit the conditions in which the various engine components must operate and to facilitate their manufacture.

10.2.3   Corrosion Resistance

10.2.3.1 *High Temperature Oxidation*

In products of combustion free from sulfur and chlorides the high temperature components, turbine blades, turbine discs and combustion chambers resist oxidation by virtue of protective films formed by selective oxidation. Depending on the alloy composition, the films are chromia, Cr_2O_3, or alumina (corundum), Al_2O_3, and alloys are distinguished by the terms, *chromia formers* and *alumina formers*.

The strongest superalloys have high aluminum-titanium ratios and high concentrations of components for solid solution strengthening. Because of requirements for γ′ stability and the need to retain alloy components in solution, the chromium contents are reduced to below the critical compositions needed to stabilize Cr_2O_3 as the protective film and its function must be replaced by a film of Al_2O_3, that is thermodynamically stable with respect to the alloy composition. However, the alloy is unable to maintain the protection unaided, because the supply of aluminum by diffusion from the interior is insufficient to replenish that which is consumed in forming and repairing the film. The deficiency is rectified by *aluminizing* the surface, i.e., diffusing aluminum into the surface to produce an aluminum-nickel alloy layer capable of supporting a protective oxide film.

10.2.3.2 *Aluminizing for Oxidation Resistance*

The simplest and most common process is *pack aluminizing*. Blades are cleaned by blasting with an abrasive and then packed and sealed in a mixture of aluminum powder, alumina, Al_2O_3, and a halide, typically ammonium chloride, NH_4Cl. The pack is heated for a few hours at a temperature in the range 750 to 1000°C, typically 900°C. The alumina has no active part and is there to prevent coalescence of the aluminum powder particles when they melt. Ammonium chloride supplies aluminum chlorides,

$$2NH_4Cl(s) + 2Al(powder) = 2AlCl(g) + 2NH_3(g) + H_2 \qquad (10.1)$$

initiating a cyclic process that transports aluminum from the powder to the alloy through the vapor phase:

$$3AlCl(g) = 2Al(\textit{at alloy surface}) + AlCl_3(g) \qquad (10.2)$$

$$AlCl_3(g) + 2Al(\textit{powder}) = 3AlCl(g) \qquad (10.3)$$

An aluminized coating forms in the absence of the halide but it is uneven and unsatisfactory.

The coating is an intermetallic compound, β-NiAl, containing up to 40 weight % aluminum, that confers good protection and it is sometimes used in that condition but it is brittle and can crack. Lively blades that experience flexing are given a subsequent diffusion anneal for a short time at a higher temperature, typically 1100°C, to disperse the aluminum further into the metal, conferring some ductility in the coating. The annealed coatings are about twice as thick with a surface aluminum content of between 20 and 30 weight %. Coatings are 25 to 75 µm thick and increase blade dimensions by about half as much.

Considerable effort has been expended in a search for improved coatings. Combination of chromium with aluminum yields only marginally better coatings but additions of silicon significantly improve the sulfidation resistance and enrichment with platinum by pre-plating the alloy surface with a thin layer of platinum prior to aluminizing enhances the high and low temperature performance of the coatings by introducing aluminum-platinum compounds.

10.2.3.3   Hot Corrosion and Sulfidation

Blade and other engine components exposed to moderate temperatures can be attacked by *hot-corrosion* and *sulfidation*, that are specific terms describing attack initiated by sodium sulfate derived from salts ingested in air contaminated with sea salt. Sea salt already contains a significant amount of sodium sulfate, Na_2SO_4 and more can be produced by interaction between sodium chloride and sulfurous oxides in the combustion products from the fuel:

$$2NaCl + SO_3 + H_2O = Na_2SO_4 + 2HCl \qquad (10.4)$$

At the highest engine temperatures, the sodium sulfate is vaporized but at temperatures in the order of 650 to 900°C, it can condense as a dew on the metal surfaces. The liquid attacks protective oxides leaving the metal vulnerable to accelerated subsequent oxidation, and can stimulate *sulfidation*, a form of corrosion in which sulfur penetrates into the metal locally precipitating chromium as chromium sulfide, Cr_2S_3.

There are two regimes of hot-corrosion, designated Type 1 and Type 2, that operate in different temperature ranges, as illustrated schematically in Figure 10.2. In Type 1 corrosion at the higher temperatures, the sodium sulfate has an acidic character imparted by the absorption of oxides of alloy components such as molybdenum and tungsten. In Type 2 corrosion at the lower temperatures it has a basic character due to the release of O^{2-} ions when sulfur is absorbed by the metal. In either condition, it can attack the amphoteric oxides, Cr_2O_3 and Al_2O_3 but Al_2O_3 is the more resistant to acid fluxing and Cr_2O_3 is the more resistant to basic fluxing.

To limit the attack, aircraft fuels in Western countries are standardized with low sulfur contents but hot corrosion is always a potential problem to a greater or lesser degree.

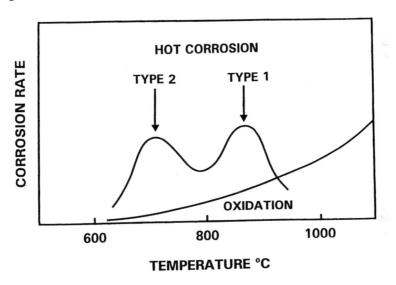

Figure 10.2
Temperature-dependence of oxidation and hot corrosion rates (schematic).

10.2.4 Engine Environment

The thermal and chemical environments within an engine depend on the functions of the various parts and on aircraft assignments, maneuvers, and flight patterns.

10.2.4.1 Factors Related to Engine Operation

The fuel is burnt with excess oxygen, so that the gaseous environment is oxidizing. The highest temperature is experienced in the high pressure turbine stage, where at full power the gas temperature approaches 1500°C and the high pressure turbines that first receive the flame may reach a peak metal temperature of 1100 or even 1150°C. The gas cools in passing through the succeeding intermediate and low pressure turbine stages, where the lowest blade temperatures are typically 700°C. There are temperature gradients within the individual blades, a topic taken up later in relation to fatigue. The rims of the turbine discs on which the blades are mounted can experience temperatures as high as 600°C and the walls of the combustion chambers can reach a temperature of 1000°C.

Compressor blades are not exposed to the flame but they are heated by a combination of the thermal ambience and the adiabatic compression of the air intake. Blades in the second section can reach temperatures between 300 and 600°C.

10.2.4.2 Flight Pattern Factors

The severity of the environments within engines depend on aircraft assignments and are conveniently considered within the following categories:

1. Long-haul civil applications, e.g., transatlantic and non-stop transcontinental flights.
2. Short-haul civil applications especially on island routes e.g., linking Hawaiian Islands.
3. Military aircraft flying low over water.

Long-haul flights impose the least demands on the corrosion resistance of engine components because they comprise long single-flight cycles of take off, cruising and landing, Full engine speed is required for take off and climb and the blade temperature rises rapidly through the ranges for hot corrosion to their peak values. On reaching the cruising altitude, typically 10000 m, the engine speed is reduced and the temperatures fall to a steady lower value where they remain until landing. After landing, the engines accelerate for a few minutes to apply reverse thrust. The temperature cycle is represented in Figure 10.3.

Aluminized high-strength superalloys alloys resist oxidation in clean air at the constant temperatures below their peak values experienced during cruise. Ingestion of air contaminated with salt is possible only during a short interval immediately after take off and during descent.

In short-haul applications, the engines experience multiple flight cycles with frequent thermal cycling through the temperature ranges for hot corrosion and if the flights are over seawater there are frequent episodes in which air contaminated with salt is ingested at low altitudes during climb and descent, creating conditions for hot corrosion.

Engines on military aircraft flying low over sea are exposed to continual thermal cycling in salt-laden air. Helicopters create spray over sea with

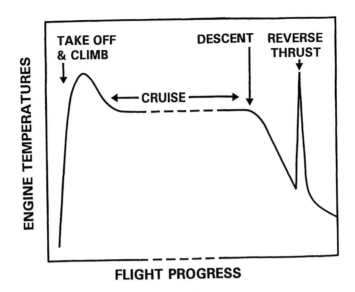

Figure 10.3
Engine temperatures during progress of a long-haul flight (schematic).

the down draught from their rotors and their engines are particularly vulnerable.

10.2.4.3   Fatigue

Flight patterns and maneuvers also influence fatigue life. Engine components, especially turbine blades and discs experience a combination of cyclic stresses. Mechanical stresses are imposed by centrifugal force and bending moment in the gas stream. Further stresses are superimposed by cyclic variations in engine temperature due to the application of power for take off, reverse thrust and maneuvers. Thus the expected fatigue lives are longer for long haul than for short haul and military assignments.

10.2.5   Materials

10.2.5.1   Turbine Blades

Turbine blades that approach temperatures of 1150 °C in civil application are usually produced from one of the high-strength multicomponent alloys with high aluminum and low chromium contents such as MAR M002 that have good oxidation resistance. Where environments or aircraft assignments impose greater exposure to hot corrosion, alloys with higher chromium contents may be preferred. High-strength alloys are difficult to fabricate and they are produced by investment casting, a method that is capable of yielding the complex shapes needed with precision. Wax patterns of the blades are assembled on a wax support that serves as a pattern for the runners to fill the mold with liquid metal. The assembly is coated with successive layers of a ceramic slurry dried in an oven to produce a self-supporting shell. The wax is melted out, leaving cavities of the required shapes. The shell is insulated and the alloy is cast into it. Some alloys can be cast in air but those containing reactive elements are cast in vacuum. Blades for low-pressure sections can be polycrystalline but blades for the high-pressure turbine section are cast by directional solidification to eliminate grain boundaries normal to the principle stress axes. This is done by establishing a thermal gradient through the metal as it solidifies in the mold, using chills and auxiliary heat sources. It can be modified by crystal selection to produce whole blades as single crystals, eliminating all grain boundaries. Some standard high-strength alumina-forming alloys, such as MAR M002 have been applied extensively for directionally solidified blades and many others have been customized for these advanced casting techniques. Two examples given in Table 10.1 are SRR 99 for directional solidification and CMSX-4 for single crystals.

Critical surfaces of the cast blades are precision machined or ground. The allowable peak temperatures for the blades are significantly lower than the flame temperatures so multiple channels are cut into the blades to serve as ducts for cooling air supplied by the compressor. The channels are cut by electro-discharge machining or lasers because these methods can be controlled to cut channels with uniform overlying metal thicknesses, minimizing the thermal gradients experienced in service to improve resistance to thermal fatigue.

10.2.5.2   Turbine Discs

Materials for turbine discs must resist oxidation at the moderately high temperatures experienced at the rims and they must be suitable for forging to shape. Hot corrosion is not usually a significant problem with turbine discs because cooling air is used to keep the contaminated gas stream away from the rims. Disc life is determined mainly by fatigue resistance and it is maximized by carefully controlling the structure. This requires alloys that are not unduly subject to segregation in the ingots cast for forging and careful control of the deformation sequence and subsequent heat treatment. The high-strength superalloys do not meet these requirements. The best-suited materials are the chromium-rich alloys, IN 718, IN 901, Waspalloy and Astralloy.

10.2.5.3   Combustion Chambers

Combustion chambers are annular structures formed from sheet. The chamber walls experience thermal cycling with peak temperatures above 1000°C. These factors impose requirements for good fabrication and welding characteristics combined with resistance to oxidation and thermal fatigue, such as C263.

TABLE 10.1
Examples of Alloys Used in Gas Turbine Engines

Alloy	Composition, weight %										
	Ni	Cr	Ti	Al	C	Mo	Co	Fe	W	Nb	Others
Mar M 002	Balance	9	1.5	5.5	0.15	—	10	—	10	—	2.5 Ta + 1.5 Hf
IN 738	Balance	16	3.5	3.5	0.11	1.8	8.5	—	2.5	0.7	1.6 Ta
Nimonic 105	Balance	15	3.8	5.0	0.15	3.3	13	—	—	—	—
IN 939	Balance	22	3.7	1.9	0.5	—	19	—	2.0	1.0	1.4 Ta
IN 718	Balance	19	0.9	0.6	0.04	3.0	—	20	—	5.2	—
Inco 901	42	13	3.0	0.3	0.04	5.7	—	Balance	—	—	—
Waspalloy	Balance	19	3.0	1.3	0.08	4.3	13.5	—	—	—	—
Astralloy	Balance	15	3.5	4.0	0.06	5.2	7.0	—	—	—	—
C 263	Balance	19	2.2	0.5	0.06	6.0	20	—	—	—	—
SRR 99	Balance	8	2.2	5.5	—	—	5	—	10	—	3 Ta
CMSX-4	Balance	6	1.0	5.6	—	0.6	9	—	6	—	7 Ta + 3 Re + 0.1 Hf

Ti = titanium, Mo = molybdenum, Co = cobalt, W = tungsten, Nb = niobium, Ta = tantalum, Hf = hafnium, Re = rhenium.

10.2.5.4 Compressor Assemblies

The first section of the compressor stage is a welded assembly comprising a drum carrying the blades. It is fabricated from a titanium alloy with 6% aluminum and 4% vanadium (listed in ASM specifications) for high strength with low mass. It is relatively cool and resists corrosion in the air it collects. As a precaution against a remote fire risk associated with the use of titanium, the stationary guide vanes between the stages are made from a martensitic stainless steel to act as fire breaks. Martensitic stainless steels resist corrosion in the atmosphere well and the guide vanes do not normally need protection but they can corrode in heavy condensates from humid air when aircraft are grounded in tropical locations, especially if the engine contributes salt contamination acquired in flight. If this is anticipated, they are protected with paint films containing aluminum powder as a pigment for sacrificial protection. To serve this purpose, the paint is made conductive by peening and heat-treating it to establish electrical contact between the aluminum particles.

The blades in the final compressor section in modern engines can reach a temperature as high as 600°C, and have to be produced from a temperature-resistant alloy such as IN 718.

10.2.5.5 Cool Components

Shafts are made from heat-treated high-carbon low-alloy nitriding steels.* They are partly protected by lubricating oils and exposed non-oiled surfaces can be protected by sacrificial or barrier paints with aluminum pigments. Corrosion problems for bearings are not so much during service as in the storage of spares. To meet stringent requirements for freedom from pitting they must be packed in wax preservatives with vapor phase inhibitors. Fan containment casings and their accessories are fabricated from aluminum alloy sheet matching that used in airframes and are typically protected by chromic acid anodizing.

10.2.6 Monitoring and Technical Development

Manufacturers give warranties on component lives but expect to exceed them. They cooperate with customers to prescribe monitoring procedures during service and make arrangements for engines to be inspected for condition on the wing. Suitable facilities are incorporated in engine design, e.g., by providing fiber optic paths to examine critical areas.

Successive engine designs are upgraded to meet increasing demands for safety, fuel economy, noise abatement, mass reduction and care of the environment. Manufacturers have strategies to anticipate and prepare for

---

* Steels containing chromium or aluminum that can be surface hardened by heating in ammonia/hydrogen mixtures.

future needs. So much investment is at stake that new materials or coatings can be incorporated in engines only after they have been proven on demonstrator rigs with full instrumentation, designed to simulate service environments.

Further Reading

Flowers, H. M., (ed.), *High Performance Materials in Aerospace*, Chapman & Hall, London, 1995.

Crane, F. A. A. and Charles, J. A., *Selection and Use of Engineering Materials*, Butterworths, Boston, 1984.

Forsythe, P. J. E., The fatigue performance of service aircraft and the relevance of laboratory data, *Journal of the Society of Environmental Engineering*, 19, 3, 1980.

The Jet Engine, 5th Edition, Rolls-Royce, Derby, U.K., 1996.

Meatham, G. W, (ed) *The Development of Gas Turbine Materials*, Applied Science Publishers, London, 1981.

Thomas, M. C., Helmink, R.C., Frasier, D. J., Whetstone, J. R., Harris, K., Erickson, G. L., Sikkenga, S. L., and Eridon, J. M., *Alison Manufacturing, Property and Turbine Engine Performance of CMSX-4 Single Crystal Airfoils in Materials for Advanced Power Engineering, Part II*, Kluwer Academic Publishers, Liege, Belgium 1994.

11

Corrosion Control
in Automobile Manufacture

11.1 Overview

Automobiles are built for an internationally competitive market that is demanding not only in vehicle performance and price but also in subjective customer perception. Corrosion resistant coatings for the body therefore have both protective and cosmetic functions. The cosmetic aspect applies both to initial showroom appearance and to premature development of blemishes. For these reasons, the quality and consistency of paint coatings has a crucial influence not only on vehicle durability but also on sales potential.

Corrosion protection of the body must be durable but not to the point where it incurs excessive costs to maintain it beyond the vehicle life expectancy due to other limiting factors, such as degradation of moving parts, obsolescent technology and changes in fashion. All of these aspects of durability are synchronized in *a design life goal* that for many automobile volume manufacturers is six or seven years. Safety concerns due to loss of structural reserves by corrosion should not be an issue within the design life, assuming normal maintenance.

Premature corrosion in the engine and associated systems can cause unreliability and expense. Corrosion can interact with the mechanical degradation of moving parts by wear, fatigue or erosion to an extent influenced by the compositions of fuels, oils and products of combustion. The protection of recirculating water-cooling systems is complicated by the use of anti-freeze additives. The integrity of exhaust systems, that have hitherto been replaceable items, is now a serious concern because of widespread adoption of catalytic converters for emission control, reduced noise tolerance and a trend by manufacturers to extend warranties.

11.2 Corrosion Protection for Automobile Bodies

11.2.1 Design Considerations

The design of an automobile body shell is inevitably a compromise to reconcile the different interests. Stylists address sales potential by designing attractive shapes and accountants monitor costs to meet price competition. Engineers develop the design into a load-bearing structure of minimum mass, composed of parts suitable for robotic assembly on a production line. The usual solution is a welded construction of panels, frames and box sections cut and press-formed from cold-rolled low-carbon steel strip, selected from a range of tempers and gauges, typically 0.5, 0.7, and 2 mm, to carry the local stresses. The overall strength and stiffness is derived from the profiles of the component parts and from cold work applied to the metal. This is the context within which protection from corrosion in service is addressed.

Basic protection is by a paint system applied over a phosphate coating that also gives the vehicle an aesthetically pleasing appearance. It is reinforced by plastic or bituminous underseal where experience shows it to be needed. This is quite adequate on most panels but attention is needed for vulnerable features revealed in service experience and destructive whole vehicle corrosion assessments using standardized tests such as exposure to salt sprays and in humidity chambers. It is a common observation that the patterns of premature corrosion in automobile bodies is model-specific. For example, one model may suffer at the skirts of doors but another is vulnerable at the joint between the front wings and bulkhead. These patterns are related to particular geometries or to methods of assembly. Some of the features that can lead to problems and measures available to counter them are itemized below.

Front and Side Panels

These areas receive the most intense exposure to grit and spray. Modern practice is to use panels pressed from steel supplied by the steel mill coated with zinc for galvanic protection; the zinc may be electrodeposited (EZ steel) as outlined in Section 6.2.6 or hot-dip coated (IZ steel) as described in Section 6.3.1 and annealed to allow zinc and iron to interact, producing a surface iron-zinc compound that provides a better base for phosphating. The decision on which kind of zinc coating to use is however usually made for the more mundane reason that EZ steel is least expensive for gauges of 2 mm and above and IZ steel for thinner gauges. Some authorities advocate building the whole shell from zinc coated steel despite the penalty in increased material cost.

Wheel Arches

Wheel arches are vulnerable to paint damage by road stones and grit. Despite this, the rear arches have smooth contours and when undersealed give little trouble. The front arches are attached to the frame supporting the engine, front wheels and front fender and may carry inset lamp fittings. The irregular contours offer traps for wheel splash that may accumulate as *mud poultices*. These poultices can remain damp, stimulating corrosion long after rain has fallen. The problem is exacerbated by the high conductivity and aggressive nature of chloride-bearing de-icing salts laid down by highway authorities in winter. Such poultices can be more active in a heated garage than in the cold open air. The solution to the problem is quite simply to deflect the splash by fitting smooth plastic internal arches over the wheels.

Joints

Joints are of various kinds, e. g. lap joints between front wheel arches and bulkhead, clinched joints between the front and back panels of doors, gaps between double metal thicknesses, etc. These joints can act not only as water traps but also as water conduits, so that the site of corrosion may not coincide with the joint. Part of the difficulty is that the joints often have to be made during the shell assembly before the application of paint. The solution is to apply beads of plastic sealer when the shell has received an undercoat of paint.

Rainways

Rainways are built into the shell to deflect rain falling on the roof clear of the doors. It can happen that due to some unforeseen circumstance the rain can be collected and inadvertently directed into a water trap, such as the gap between the hood and bulkhead. This is a matter for minor modification.

11.2.2 Overview of Paint-Shop Operations

A body shell presented for painting is a large intricate work-piece presenting surfaces with many different orientations, cut edges and internal and re-entrant features to which it is difficult to gain access. The paint shop has an input in the design to ensure that all parts of the shell are accessible to the entry and drainage of process liquors. Furthermore, as produced, the surfaces of the body shells are unsuitable to receive paint because they are contaminated with lubricants and detritus from the various fabrication operations. Cleaning and phosphating to provide the key for the paint are therefore an integral part of the painting operations. Failure to fulfil these preparatory tasks meticulously is not necessarily apparent in the superficial appearance of the finish but is manifest by premature corrosion in service.

A paint shop receives the bare metal shells from the vehicle build production line, typically at a rate of 30 to 60 per hour and painting and related operations are organized to match. Pre-cleaning, phosphating, painting, sealing, paint baking, and associated operations are arranged in sequence in the same production line, shielded by tunnels to contain liquor splashes. The body shells, complete with doors, hoods and trunk lids braced slightly open are suspended from an overhead conveyor that carries them through and between all of the processes throughout the line.

The paint line represents a capital investment of more than $100 M and must be adaptable to accept different automobile models in any order, to anticipate future models and be capable of treating steel, zinc and aluminum surfaces. The throughput must be variable to synchronize with the output of the vehicle build lines and since this influences the treatment times in the various processes, consistent performance and quality is maintained by adjusting other parameters, using arrangements to set, control and monitor them automatically.

11.2.3   Cleaning and Pretreatment of Body Shells

The body shells are manually wiped with a safety-approved hydrocarbon solvent to remove grease and loose contamination carried over from press-forming and vehicle assembly; this and joint-sealing operations after application of the paint undercoat are the only labor intensive part of the whole sequence. The shells are then cleaned in an alkaline solution that is essentially 0.2 to 0.4 M sodium or potassium hydroxide with a surfactant, usually supplied as a proprietary formulation. To reach all parts of the intricate assembly, including undersides and box sections, it is applied in three sequential stages: (1) a deluge at ambient temperature to reduce contamination of subsequent stages and to dislodge swarf, (2) a high pressure (100 kN) spray at 50°C from strategically placed multiple nozzles, and (3) complete immersion of the body shell in the solution, also at 50°C. There follows two rinses in water to prevent transfer of alkali to the following acidic phosphating process.

11.2.4   Phosphating

The phosphate coatings are produced by applying the principles described in Section 6.4.1 on a large scale. The body shells are completely immersed in the phosphating liquor for typically for 3 to 5 minutes, entering and leaving it on the conveyor. The capacity of the phosphating liquor tank, about 100000 liters, gives an impression of the large scale of the operation. On emerging, the shells have a smooth matt gray appearance. They are thoroughly rinsed twice, once by complete immersion in water to remove all soluble materials that could subsequently cause the paint to blister and again in a 0.005 M chromic acid solution to passivate metal exposed at any microscopic gaps and to impregnate the coating with the active corrosion inhibitor, chromium (VI). Even although the next operation is coating with a water-borne paint the phosphated shells are dried in hot air as a precaution against rusting during temporary holdups that would adversely affect paint performance.

The solutions are usually proprietary formulations and the coatings produced are mixed hydrated tertiary phosphates in which, zinc, manganese, and nickel and iron cations all participate. The relative proportions of the cations are chosen to meet the requirements that the coatings spread rapidly over the surface and that their chemical and physical natures provide the optimum base for the paint coatings. Figures 11.1 and 11.2 show the appearances of the surfaces of satisfactory coatings at a magnification of × 1125, illustrating their crystalline structure.

The physical features that characterize good coatings are:

1. The coatings are thin and cover the metal surface completely. If the coatings are too thick, they detract from the smooth finish required for the paint. The coating thickness is expressed conventionally as a *coating weight*, typically within the range, 2.2 to 3.0 g m^{-2} on bare steel and 2.5 to 3.5 g m^{-2} on zinc coated steel.

2. The crystals are equiaxed, regular and of a uniform optimum size. Crystals in satisfactory coatings have maximum dimensions as viewed on the surface of 2 to 8 µm on bare steel and 4 to 10 µm on zinc coated steel. The higher range for zinc allows for the acicular morphology illustrated in Figure 11.2. If the crystal size is larger, the coating can tend to delaminate, undermining the adhesion of the paint. If it is smaller, the coating does not offer such a firm base for paint.

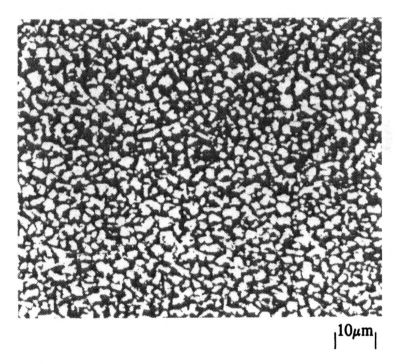

|10µm|

Figure 11.1
Structure of phosphate coating on cold-rolled steel, showing equiaxed crystals.

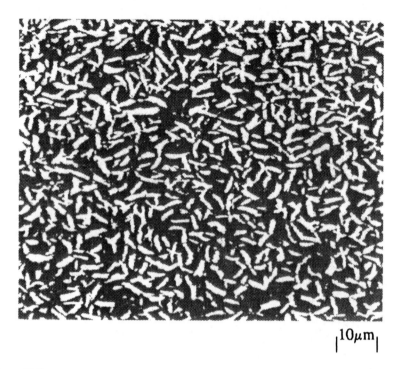

$_|10\mu m_|$

Figure 11.2
Structure of phosphate coating on zinc-coated steel, showing acicular crystals.

The phosphate liquor is formulated to suit mixed metal body shells that may contain steel, zinc and aluminum components. The combination of zinc and manganese cations in the formulation produces thin, smooth coatings on steel. The contribution of the nickel cation is to control the growth of phosphate crystals on zinc. The formulation may contain a small quantity of sodium fluoride to limit contamination of the liquor with aluminum, because an aluminum content > 0.3 g dm^{-3} inhibits the deposition of zinc tertiary phosphate. It eliminates aluminum by complexing it as AlF_6^{3-} that precipitates from the solution as cryolite, Na_3AlF_6. The small uniform crystal size required is promoted by nuclei provided by a titanium based surface conditioner added to the second rinse following alkali cleaning.

Strict controls are needed to maintain consistent good coatings. Daily test panels representing the materials in the body shells are processed and used to monitor coating weight and crystal size. The phosphating liquor is continuously circulated through heat-exchangers to control temperature and filters to remove sludge. Excessive ionic content in the water supply, especially in the rinse water, can reduce the corrosion resistance of the coating and paint system applied to it; this requires particular attention if the water is "hard", with $[HCO_3^-] > 4 \times 10^{-3}$ mol dm^{-3} or if the sulfate and chloride concentrations are higher than usual; these characteristics are related to the hardness of the water discussed in Section 2.2.9.2. It is usually dealt with by matching the composition of the water to the quality of the coating and de-ionizing only a sufficient fraction of the water to meet the required standards, using ion exchange resins.

11.2.5 Application of Paint

The phosphate coating keys the paint to the metal and prevents undermining by corrosion creep but it is the paint system and associated operations that provides protection. The following sequence of operations is typical:

1. A priming coat is applied by electropainting.
2. Crevices are sealed.
3. Undercoat, color and gloss coats are applied by electrostatic spray systems.
4. Supplementary protection is applied to critical areas.

11.2.5.1 *Primer Coat and Sealing*

The primer coat is applied by electrodeposition of a water-borne paint of the kind described in Section 6.5.1.3. An electrical contact is attached to every body shell moving on the conveyor as it descends into a tank containing the paint. The shell is completely immersed and a negative DC potential of the order of 200 V is applied. The current entering a medium sized body shell is about 900 A, that builds a paint film 23 to 35 μm thick in the few minutes allowed in the production line program. Although the insulating properties of the paint tends to promote an even coating by diverting current away from thickly coated areas, the throwing power is finite, extending to about 25 cm, and strategically placed cathodes are needed to obtain an even coating on such a large convoluted workpiece as a body shell. The paint is coherent and free from noxious solvents. It is rinsed to remove loose particles, dried and baked by passing through a continuous oven. The electrodeposited coat does not penetrate crevices > 0.4 mm deep and seams, hemming flanges under wheel arches and in the floor are sealed with a mastic filler after baking.

11.2.5.2 *Undercoating Color and Gloss Coats*

The undercoat, color and gloss coats are applied sequentially by dedicated electrostatic spray systems in clean dust-free spray booths. The paint is delivered through *spinning bells*, i.e., spray heads revolving at 400 to 500 revolutions per second that atomize the paint and charge it to a high DC potential, e.g., 90000 V. The charged paint particles are attracted to the earthed body shell, minimizing overspray. The spinning bells move on computer controlled tracking systems, executing predetermined paths following the body shell contours.

The undercoat is applied, baked in a continuous oven, manually sanded to remove small blemishes, automatically dusted down, and passed to the color spray booth. The color coat is automatically selected by a computer programmed with order information. It is applied in its own dedicated spray booth by automatic electrostatic spray equipment, supplemented by manual spraying where needed. The body shell with its color coat is passed to an adjacent spray booth to receive the gloss coat, also applied by electrostatic spray. Finally, the color and gloss coats are baked.

11.2.5.4 Supplementary Protection

PVC based underseal is applied over the paint inside the wheel arches and under the sills as defenses against abrasive wheel splash. As much as 1.5 kg of warmed fluid wax with a corrosion inhibitor is injected into underbody box sections, where it solidifies to form a thick coating, reinforcing the protection against water accumulated from road wash. The finish is inspected in bright light and minor blemishes touched in.

11.2.6 Whole-Body Testing

No matter how carefully a body shell is designed, there are sometimes areas vulnerable to corrosion that come to light only when it is built and whole body testing is needed. As might be expected, manufacturers test bodies in conditions simulating extremes encountered in service, notably wash from highway de-icing salts, high humidity, and exposure to UV radiation. The tests are standardized and can include the following procedures:

1. Spraying the whole automobile with 5% sodium chloride solution with frequent intermittent drying cycles.
2. Storing the vehicle continuously in air with 100% relative humidity at 30°C.
3. Continuous exposure of a vehicle coated with primer coat only to artificial UV radiation at a prescribed intensity.

Survival to a prescribed acceptable degree of degradation for 960 h is usually required.

11.3 Corrosion Protection for Engines

11.3.1 Exhaust Systems

Exhaust systems were formerly maintenance items, easily replaced when perforated by corrosion but the advent of catalytic converters and a desire by manufacturers to give extended warranties puts them in a new category as long service items. Abatement of noise from perforations is another factor. Therefore, corrosion problems are now addressed more urgently.

A catalytic converter is an expensive and delicate device, easily damaged mechanically or by contamination. The active agents are contained in a ceramic foam supported on anti-vibration mountings. Installation of the converter has repercussions in both the front section of the exhaust between the engine and the converter and the rear section leading from the converter to the tailpipe.

The environment in the exhaust front section is more aggressive because it must run hotter to admit gases to the converter at a high enough temperature and at the same time the consequences of corrosion are more serious because oxide flakes can block the catalyst and perforation can allow combustion gases to bypass it. For these reasons, the front end and the catalyst containment casing are formed from AISI 409 stainless steel, listed in Table 8.3.

During short trips, the effect of the converter increases the condensation of acidified water in the exhaust rear section and mufflers associated with it. These items must be well protected; they are made from seam-welded mild steel protected by hot-dip aluminizing.

11.3.2   Cooling Systems

The cooling system in an automobile is a mixed metal closed circuit. The water ways in the engine block are through the iron or aluminum-silicon alloy from which the block is cast, the heat exchangers may be made from aluminum or copper sheet, thermostats can be of soldered copper bellows and the whole system is connected by rubber hoses. The system differs from closed circuit mixed metal systems described later in Section 13.5.4 in that the coolant in winter is not water but an antifreeze mixture of water and typically 25% ethylene glycol, $(CH_2OH)_2$.

The presence of glycol precludes the use of oxidizing inhibitors such as chromates and nitrites that attack it. The possibility of attack on rubber hoses eliminates some others. A mixture of inhibitors is needed to cope with the mixed metals system. A common system is 1% borax, $Na_2B_4O_7 \cdot 10H_2O$, to act as a mild alkaline buffer, pH 9, to passivate iron and steel with 0.1% mercaptobenzothiazole to inhibit cuprosolvency that can deposit copper on steel causing indirect galvanic stimulation.

The inhibitors are included in antifreeze sold at gas stations to make up the cooling fluid at the beginning of a winter. The system is at its most vulnerable at the end of the winter, when owners can change the coolant for water that is not inhibited. For this reason there is some merit in using antifreeze in summer as well as winter.

11.3.3   Moving Parts

Moving steel parts are protected by lubricating oil but with some reservations. The oil is not without some effects of its own and it has been shown to influence fatigue under cyclic loading. There is a form of damage sometimes found on mating surfaces of cams and tappets that is called pitting but is in fact surface fatigue damage. Automotive engine oil is formulated from a hydrocarbon base stock with additives for various purposes. To justify suspicion that surface fatigue was associated with oil formulation, laboratory fatigue tests on rotating steel samples were conducted in the hydrocarbon base stock of an automobile engine oil with separate additions of an alkaline detergent, a soot dispersant and an anti-wear agent.*
The tests showed that fatigue is sensitive to the composition of an oil environment as it is for aqueous media. The dispersant has no significant effect but the detergent improves the fatigue life and the anti-wear agent diminishes it, as shown in Figure 11.3. The detergent probably acts as a scavenger for unidentified acidic agents in the base stock persisting through oil

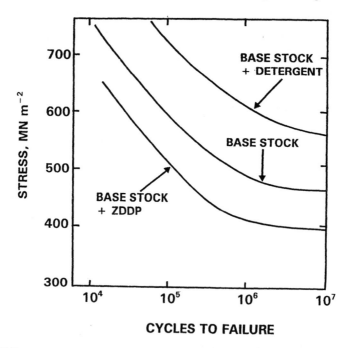

Figure 11.3
Influence of additives on the fatigue life of a case-hardened steel in an automotive oil. Anti-wear additive: 1.5 volume % zinc dipentyl dithio phosphate (ZDDP). Detergent: 5 volume % calcium sulphonate overbased with calcium carbonate. (Reproduced with permission of NACE, Houston, TX.)

---

* C. Chandler and D. E. J. Talbot, Corrosion fatigue of a case hardening steel in lubricating oil, *Materials Performance*, p. 34, 1984.

refining, a view supported by observations that fatigue life in the base stock oil is diminished if components are added simulating contamination of the oil by blow-by combustion products from the fuel, e.g., water, sulfuric acid, and chlorides, and restored if detergent is added. The effect of the anti-wear agent is probably associated with its film-forming capability that is responsible for its anti-wear function, perhaps introducing some form of cell at cracks in the film.

11.4 Bright Trim

Bright trim is not as widely used on automobiles as it was some years ago. It was popular for decoration and gave an integrated appearance to disparate accessories like steel fenders and diecast zinc door handles. Replacement of many of these items with plastics and change of fashion have reduced the applications of metal trim but it is still used to a significant extent for radiator grills, lamp bezels, wheel embellishment, and side trim; even so, plated plastics have made inroads on metal trim except for up-market automobiles. The standard for acceptable appearance is electrodeposited bright nickel/chromium systems or matching alternatives. One is polished ferritic stainless steels that need no explanation, except to note the expense. The other is brightened and anodized aluminum alloys.

11.4.1 Electrodeposited Nickel Chromium Systems

In these systems, corrosion protection is afforded by one or two thick coats of nickel. Although protective, nickel tends to dull on outside exposure and is notoriously prone to tarnishing in urban atmospheres polluted by sulfur dioxide. A very thin overlay of chromium electrodeposited on the nickel confers brilliance and tarnish resistance. Recalling discussions in Section 6.2.2, nickel deposited from standard acidic baths, developed from the Watts solution, do not adhere to steel or zinc, and must be preceded by copper strike coats deposited from alkaline solutions.

A problem in the design of the coating system is that the passive chromium is a galvanic stimulant to attack on the nickel. Bright chromium coatings are porous with exposed nickel sites at the bases of the pores that become anodes in a nickel/passive chromium cell. Formerly, only one chromium coating was applied and it was not uncommon to see scattered small brown spots of corrosion product on nickel/chromium plated fenders. The problem can be addressed by applying two chromium coatings; one is a standard bright porous coat and the other is a coating applied by a proprietary process that causes the plate to crack on a microscale. The pores and cracks do not coincide and so the coatings cover each others' defects.

A further feature is that both nickel and copper stimulate galvanic attack on a steel substrate, if exposed. To ameliorate the consequence of breaching the chromium coating, two nickel coats can be applied with small additions of other metals to make the outer nickel coat cathodically protective to the inner coat. This deflects any corrosion attack along the interface between the two nickel coats so that it does not reach the steel.

A typical complete system is:

1. A strike copper deposit from dilute alkaline cyanide solution.
2. A second thicker copper deposit from a more concentrated cyanide solution.
3. A deposit of pure nickel from a standard acidic solution.
4. A second deposit of nickel, cathodically protective to the first.
5. A deposit of porous chromium.
6. A deposit of microcracked chromium.

11.4.2 Anodized Aluminum

Acceptance standards for aluminum alloys for bright automobile trim are high and they are critical products based on relatively high purity, > 99.8%, aluminum to avoid dulling due to elements that yield intermetallic compounds, especially iron. A typical alloy contains 1% Mg with 0.25% Cu to give some ability to strengthen the otherwise soft pure metal by work hardening it. The metal must be homogeneous otherwise the differential brightness obtained over an inhomogeneous structure gives a banding effect, *bright streaking*. To avoid it, the cast ingots from which wrought products are fabricated must be held for a long time e.g., 24 h at high temperature, e.g., > 600°C to disperse segregates by diffusion. Another requirement for anodizing is efficient filtering of the liquid metal before casting to remove oxide inclusions that lead to disfigurement by *linear defect* that appears as scattered, short, fine lines parallel to the direction of fabrication.

Material used for car trim is chemically brightened and anodized in sulfuric acid, using low acid concentration, cool temperature and high current densities to produce anodic films that are bright yet hard enough to resist abrasion, as explained in Section 6.4.2.2.

Further Reading

Crane, F. A. A. and Charles, J. A., *Selection and Use of Engineering Materials*, Butterworths, Boston, 1984.

Fenton, J., *Handbook of Vehicle Design Analysis*, Institute of Mechanical Engineers, London, 1996.

Ettis, F., *Automotive Paints and Coatings*, VCH, New York, 1995

Freeman, D. B., *Phosphating and Metal Pretreatment*, Woodhead Faulkner, Cambridge, 1986.

12

Control of Corrosion in Food Processing and Distribution

12.1 General Considerations

Commercial food processing and distribution are competitive enterprises that must preserve the value of capital invested in plant and minimize the costs of products and packaging. To that extent, their needs to control materials degradation are similar to those of other industries but there are additional factors that introduce differences in approach:

1. Supply and distribution of food is subject to scrutiny by public health authorities and sensitive to consumer confidence.
2. Food product environments have special characteristics as corrosion environments.
3. The life expectancy of plant is many years but the life required of metallic packing depends on short-term retail turnover.

12.1.1 Public Health

Surfaces specified for contact with food products must not introduce toxic substances nor influence flavor. Materials that fulfil these requirements are glass, earthenware, some plastics, and a restricted range of metals. Glass is sensitive to shock and fragments are sharp. Inexpensive plastics are not heat resistant and in thin gauge they are permeable to gases. Metals do not have these disadvantages.

12.1.1.1 Possible Toxic Effects

Several metals are capable of resisting corrosion in food environments. Of the candidate metallic materials, long experience of tin-coated cans and of aluminum cooking utensils has established the non-toxic natures of tin and of aluminum in both the short and long term. Coated steel is often used for processing plant and for cans but there is little health hazard from iron exposed locally by breakdown of thin protective coatings. Stainless steels have been widely applied for equipment in large-scale food processing, so

339

that nickel, chromium and molybdenum as alloy components are considered safe; the same is true for magnesium as an alloying element in aluminum alloy beverage cans. Copper was used for food vessels before the advent of stainless steels and is now used extensively as piping to distribute domestic water supplies but it can have indirect effects, such as catalytic activity accelerating rancid degeneration of oils and fats. Metals such as lead and cadmium are precursors of very toxic substances and are prohibited from contact with food by health and safety regulations.

12.1.1.2 Hygiene

Food contains beneficial bacteria but the entry of harmful bacteria that cause food poisoning from external sources must be prevented not only by regular removal of food residues and disinfection but also by designing product contact surfaces so that they do not offer sites for these harmful bacteria to colonize, such as accumulated corrosion products and degraded protective coatings. In this respect, austenitic stainless steels and aluminum alloys are well-suited to the construction of process plant and permanent containers, such as beer kegs. For service in most food environments these metals require no coatings; they are strong, ductile, weldable and have acceptable casting characteristics. They can be formed into large complex components with smooth easily cleaned surfaces, joined by welding to eliminate crevices.

An obvious aspect of hygiene is sterilization. A common disinfectant contains sodium hypochlorite, $NaClO$, and is manufactured by chlorinating sodium hydroxide:

$$Cl_2 + 2NaOH = NaClO + NaCl + H_2O \tag{12.1}$$

The product of the oxidizing reaction that kills microorganisms is sodium chloride:

$$NaClO \rightarrow NaCl + [O] \tag{12.2}$$

where the symbol [O] indicates nascent oxygen that kills biological material by oxidation.

Sodium chloride and sodium hypochlorite are easily washed away from food contact surfaces by rinsing with water but none must remain on any internal or external part of the metal surface, because residual chlorides and hypochlorites concentrated by evaporation can induce pitting corrosion, as explained in Section 8.3.1.4. The chloride ions are the primary pitting agent and the hypochlorite ion can impose a redox potential that can be more positive than prevailing pitting potentials for stainless steels by the reaction:

$$ClO^- + H_2O + e^- = Cl^- + 2OH^- \tag{12.3}$$

for which: $E^{\ominus} = +0.90$ V (SHE)

12.1.2   Food Product Environments

Foods are usually associated with slightly acidic water, with pH values mainly in the range pH 3 to 8, as illustrated by examples given in Table 12.1. The generally satisfactory behavior of aluminum, tin and stainless steels within this pH range is consistent with the information given for aluminum and tin in Figures 3.4 and 3.7 and for stainless steels in Figures 8.6 and 8.7.

Features of food and associated materials that distinguish them from many other corrosion environments are the activities of microorganisms, especially bacteria, the presence of emulsions and interplay between organic and inorganic components.

TABLE 12.1
pH Ranges for Some Common Foodstuffs

Fruits		Vegetables	
Apples	3.0–3.5	Potatoes	5.5–6.0
Oranges	3.0–4.0	Cabbage	5.0–5.5
Lemons	2.0–2.5	Carrots	5.0–5.5
Tomatoes	4.0–4.5	Beet	5.0–5.5
Raspberries	3.0–3.5	Pickles	3.0–3.5
Dairy Products		**Beverages**	
Cow's Milk	6.0–6.5	Beers	4.0–5.0
Butter	6.0–6.5	Fruit drinks	2.0–4.0
Cheeses	5.0–6.5	Wines	3.0–4.0
Meat and Fish		**Bakery**	
Fish	6.0	Bread	5.0–6.0
Meats	7.0	Wheat Flour	5.5–6.5

12.1.2.1   Bacteria in the Environment

Bacteria are primitive single-cells bounded by rigid walls, typically spheres 0.5–1 µm in diameter or rods 1–5 µm long, that live in aqueous media. They may exist as isolated individuals or as clumps or sheets. They cannot produce their nutrients by photosynthesis but must obtain them directly from solution, transforming them by their metabolism into other substances returned to solution. The range of nutrients needed includes a source of energy such as glucose and small quantities of minerals for the cell structure, such as nitrogen, phosphorous, and sulfur. The pH of the medium and concentrations of ions e.g., Cl^- within it are important factors affecting the growth, activity, and death of the microorganisms; for example, a bacterium associated with milk, *Lactococcus lactis*, grows best at pH 3–4 and will not grow in a media of pH > 5.

Bacteria are agents responsible for many different chemical effects corresponding to the vast number of varieties. Some reduce sulfates to sulfides and others oxidize sulfides to sulfates, some thrive in chloride media and others do not, some reduce nitrates, some utilize hydrogen, etc. These chemical transformations can occur quite rapidly because:

1. Bacteria reproduce by successive binary division i.e., single cells divide in two, so that they multiply rapidly. A single cell can potentially produce millions of progeny per day, although division stops if nutrients are exhausted, waste products accumulate or all of the available space is occupied.

2. A colony of bacteria has a very high surface to volume ratio because of the minute size of the individuals.

Bacteria are ubiquitous and according to type they are responsible for decay, putrefaction, soil fertility, some diseases of plants and animals, etc. The effects are produced only while the microorganisms are growing but they can lie dormant for a long time. Bacteria can intervene in corrosion and protection in various ways. The following briefly summarizes some general effects.

Environmental Changes

Bacteria can change the environment by replacing one substance with another. For example, *Streptococcus lactis* sours milk by converting lactose to lactic acid, reducing the pH.

Local Action Cells

By their physical presence in clumps or films, bacteria can screen metal surfaces, stimulating differential aeration and concentration cells e.g., through local oxygen starvation.

Depolarization

Some sulfate-reducing bacteria produce an enzyme, *hydrogenase* that enables them to utilize hydrogen in a process summarized by the equation:

$$SO_4^{2-} + 4H_2 \rightarrow S^{2-} + 4H_2O \tag{12.4}$$

This can accelerate corrosion by depolarizing the hydrogen evolution cathodic reaction if it is under rate-control by a film of hydrogen at the metal surface in anaerobic conditions.

Degradation of Protective Mechanisms

Foods can contain ascorbic acid (Vitamin C) and folic acid, both of which are corrosion inhibitors for iron and other metallic materials but they are susceptible to depletion in various ways, one of which is by bacterial metabolism. Bacteria can also degrade protective coatings such as those containing cellulose materials.

12.1.2.2 *Emulsified Environments*

Many foods subject to industrial processing, such as milk, milk products and soft margarine, are two-phase emulsions containing small micelles of fats and oils in water. They can sustain electrochemical activity in the water phase but they also have characteristics conferred by the micelles that carry electric charges. A further factor is that in some food processing the emulsions can separate into bulk phases, as in curds and whey.

12.1.2.3 *Interaction Between Inorganic and Organic Solutes*

Food solutions contain organic solutes that can modify the behavior of inorganic ions. One example, quoted earlier in Section 4.1.3.6 concerns polarity reversal of iron-tin bimetallic couples. From their relative standard electrode potentials, given in Table 3.3, tin might be expected to stimulate corrosion on iron in a bimetallic couple and this is found to be true for some purely *inorganic* solutions such as sodium chloride solutions but in many solutions of *organic* acids found in foods such as fruit juices, milk and milk products the polarity of the couple is reversed so that tin coatings galvanically protect steel exposed at defects. Another example, described later relates to intensification of pitting corrosion of stainless steels due to the simultaneous presence of chloride ions that stimulate corrosion and organic solutes such as folic and ascorbic acids that inhibit it.

12.1.2.4 *Corrosion Control in Practice*

Corrosion presents few problems during short-term contact of foods with suitable metals as in cooking and associated preparation provided practices are hygienic. Aluminum, the aluminum-manganese alloy, AA 1100, and AISI 304 stainless steel are standard materials for utensils and contact surfaces. Martensitic stainless steels, e.g., AISI 431, are standard for knives. Steel coated with an iron-tin alloy layer gives good service as ovenware.

Proactive corrosion control, especially in materials selection, is required in commercial processing and for long term contact in packaging. Commercial enterprises include dairying, brewing, baking, and production of sugar, margarine, soft drinks, and pre-cooked convenience foods. Dairying and brewing are featured later in this chapter because between them they exemplify the contexts within which corrosion control must be exercised. Canning is a means of distribution common to most of them and is responsible for by far the greatest area of food/metal contact. Tin-coated steel is the dominant material but more is involved in its production and use than simply applying the coating to steel. It is produced by special technologies to secure corrosion resistance and to enable it to survive the severe deformation required to fabricate cans meeting public health standards at a cost that the market will bear.

12.2 The Application of Tinplate for Food and Beverage Cans

12.2.1 Historical

Tin is one of the oldest available metals and centuries of use have established its non-toxic nature. Its inherent corrosion resistance in neutral and mildly acidic aqueous media and its low melting point, 232°C, facilitating hot-dip coating of steel made it a natural choice as a protective coating for steel food-handling equipment and containers. Historically, single steel sheets or continuous steel strip were passed through a flux to remove the oxide barrier and thence into a bath of molten tin and the surplus tin was wiped off by pads on exit. Cans made from this material had proved their value in safely preserving perishable foods some sixty or seventy years ago, when household refrigeration was not generally available. The first successful mass-produced can was a three-piece can in which the body was formed by rolling a tin-coated steel sheet blank into a ribbed cylinder with an overlap side-seam sealed by a lead-tin soldered joint. The circular base was clinch-sealed to the cylinder and the top was also clinch-sealed to it, after filling the can.

Continuous electrodeposition replaced hot-dipping because thinner coatings can be applied, economizing in tin, which has a high and volatile price, currently about $6000/tonne. Techniques for steel strip production, for electrodeposition, and for the manufacture of cans subsequently evolved to support a massive increase in production, the introduction of the pressurized beverage can and competition from other container materials.

The preservative function of cans has been overtaken by their convenience as food and beverage dispensers, in supermarket shopping. Consequently, can production has attained a volume in which the can is the most prolific human artifact ever produced. This has created intense competition, not only between manufacturers of tinplate but also between tinned steel and other materials, notably chromized steel, aluminum alloys and polymers. Nevertheless, production of tinplate for cans remains buoyant. Currently, billions of cans are produced annually from tinplate manufactured in the U.S. alone.

12.2.2 Modern Tinplate Cans

Modern tinplate for cans is produced from steel strip with tin applied by continuous electrodeposition. There are three can types: (1) the three-piece can evolved from the original concept, (2) the two-piece draw/redraw (DRD) can, and (3) the draw/wall-ironed (DRW) can, in which the terminology refers to the methods of fabrication. Tinplate for these cans is customized to economize in material costs, consistent with its fitness for purpose, including corrosion resistance, adequate strength and amenability to can fabrication and filling.

The mass of tin on each side of the tinplate is 1.4–11.2 g/m^2 of steel surface, corresponding to tin thicknesses of 0.2–1.5 µm. As deposited, the surface of the electrolytic tin is dull gray and standard practice is to melt the tin momentarily to produce a bright finish ("flow melting"). This operation consumes some of the tin by introducing an alloy layer of the compound FeSn$_2$ between the steel and the remaining free tin of the coating. Tinplate with < 2.8 g/m^2 of tin per surface, corresponding to a tin thickness of 0.4 µm is designated "low tin-coated" (LTS) steel. Continuing competitive pressure forces can manufacturers to use the thinnest tin coatings where possible, for which all of the tin is consumed in flash melting, so that the protective layer is wholly FeSn$_2$.

All tinplate is passivated in aqueous Cr(VI) solutions, usually cathodically, i.e., by application of a negative potential to the strip. This improves the corrosion resistance and provides a key for lacquer to improve it further. For the thinnest tin coatings on steel, the passivated surface and lacquer are the main defence against corrosion. The lacquer coat is applied by spray to the inside surfaces of finished cans, because it would not survive the severe deformation of the steel in the can forming operations.

12.2.2.1 Three-Piece Cans

Three-piece cans remain in service for non-pressurized food contents. The high volume of production, the drive for economy in tin and concern for health hazards from the lead content in solder led to the replacement of soldering by overlap welding to seal the side-seams. This operation is done automatically and a stripe of lacquer is applied to cover the narrow heat-affected zone, 1 to 2 mm wide, where protection is broken down by welding.

12.2.2.2 Two-Piece Draw/Redraw Cans (DRD)

An alternative to the three-piece can is the draw/redraw can. Deep-drawing is a press-forming operation for producing cylindrical hollow ware from flat metal sheet. A flat circular blank is sheared from the sheet and gripped by a blank-holder. It is pressed by a punch through a die, the *draw-ring*, of the diameter required for the hollow shape, while partially restrained by pressure applied to the blank-holder. This pressure is adjusted to allow the whole of the blank to enter the die, with sufficient restraint to avoid wrinkling. The blank pressure, punch and draw-ring radii, clearance between the punch and die and the friction between the base of the punch and the metal are all interacting dependent parameters, which must be carefully optimized to ensure that a smooth shape is produced without fracture. The cylindrical walls of the hollow shape are strengthened by work-hardening as the blank is drawn into the die and over the die-ring, enabling it to resist fracture under the tensile stress imposed by the load on the punch. Friction is needed at the base of the punch to restrain the unhardened metal below it from slipping into the cylindrical walls where it can fracture under the applied load. Successive deep-drawing operations through dies of progressively smaller diameter are needed to produce the cans; hence the term draw/redraw. The can lid is formed separately and is crimp-sealed to the can after filling. The corrosion protection conferred by the tin coating must survive this severe fabrication sequence.

12.2.2.3 Draw/Wall-Iron (DWI) Cans

The draw/wall-iron can was developed to compete with aluminum alloys for pressurized beverages containers, which have become the highest growth market for cans, due to their popularity for dispensing drinks.

Beverage containers are designed to have the requisite strength with the absolute minimum of material. The rigidity of the very thin-walled can is due to the internal gas pressure from the contents. The cylindrical walls can support the internal pressure but for matching support, the base of the can must be of thicker metal and formed into an internal dome. A further consideration is a specified column strength of the open can, about 91 kg, to resist the force applied to the can during filling.

The objective of the forming operations is to produce thin-walled cans of complex shape with a high aspect ratio. A shallow cylindrical hollow shape is first drawn from a flat blank It has the same internal diameter but a lower aspect ratio and thicker walls than are required in the final product. The basic form of the can is then developed by wall-ironing the cup in which it is forced through a series of ironing dies of progressively smaller diameter while supported internally by a punch. The effect is to reduce the wall thickness, stretching the cylindrical walls to their full height and work-hardening the metal. Other operations are integrated with the wall-ironing operation to impart other required features to the can shape, e.g., the base of the can is domed by pressing it against a convex bottom end tool, mating with a matching concavity in the base of the punch.

12.2.3 Steel Base for Tinplate Manufacture

12.2.3.1 Steel Grades for Three-Piece Cans and Non-Critical Parts

Competing aluminum alloys are well-suited to draw/redraw and draw/wall iron techniques and have tear characteristics needed for the easy-open end but their material cost is high and their economic use depends on a premium price for the lid material and the return of used cans for recycling. Tinplate has an immediate advantage in material cost but special grades are needed to match the performance of aluminum alloys in this field.

Steel normally contains carbon (C), manganese (Mn), phosphorous (P), sulfur (S) and silicon (Si), derived from its iron precursor produced in the blast-furnace. Steelmaking processes are available, which can reduce any or all of these elements to very low contents, if required, but at an appropriate cost. Steel specifications for tinplate manufacture typically include the following maximum composition limits:

C 0.13%, Mn 0.6%, P 0.02%, S 0.05%, Si 0.02%

Sulfur is deleterious in steels but sulfur contents of up to 0.02 % are normal in low carbon steels for general purposes, because ultra-low sulfur contents incur extra cost in steelmaking. The function of the manganese is to scavenge the sulfur present in the form of manganese sulfide to avoid hot embrittlement due to iron sulfide. Steels with normal sulfur and manganese contents are suitable for forming three-piece cans but residual manganese in solution in the metal impairs the deep-drawing characteristics of the steel required for drawn and wall-ironed cans.

12.2.3.2 Steel Grades for Drawn and Wall-Ironed Cans

The composition limits given in Section 12.2.3.1 are maxima and tinplate manufacturers are at liberty to reduce them. If the manganese content is reduced to > 0.2%, the deep-drawing characteristics of the steel improve significantly. This requires a sulfur content of > 0.005%. In Japan, some steel is made with an ultra-low carbon content, 0.005%, as well as an ultra-low sulfur content, 0.003%. The particular compositions produced for draw/redraw and draw/wall-iron cans are probably subject to commercial confidentiality. Brochures describe deep-drawing quality steels as "low in metalloids and residual elements".

12.2.3.3 Steel Form, Thickness, and Tempers

Tinplate is produced in nominal thicknesses from 0.16–0.60 mm. It is supplied to can makers, in a variety of aesthetic finishes, as cut sheets or as coils of continuous strip up to a meter wide, weighing several tonnes.

To suit various applications, the steel base of tinplate is manufactured to different *tempers* ranging from soft deep-drawing steels to harder steels needed for pressure packs such as carbonated drinks. Temper is designated by a symbol, such as T1 (soft) to T4 (hard and strong). (ISO R1111 Parts 1 & 2 and Euronorms 145-78 & 146-80). The methods by which different tempers are produced are described in Section 12.2.4.2.

12.2.4 The Manufacture of Tinplate

Successful application of the tin coating on steel and the ability of the material to withstand the deformation in can manufacture depend critically on all aspects of the steelmaking and processing operations. These are extensive topics dealt with in specialized texts and the following description is only a brief summary.

12.2.4.1 Steelmaking and Ingot Casting

Most tinplate is made from steel produced from blast furnace iron, which contains quantities of carbon, manganese, phosphorous, sulfur, and silicon. Carbon is usually required in the steel but in very much smaller quantities (0.01 to 0.6%) than are present in the iron (typically 4%). Phosphorous and sulfur are deleterious and the quantities present are reduced to as low a level as is cost-effective for the intended application of the steel. For non-critical applications, 0.05% sulfur and 0.02% phosphorous are acceptable.

Carbon, phosphorous and sulfur contents are reduced to values suitable for general steels by *pneumatic converter processes,* conducted in 100 to 300 tonne vessels in which oxygen gas is blown into the iron either at the bottom, (QBOP process, used in the U.S.A.) or at the top (BOF process, used in Europe and Japan). When the desired composition is reached, the steel is poured into a ladle, where the oxygen remaining in the steel is removed by adding a calculated quantity of aluminum metal. To attain the low sulfur contents for deep-drawing cans, as described in Section 12.2.3.2, additional treatment is needed. This can be done by injecting magnesium metal into the steel while it is covered by a slag rich in lime or by a *secondary steelmaking process,* in which a vacuum is applied to the ladle and at the same time, argon and lime are blown in at the bottom.

The finished steel is cast into a rectangular section for rolling to strip. Almost all steel for tinplate production is solidified as a *continuously cast strand,* typically 20 cm thick. In this method, the liquid steel from the ladle is poured into a short water-cooled mold open at the bottom, from which the solid metal is withdrawn.

12.2.4.2 Strip Rolling and Annealing Sequences

Hot-Rolling

The continuously cast strand is cut into lengths, each of typically 10 tonne mass, preheated to 1100°C and hot-rolled to typically 2-mm thick hot-line gauge. The purposes in rolling the metal hot are (1) to reduce the power required, and (2) to break up the crystal structure and to disperse segregate characteristic of cast metal. The reduction in thickness produces a corresponding increase in length and on completion of hot rolling, the material is coiled to handle it for further processing. A most important consideration is that the hot-finishing temperature, i.e., the temperature of the final gauge strip is kept above 900°C. Below this temperature, a phase change occurs in the steel; the high temperature form of the iron-carbon alloy, *austenite,* changes to the low temperature form *ferrite*. If hot-rolling were to continue when ferrite is present, the steel would develop anisotropic properties and would not deform uniformly when subsequently drawn into cans.

Pickling

The metal is cold-rolled from the hot-line gauge to the final gauge, which for tinplate is one of several standard gauges in the range, 0.60 to 0.16 mm. The reason for cold rolling is to gain accurate control of the final gauge and to produce a smooth bright surface finish.

The first task is to remove surface oxides, which are inevitably present when hot steel is exposed to the atmosphere, as in hot-rolling. This is done by passing the steel strip through dilute sulfuric acid to dissolve the oxides, an operation called *pickling*.

$$Fe_3O_4 + 4H_2SO_4 = FeSO_4 + Fe_2(SO_4)_3 + 4H_2O$$

Although this is an apparently trivial process, it requires expensive complex plant with extensive floor space, because of the need to handle strip and to operate continuously. A typical plant that can process a 10 tonne coil every five minutes includes:

1. Arrangements to unwind the steel supplied as coils.

2. A welding station to join coils together as they follow in sequence, with a looping pit to allow the ends of the coils to stop for welding, without interrupting the steady flow of material in the rest of the line.

3. The active part of the line, in this case, several shallow tanks of 10% sulfuric acid in series, followed by a water spray, a hot water wash tank and a hot-air dryer to remove the acid from the steel before recoiling. To use the acid economically, it flows counter-current to the strip, i.e., fresh acid is delivered to the last tank and is transferred from tank to tank until it reaches the first tank, from which it is discharged.

4. Arrangements to rewind the strip and shear it into separate coils for onward transmission to the next process. Another looping pit is inserted to allow the strip to be stopped momentarily so that it can be sheared.

Cold-Rolling

The strip is cold-rolled in one operation, using a 5-stand 4-high tandem cold mill. *5-stand* means that the strip is uncoiled, passed through five pairs of steel rolls in sequence, and recoiled; the speed of the strip increases between stands and the roll speeds are adjusted to suit. *4-high* means that each pair of *work rolls* contacting the steel strip is backed by a second pair of rolls for support, permitting the use of small diameter work rolls, which produce even deformation through the metal thickness.

The metal slips forward on the rolls because the exit speed exceeds the entry speed as the thickness is reduced and the work generates heat. Mineral oil is supplied to the roll gap as a lubricant and coolant and must be meticulously removed before the strip is processed further. If this were neglected, the next operation, annealing, would stain the metal, interfering with the eventual application of the tin coating. To accomplish this, the strip is passed through an alkaline cleaning line, scrubbed, hot-rinsed, and dried, using a continuous system similar in concept to that described for pickling. It is then either sent directly for continuous annealing or re-coiled for batch annealing, as described below.

Annealing and Temper-Rolling

The cold-rolled metal is unsuitable for can production because it is work-hardened and so it must be annealed. Annealing is a controlled heating process, which relaxes the deformed structure responsible for work-hardening and restores ductility. One or the other of two different annealing techniques is used, depending on the requirements of the end application. These are (1) batch annealing and (2) continuous annealing.

Batch Annealing

Batch annealing is generally used for softer and more ductile products and is ideal for drawn can bodies. The coils are stacked typically 4-high on an insulated base and separated by convector plates to facilitate even heat distribution. A cover is placed over the stack and filled with a mixture of hydrogen and nitrogen, providing an inert gas mixture to shield the metal from oxidation in the atmosphere. This particular gas mixture can be obtained cheaply by thermal decomposition of ammonia. A furnace hood is lowered over the shielded coils and the covered stack of coils is heated by gas burners. The stack is heated to 600 to 640°C and allowed to cool in the inert gas; the cycle takes 48 hours.

Continuous Annealing

In continuous annealing, the steel strip moves through the process in an uninterrupted flow, accomplished by welding the ends of successive coils together and feeding the strip into a looping tower. This allows the strip to be stopped temporarily to permit welding while paying out strip from the buffer stock in the looping tower at a constant speed through the subsequent heating and cooling sections. The strip is heated to 650°C in an inert gas atmosphere but the duration of heating is much shorter than in batch annealing and any given point on the strip passes through the heating and cooling cycle in < 1 ½ minutes.

Continuously annealed strip is stiffer and harder than its batch annealed counterparts and is used for three-piece can ends and bodies and for some two-piece cans. The extra strength of continuously annealed strip allows some reduction in the gauge used for cans without loss of performance.

Temper-Rolling

Another method for increasing the strength of the strip is to impart a controlled degree of work-hardening by cold-rolling it after annealing. Such strip is known as double reduced (DR) strip. The increased strength is obtained at the cost of some loss of ductility. For applications which do not require the metal to withstand severe deformation, this procedure also permits a reduction in gauge.

12.2.4.3 Electrodeposition, Flow-Melting, and Passivation

In the manufacture of tinplate, the acid sulfate and halide baths are used extensively and the fluoborate bath is used only occasionally. The three operations, electrodeposition, flow-melting and passivation are carried out continuously in sequence in the same strip line. As with other operations described earlier, the line has looping storage devices to allow continuous operation while the ends of successive coils are welded together for entry and finished tinplate is re-coiled on exit.

Before entering the tinning section, the strip is meticulously cleaned and pickled to present a good surface for tin-coating. The tinning section consists of a series of tanks holding the electroplating bath, through which the strip passes at a typical speed of 6 m s^{-1}. The anodes are bars of high purity tin and the cathode is, of course, the strip itself.

If a bright finish is required, as it usually is, the tin-coated strip passes through a heated section, raising the temperature of the strip momentarily above the melting point of tin, 232°C, to flow-melt the tin coating. The final stage is the passivation treatment, in which the strip is passed through an aqueous Cr^{VI} solution, usually cathodically, i.e., by applying a negative potential to the strip. If required, an additional plating section may be incorporated in the line to produce a complementary product, "tin-free steel" described in Section 12.2.5.

Before recoiling for delivery to can fabricators, the strip is inspected and passed through an automatic on-line monitoring station to detect "pin-holes" in the coating that would compromise its corrosion resistance and deviations from the prescribed gauge of the strip and the mass of tin coating.

12.2.5 Tin-Free Steel for Packaging

Tinplate is still the principal material used for packaging processed food and other products but alternative coatings are becoming available. Since very thin tin coatings are completely converted to FeSn$_2$ on flow-melting, requiring chromizing passivation and lacquering for corrosion resistance, it is natural to consider eliminating the electrodeposition of tin and chromizing the steel directly to provide a suitable substrate for lacquer.

This reasoning stimulated developments which led to electrodeposition of chromium-bearing coatings. The dominant process is a proprietary process delivering a chromium coating consisting of chromium metal and nontoxic chromium^{III} oxide in the ratio 2 or 3:1 The chromium metal is adjacent to the steel with the oxide in a layer above it, forming the key for lacquer. An advantage of the process is that the steel base for the coating and the processing requirements are similar to those for tinplate, so that passivated tinplate and chromium-coated steel can be produced as complementary products in the same plant.

The electroplating solution is derived from industrial chromium metal plating solutions based on chromic acid anhydride, $Cr^{VI}O_3$. It is more dilute and is operated at a much higher current density. Although well-suited to its purpose for can stock, the deposit is different from chromium metal deposits produced for decoration or industrial hard-facing.

A typical formulation with operating parameters is:

Chromic acid anhydride, CrO_3	50–180	g dm^{-3}
Sulfuric acid, H_2SO_4	1	g dm^{-3}
Fluoboric acid, HBF_4	0.8	g dm^{-3}
Temperature	40–55	°C
Current density	50–100	A/dm^{-2}

A continuous strip speed of 6 m s^{-1} can be accommodated, allowing a process time of 1 to 8 s. The coatings are 0.01 to 0.03 μm thick, which is much thinner than the 0.4 μm thick low-tin coatings they replace. The corrosion resistance of the more economic $FeSn_2$ and tin-free coatings is not so good as for metallic tin but it is sufficient for many food products.

12.3 Dairy Industries

Dairy industries treat and distribute milk, use it as the feedstock for other foods, principally butter, cheese and yoghurt and extract residual values as by-products. Corrosion control of the plant depends on the natures of these products and on the processes by which they are produced.

12.3.1 Milk and Its Derivatives

12.3.1.2 Constitution

Cows' milk is a complex aqueous medium containing fats, proteins, and carbohydrates, with small but important contents of minerals, vitamins, enzymes, and bacteria; the composition varies and Table 12.2 is a simplified example. The *fats* are present as an emulsion of minute globules. *Casein*, the most abundant protein, is combined with calcium and is dispersed in the liquid phase as a colloidal suspension of gelatinous particles. *Lactose* is a sugar in aqueous solution and is the nutrient for bacteria in fermented products. The bacteria *Lactococcus lactis* and *Lactococcus cremoris*, naturally present in milk, progressively convert it to lactic acid during souring. *Calcium and phosphorous* are present as minerals vital for nutrition and bone formation. *Vitamins A and D* are fat soluble and *Vitamins B_1, B_2, B_{12}, C, and M* are water soluble. *Vitamins B_1 and C* are unstable when the milk is heated.

TABLE 12.2
Representative Composition of Natural Milk

Category	Weight %	Principal Constituents
Fats	3.8	Trihydric glycerol esters Phospholipids Other fatty acid derivatives
Proteins	3.3	Casein Lactoalbumin Lactoglobulin
Carbohydrate	4.7	Lactose
Mineral ions	0.75	Ca^{2+}, Mg^{2+}, Na^+, K^+, PO_4^{3-}, Cl^-, $C_6H_5O_7^{3-}$ (citrate)
Vitamins	Traces	A (retinol) B_1 (thiamin) B_2 (riboflavin) B_{12} (cyanocobalamin) C (*l*-ascorbic acid) D (a steroid derivative) M (folic acid)
Enzymes Bacteria Water	 Balance	

12.3.1.1 Processing

Operations using milk as a feedstock are summarized in the typical flow sheet given in Figure 12.1. They start by fractionating the milk by one or other of two methods:

1. Centrifugal separation based on differences in density of milk components. This yields a fat-rich fraction, cream, and fat-free *separated milk*.

2. Curdling induced by adding *Chymosin* at 32 °C, a synthetic copy of the active component in natural rennet. This breaks down casein, yielding a colloidal mass, *curds* and a watery liquid, *whey*.

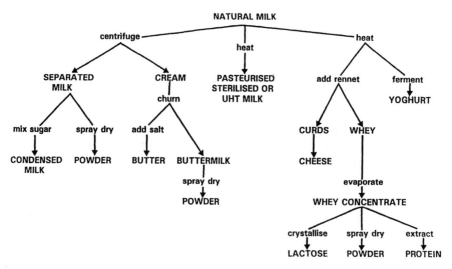

Figure 12.1
Simplified flowsheet for milk processing.

Traditional Dairy Products

The fat-rich fraction from centrifugal separation contains 50% fat with 50% of retained liquid and the fat soluble vitamins; it is marketed as creams or churned into butter releasing the retained liquid as buttermilk. Curds contain most of the fat, casein, colloidal calcium phosphate, and bacteria; they are extracted from the whey and pressed and processed to cheese. These traditional dairy activities are not particularly demanding on the corrosion resistance of metallic contact materials.

Evaporation of Separated Liquors

Separated milk can be sold as a product, concentrated by evaporation and sweetened for sale as condensed milk or spray-dried to powder. Buttermilk is spray-dried to powder.

Recovery of Proteins from Whey

Whey retains the proteins lactalbumin and globulin in colloidal solution, the lactose, the water-soluble vitamins l-ascorbic acid (vitamin C), folic acid (vitamin M), thiamin (vitamin B_1), riboflavin (vitamin B_2), and vitamin B_{12} and the minerals as listed in Table 12.3. It can be spray dried and disposed as low value material for such uses as animal feed. A better alternative is to recover the high quality protein content and the lactose. In one process the whey is treated with hydrochloric acid, concentrated by reverse osmosis and filtered through a membrane with pores of the order of 1 μm.

TABLE 12.3
Representative Composition of Whey

Category	Weight %	Principal Constituents
Fats	0.01	—
Proteins	0.6–0.7	Lactoalbumin
		Lactoglobulin
Carbohydrate	4–5	Lactose
Mineral ions	0.3	$Ca^{2+}, Mg^{2+}, Na^+, K^+, PO_4^{3-},$
		$Cl^-, C_6H_5O_7^{3-}$ (citrate)
Vitamins	Traces	B_1 (thiamin)
		B_2 (riboflavin)
		B_{12} (cyanocobalamin)
		C (*l*-ascorbic acid)
		M (folic acid)
Enzymes		
Bacteria		
Water	Balance	

12.3.2 Materials Used in the Dairy Industry

To limit contamination from external bacteria, product contact surfaces of processing equipment must be hygienic. Materials, surfaces, construction of plant, cleanability and inspection are considered in publications by the International Dairy Federation.* The recommendations include the use of surfaces that are self-resistant to corrosion and robust enough to withstand cleaning and sterilization. Organic coatings such as paints are not approved because of degradation by bacterial attack and difficulty in cleaning.

12.3.2.1 Surfaces in Contact with the Product

Metal surfaces in contact with the product must satisfy the following requirements: (1) low toxicity and physiological indifference, (2) corrosion-resistance to the products and to cleaning and disinfecting solutions, and (3) readily availabile in standard forms at acceptable cost.

Associated materials contacting metals, including rubbers, plastics, ceramics, and carbon for gaskets, rotary seals, bearings, and other adjuncts are subject to the same provisos. Care is required to ensure that seals and gaskets fit well and do not offer sites for crevice corrosion.

Numerous materials have been suggested but austenitic stainless steels are in general use for the reasons given in Section 12.1.1.2. AISI 304 type steels are standard materials in general use for milder environments, such as milk storage vessels and separators; the stabilized versions, AISI 321, AISI 347 or AISI 304L, listed in Table 8.3, are clearly preferable for on-site welded structures where weld decay might otherwise be expected.

In filtration, evaporation and spray-drying plants, various combinations of higher temperatures, higher pressures, diminished solubility for oxygen, higher concentrations of e.g., chlorides in feed fluids, oxygen

---

* *International Dairy Federation Bulletin*, 1987, No. 218.

starvation under deposits can produce environments that are too aggressive for AISI 304 steels and must be resisted by using the more expensive molybdenum-bearing austenitic steels, AISI 316 or AISI 316L.

One particular environment that is too aggressive for even AISI 316 is the intermediate liquor obtained after treating whey with hydrochloric acid to recover the proteins. Stainless steels do not exhibit an active range at the prevailing pH, 4 to 5, but there is a chloride concentration of 0.07 M that introduces pitting potentials below the potential for dissolved oxygen in equilibrium with air. The problem is made worse by the presence of folic acid in whey that obstructs repassivation of pits; it does so by increasing the potential difference between active pit sites and the general passivated surface. This is manifest in marked hysteresis in the polarization characteristics, illustrated and explained in Figure 12.2. One approach is to replace AISI 316 with duplex stainless steels that do not exhibit this hysteresis; another is to use an alternative process that does not introduce chloride ions.

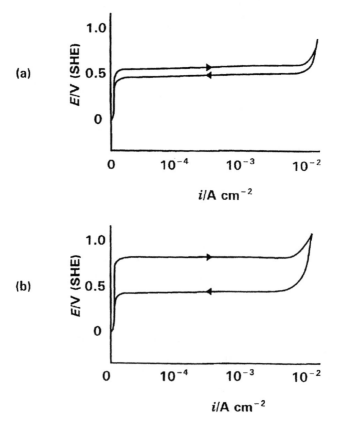

Figure 12.2
Polarization characteristics of AISI 316 stainless steel in 0.07 M sodium chloride solutions in (a) water, (b) whey, both of pH4, for rising (→) and falling (←) potentials. The steel exhibits a higher pitting potential with more hysteresis in the whey solution than in the corresponding water solution.

12.3.2.2 Surfaces Not in Contact with the Product

Stainless steels are also preferred over less expensive materials for surfaces not in contact with the product. For example, the enclosed space between double walls in jacketed and insulated equipment must be sealed against entry of water vapor and bacteria. Mechanical joints do not provide such reliable seals as welding and this implies that the outer wall is of the same metal as the inner wall in contact with the product. If, for economy, other metals are used in proximity to stainless steels, precautions must be taken to avoid galvanic stimulation. Passivated stainless steels are more noble than most common metals and stimulate attack on them. For example, uninformed use of other steels, such as plain carbon steels for less critical parts contacting stainless steels is a false economy.

12.4 Brewing

12.4.1 The Brewing Process

The raw materials for beer are barley, hops, yeast and water. Important by-products are surplus yeast processed for health foods and products, spent barley sold for animal feed and spent hops used for fertilizer. A flow sheet for a typical large brewery is given in Figure 12.3. The barley is *malted*, i.e., allowed to germinate, during which it secretes an enzyme, diastase, that is needed for conversion of starch in the grains to the sugar, *maltose* in the next stage. The malt is pulverized to grist that is accumulated for use in *grist hoppers*, from which it is fed as required. The grist is *mashed*, i.e., steeped in hot water for an hour or two in *mash tuns*, to effect the conversion of starch to maltose, producing a sweet liquid called *wort*. The water for the process is filtered through activated carbon to remove pesticides and its content of ions such as nitrates is reduced by reverse osmosis or ion exchange resins.

The wort is drained off into a large vessel called the *copper* and the spent grains are retrieved and conveyed to large hoppers, awaiting transport to manufacturers of animal feed. Hops are added to the wort in the copper and the mixture is boiled for an hour or two by heat supplied through steam coils. This serves two purposes; the boiling stops the starch to sugar reaction and infusion of hops gives the beer its characteristic flavor. It is discharged into a *whirlpool* separator that retrieves the spent hops by suction to the center of the vortex. The separated liquid wort is cooled through a heat-exchanger to ambient temperature, aerated to sustain the growth of yeast and transferred to fermenting vessels, where the yeast is added and the brew is allowed to ferment at about 17°C for 40 to 60 hours.

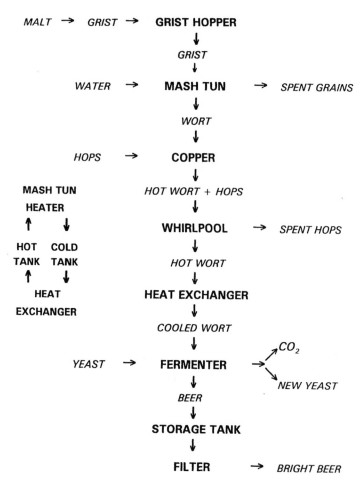

Figure 12.3
Schematic brewery flow sheet. Inset: heat-recovery circuit.

This generates the alcohol and liberates carbon dioxide gas that is recovered, purified and liquified for use in pressuring beer in cans or barrels. The beer is cooled and the yeast is removed and recovered by centrifuging. Part of the yeast is recycled for use in subsequent brews and the surplus is sold as a valuable protein rich ingredient for human and animal foods. The beer is allowed to mature in tanks, then filtered yielding bright beer that is stored in holding tanks from which it is drawn as required, pasteurized by heating and distributed in cans, bottles, barrels or mobile tanks.

An essential adjunct to the process is a water circulation system to conserve energy by recycling heat. Cooling water, running countercurrent to the hot wort in the heat exchanger is pumped to heating coils in the mash tuns from which it is returned to the heat exchanger in a closed circuit. The system contains a large quantity of water that is accumulated in thermally lagged hot tanks (*the "hot liquor tanks"*) at a temperature of 80°C; after heating the mash tuns it is again accumulated in cold tanks (*the "cold liquor tanks"*). The hot tanks are potential sites for the stress-corrosion cracking problem described in the next section.

12.4.2 Materials Used for Brewing Plant

Originally, breweries were small and their operations were carried out in wooden vessels. The first metal replacements were cast iron for the mash tuns and copper for boiling wort — hence the name, copper, for the vessel. When stainless steels became available with their superior hygienic surface qualities, they progressively supplanted all other materials for contact with the beer and its precursors and also for many non-contact surfaces.

Plant design is regulated in much the same way as in the dairy industry; materials for product contact surfaces are required to be nontoxic, self-resistant to corrosion and able to withstand cleaning and sterilization; construction is by welding to eliminate crevices and welds in contact with the product are usually subject to 100% radiographic inspection.

12.4.2.1 General Corrosion Resistance

Beers and their precursors are in the pH range 4 to 5 and AISI 304 stainless steel and its low carbon and stabilized variants are resistant to them if they are cold and fully-aerated. Hence welded sheet and plate materials of these steels are used for grist hoppers, fermenting vessels, holding tanks, and associated pipes.

Hot or boiling liquids are more aggressive, both because of the higher temperatures and because they are depleted in oxygen. Consequently, the more strongly passivating (and more expensive) molybdenum-bearing steel, AISI 316, is used for the mash tuns, copper, whirlpool separators, and heat exchanger plates in the wort cooler. These vessels are lagged with mineral wool to conserve heat and AISI 304 is satisfactory for the outer casings holding the lagging in place.

12.4.2.2 Stress-Corrosion Cracking

Several cases of stress-corrosion cracking have been observed in brewing vessels handling hot liquids and, as a consequence, precautions against it are taken in design and operation. There are several sources of stress including the weight of large masses of liquid and contraction stresses after welding. Two distinct sources of the specific agent, chloride ions, have been found, one external to the vessel and the other internal.

The external source is chloride leached from the lagging by condensing water vapor; this is now controlled by applying a coat of proprietary anti-chloride barrier paint on the external surfaces of all vessels to be insulated, specifying low chloride contents in the lagging material and interposing vapor sealing barriers of aluminum foil between successive layers of lagging.

The internal source is chloride content of the water used for brewing or residues from disinfectants and from acids used in descaling. Though small, these chloride contents can become concentrated locally by thermal cycling or evaporation, an effect well-recognized in certain other cases of stress-corrosion cracking. This source is controlled by de-ionizing the water and prohibiting the use of hypochlorites as disinfectants.

Despite these precautions, there remained a risk of stress-corrosion cracking in hot liquor tanks constructed of AISI 316, induced by chloride leached from lagging. The risk is eliminated by constructing them from an expensive duplex steel, in which the less susceptible ferrite phase acts as a barrier to stress corrosion.

12.4.2.3 Biologically Promoted Corrosion in Spent Grain Hoppers

Spent grain hoppers are not directly associated with the beer production sequence and are simply used to accumulate spent grains awaiting disposal. They are essentially large cylindrical bins made from plain carbon steel, set at a height suitable for discharging the grains into trucks.

The spent grains are moist and contain sugar residue forming a solution that drains down and accumulates at the bottom of the hopper, where bacteria can convert the sugar into an acidic medium that eats into the structural reserves of the steel. In one occurrence, the base of a hopper was so weakened that the bottom dropped away under the heavy load of grains inside but fortunately it caused no injury.

12.4.2.4 Cleaning and Sterilization

The system is cleaned periodically by flushing with 0.5 to 1 M sodium hydroxide solution to loosen and scavenge organic deposits; it is neutralized with a benign dilute acid and thoroughly rinsed.

Sterilization with hypochlorite solution is discouraged to reduce risks of stress-corrosion cracking from an *internal* chloride source. Sterilization by live steam is an unacceptable alternative, because it can leach chlorides from lagging, increasing the risk of stress-corrosion cracking from an *external* chloride source. Present practice is to use chemical sterilization based on non-chloride reagents, such as peracetic acid, that degrade to safe reaction products.

12.4.3 Beer Barrels, Casks, and Kegs

Although for convenience a large proportion of the beer currently brewed is distributed in cans, many discerning consumers, especially in Europe, prefer it drawn from barrels. Casks and kegs are large unpressurized and small pressurized barrels, respectively. The introduction of modern barrels provides a useful insight into the relative merits of two corrosion resistant metals, a stainless steel, AISI 304 and an aluminum alloy, AA 6061.

12.4.3.1 *Historical*

Traditionally, beer in bulk was distributed in barrels made from wooden staves bound by steel hoops, produced by a labor intensive highly skilled craft. Some thirty or forty years ago, they were replaced by metal casks and kegs whose production could be mechanized and that can better contain the carbon dioxide pressure in modern effervescent beers. Austenitic stainless steels, with which the industry was familiar, were a natural initial choice but at the time, techniques had not been developed to exploit their full strength and consequently the barrels were thick-walled and heavy. They were replaced by lighter age-hardening aluminum alloys but stainless steel barrels were subsequently reintroduced when techniques for producing and fabricating them were improved.

12.4.3.2 *Design*

Barrels must be corrosion resistant to cold beers, robust to withstand rough handling and volumetrically stable under internal pressure from dissolved carbon dioxide and sometimes also nitrogen introduced under pressure to generate effervescence. The brewing industry imposes standard drop and pressure tests to ensure these attributes. The barrels are expected to last for a very long time and there are now millions of both aluminum alloy and stainless steel barrels in service. Since every one costs about $60 to produce, the capital invested is very considerable and the refurbishment of existing barrels, at intervals of about ten years, is an ongoing project, as essential as the production of new ones.

The basic form of a metal barrel is illustrated schematically in Figure 12.4. It is formed by deep drawing the top and bottom halves from rolled sheet. They are trimmed to size and joined by automatic electric arc (TIG) welding. Before welding, while the internal walls are easily accessible, grease and detritus from deep drawing is removed as an essential preliminary for application of a corrosion resistant coating to the inside of the finished

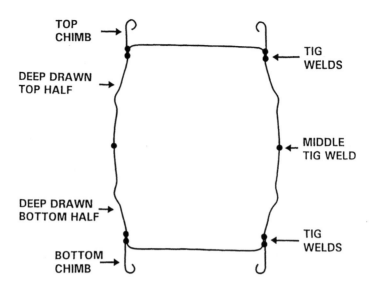

Figure 12.4
Typical construction of a metal beer barrel. The top and bottom halves are deep drawn from sheet and joined by tungsten inert gas electric arc (TIG) welding. The chimbs are each attached by two parallel TIG welds.

barrel. The *chimbs* that support the barrel are shaped from sheet by roll-forming and these too are attached by welding. Appurtenances, such as bosses to re-enforce orifices for filling and dispensing, are attached in the same way.

12.4.3.3 Materials

Aluminum Alloys

Medium strength aluminum-magnesium-silicon alloys, as listed in Table 9.2, e.g., AA 6061 are used because they are amenable to deep-drawing and the higher strength alloys in the AA 2000 and AA 7000 series have lower inherent corrosion resistance and provide less satisfactory substrates for protective coatings. Barrels are fabricated as described above and then solution-treated, water-quenched and age-hardened as appropriate for the alloy. Sheet 2.6 mm thick develops sufficient strength for 50 liter barrels. Degreasing before welding is by alkaline cleaning as described in Section 6.1.2.2. The internal surfaces may be anodized or a protective chromate coating is applied by spray, using orifices in the barrels for access. A final coat of lacquer may be applied by spray. Originally, the alloy was supplied clad with a layer of pure aluminum on the side corresponding to the inside wall of the barrel to enhance the corrosion resistance but for economy this practice has been discontinued. It does however place greater emphasis on maintaining the integrity of the protective coating. This requires close control of the heat-treatments because if the full hardness is not developed, the barrels are vulnerable to denting that cracks the coating, allowing the exposed metal to corrode, producing nodules of corrosion product.

The refurbishment of existing aluminum alloy barrels is costly, i.e., $15 to $25 each and this has contributed to the re-introduction of stainless steel barrels. On receipt, an aluminum alloy barrel is cleaned to the bare metal in nitric acid and rinsed. Dents are removed by shock waves generated by hammering them with the barrel filled with water. Repairs are made by welding where needed and the barrel is re-heat-treated. Finally new protective coatings are applied, using the same procedures as for new barrels.

Austenitic Stainless Steels

Austenitic stainless steel barrels are shaped and welded in essentially the same way as for aluminum alloy barrels. Cleaning grease and detritus from the deep-drawn material is rather more difficult because of the inherent resistance of the material to dissolution. An aqueous solution containing 10% nitric + 2% hydrofluoric acids at 60°C is applied by spray and rinsed off.

AISI 304 can resist cold beers almost indefinitely without further protection but this advantage over aluminum is offset by greater difficulty in meeting other criteria. The sheet thickness is restricted to about 1.5 mm to minimize weight and material cost and the steel develops its strength by work-hardening during the deep-drawing operation. This requires careful attention to the drawing operation because the work-hardening is not uniform but depends on the degree of deformation the metal receives from place to place. It is particularly important to ensure that the strength is adequate where needed to prevent elastic distortion of the barrel shape under internal pressure, that would offend the criterion for a stable volume.

Though unlikely, some corrosion problems are possible. Regular AISI 304 is preferred over the low carbon version, AISI 304L, listed in Table 8.3, because it is slightly stronger and has better deep-drawing characteristics. Sensitization to intergranular corrosion is not usually a problem because such thin sheet cools rapidly after welding but instances have been observed when welding techniques have been unsatisfactory. The strength imparted by work-hardening is accompanied by internal stress which could threaten stress-corrosion cracking in the unlikely event of exposure to an aqueous environment containing chlorides.

The periodic refurbishment of stainless steel barrels is less costly than for aluminum alloy barrels. All that is required is to clean them in a nitric acid/hydrofluoric acid spray and to make repairs by welding. No heat treatment or renewal of a protective coating is needed.

Further Reading

Morgan, E., *Tinplate and Modern Canmaking Technology,* Pergamon Press, London, 1985.

Hosford, W. F. and Duncan, J. L., The aluminum beverage can, *Scientific American,* 1994, p.48.

Porter, J. W. G., *Milk and Dairy Foods,* Oxford University Press, London, 1975.

Little, B., Wagner, P., and Mansfield, F., Microbiologically influenced corrosion of metals and alloys, *International Mat. Reviews,* 36, 253, 1991.

Foster et al., *Dairy Microbiology,* Prentice-Hall Inc., London: 1957.

Sekine et al., *The Corrosion Inhibition of Mild Steel by Ascorbic and Folic Acids,* Pergamon Press, London: 1988.

Briggs, D. E., Hough, J. F., Stevens, R., and Young, C. W., *Malting and Brewing Science. Vol 1: Malt and Sweet Wort,* Chapman and Hall, New York, 1981; *Vol 2: Hops Wort and Beer,* Chapman and Hall, New York, 1982.

Moll, M. (ed.), *Beers and Coolers Definition Manufacture and Composition,* (translated from French by Wainwright, T.) Intercept Ltd, Andover U.K., 1991.

Guidelines for Use of Aluminum with Food and Chemicals, Aluminum Association, Washington, D.C., 1983.

13

Control of Corrosion in Building Construction

13.1 Introduction

Grand buildings in historic cities like Rome, London, and Venice were constructed under the private initiatives of ecclesiastical, royal, municipal, or rich mercantile patronage; the contemporary buildings for dwellings, farms, and trade were built for relatively small populations, using available local resources. All were built without the concept of economic life spans, by craftsmen using natural materials, such as stone, well-seasoned woods, and baked clay bricks that had acclimatized to the local environments. Examples are found in most older cities including, Philadelphia, Boston, and New Orleans. The good condition of many of them testifies to the resilient qualities of these materials.

New buildings are entirely different in concept; they are mostly the products of large commercial enterprises developing and redeveloping built environments to satisfy the diverse needs of large populations in industrial societies. The buildings are constructed to meet definite design life goals specified by clients according to use. They can be characterized by three general time scales:

1. **Indefinite life** — a concept that is confined to prestige buildings and vital civil engineering works. Examples are buildings of national significance, such as legislative assemblies, centers of culture, and strategically important bridges.

2. **60 to 120 year life** — applies to most building of semi-permanent character such as city center development, residential estates, and freeway structures.

3. **20 year life** — is characteristic for buildings that are sensitive to redefinition of land use, replacement of declining technologies and changing lifestyles. Examples are supermarkets, light-industrial units, warehouses, leisure premises, and gas stations.

Construction is driven by funding within budgets and clients are often inclined to prefer expenditure that enhances outward appearance and internal amenities over that which contributes to maintaining the integrity of the building shell. In addition, competitive tendering induces constructors to economize wherever possible.

Modern construction relies on the application of metals. Some uses are overt, such as external cladding and siding, roofing, plumbing, and central heating installations but the most vital roles are concealed in their application as load-bearing members in steel-reinforced concrete and in steel framing for high-rise structures. In selecting metals to retain their integrity against corrosion, different general approaches apply to rural, urban, industrial, marine, or tropical locations and other specific environmental factors may need to be addressed, including the chemical nature of the soil and subsoil, mean temperature and temperature variations, humidity, rainfall, and water quality.

Taking design life, economic, and environmental considerations into account, there is not a universal set of standards for corrosion control but a menu of options from which to select to suit particular projects.

13.2 Structures

Three methods of construction predominate:

1. Erection of a frame with pre-cast reinforced concrete sections connected mechanically to support the structure that is subsequently filled with masonry or clad with reinforced concrete, glass or metal panels.
2. Erection of a frame with heavy steel sections, welded or bolted together to support the structure, clad with concrete or metal panels.
3. Traditional methods with bricks or cement blocks bonded together with mortar, where the walls both bear the load of the structure and contain the internal environment.

Traditional building is suited to low-rise, often residential, development with a long design life to cover financing by savings and loans. Framed buildings are better suited to commercial and industrial use. An important consideration is speed of erection and commercial clients often prefer "fast-track" steel framed buildings because they can be completed weeks ahead of equivalent buildings constructed with reinforced concrete.

13.2.1   Steel Bar for Reinforced Concrete Frames

13.2.1.1   Reinforced Concrete

Concrete is a macroscopic composite material in which an aggregate of carefully sized gravel and sand is bound together by a matrix of hardened cement. Cement is a calcined mixture of limestone and clay containing calcium oxide, CaO, that sets to a hard mass when mixed with water. This is due to the formation of a hydrated cement gel in which calcium hydroxide, $Ca(OH)_2$ is present.

The theoretical quantity of water for complete hydration of the material is insufficient to confer sufficient lubrication in the concrete to enable it to be shaken or vibrated into place and in practice additional water is required. The surplus eventually dries out, leaving pores distributed through the concrete after setting, some of which are interconnected.

The composite is strong in compression but weak in tension and can fracture if the tensile components of resolved stresses imposed on it are excessive. This weakness is remedied by laying the concrete around 0.4% carbon steel rods to which the tensile component is transferred. Although the steel is not exposed to the environment, it is susceptible to corrosion in water penetrating from the environment through the concrete pore structure.

13.2.1.2. Chemical Environment in Concrete

Calcium hydroxide is strongly alkaline and has a significant solubility in water, i.e., 2 g dm^{-3} at 20°C and the interior of moist newly hardened concrete has a pH in the range 12.5 to 13.5, depending on the quality of the cement. Reference to the iron-water Pourbaix diagrams, given in Figures 3.2 and 3.3 indicates that this environment is well-suited to protect steel reinforcement bar by passivation. There are however, two agencies that can degrade this environment during service and depassivate the steel initiating corrosion, the penetration of chloride ions through the concrete and loss of alkalinity due to the absorption of carbon dioxide. It might be supposed that sulfur dioxide in industrial or otherwise polluted atmospheres could also enter the concrete and attack the steel but this does not happen. Instead, the sulfur dioxide is dissolved in water and falls as acid rain that is neutralized by attacking the cement itself at the concrete surface. Old concrete in industrial atmospheres often shows this form of attack as aggregate standing proud of the surface.

The mode of failure is by cracking and *spalling* of the overlying concrete under tensile stresses induced by a large expansion in volume that occurs when the steel transforms into corrosion products. There is rarely a risk of reinforcement failure because damage to the concrete is so obvious that the problem is identified long before there is significant loss of structural reserves in the steel.

Chloride Penetration

Until the late 1970s, it was permissible to add calcium chloride to concrete to promote setting in cold weather and, although this practice is no longer acceptable, some buildings with this internal source of chloride ions still exist. External sources of chloride include marine atmospheres and highway de-icing salts. The chloride ions can migrate through water penetrating the porosity in the concrete to the surfaces of the steel reinforcement bar. The permeation can be estimated, using models based on standard diffusion theory; in practice for a given environment the rate decreases with a parabolic or a cubic time constant.

Loss of Alkalinity

Carbon dioxide from the atmosphere is carried by water through the pores in the concrete, reducing the pH by the reaction:

$$Ca(OH)_2 + CO_2 = CaCO_3 + H_2O \qquad (13.1)$$

A plane of reduced alkalinity advances into the concrete from the surface and initiates corrosion when it reaches the steel. The effect is called *carbonation* and it proceeds at a significant rate for atmospheric relative humidities in the range 50 to 70%. At lower humidities, there is insufficient water in the pores to sustain the process and at higher humidities the water content is so high that it blocks the ingress of carbon dioxide. The damage caused by carbonation is entirely due to a reduction in the ability of the concrete to protect steel; it does not harm the concrete itself and in fact it slightly increases its strength.

13.2.1.3 Protective Measures Applied to the Concrete

The onset of damage depends on the time taken for chloride or carbon dioxide to penetrate to the steel and, other factors being equal, this depends on the minimum depth of concrete cover over the outermost reinforcement bars. This ranges from 15 mm for thin concrete sections in benign environments to 75 mm for marine exposure.

There is a drive to modify the pore structure to slow the ingress of carbon dioxide and chloride by two expedients:

1. Reducing the quantity of surplus water in the cement mix by adding plasticizers such as melamine or naphthalenes both to reduce the overall volume of porosity in the hydrated cement gel and to increase the proportion that is discrete rather than interconnected.

2. Replacement of some of the cement in the concrete mixture with certain grades of ground granulated blast furnace slag or fly ash. The advantage that this has in controlling corrosion of the steel is offset by some loss in strength of the concrete. The practice is environmentally beneficial because it recycles a waste product.

Another expedient is to apply paint coatings to the concrete surface. The paint is specially formulated to have a structure that serves as a filter, preventing the ingress of carbon dioxide but allowing the escape of water from the concrete that would otherwise accumulate at the interface and break the bond between paint and concrete. A typical paint is based on a polyurethane binding medium.

13.2.1.4 Protective Measures Applied to the Steel

Using good quality concrete to avert the degradation mechanisms described above, bare steel reinforcement bar gives a good life, relying on the protection of the alkaline environment and it is satisfactory for most building projects. Other options are available to meet particular conditions; these are zinc-coated steel, epoxy resin-coated steel, and AISI 304 and AISI 316 stainless steels. All of them incur additional expense that must be justified.

Zinc Coated Steel

Zinc galvanically protects steel but reference to the zinc-water Pourbaix diagram, given in Figure 3.5, might suggest that because zinc is unstable at pH > 10.5, it would be rapidly attacked in moist concrete where the pH is 12.5 to 13.5. In practice, the zinc is not attacked and it protects the steel very well from attack due to loss of alkalinity following carbonation. This illustrates the caution needed in making predictions from Pourbaix diagrams that may not include all of the pertinent information. In the present context, the basic diagram does not include other species in concrete, notably calcium ions. The effect of the calcium is to form the insoluble product calcium zincate, $CaZnO_2$, that passivates the zinc surface. Protection against chloride depassivation is more equivocal. One school of thought considers that zinc confers protection against external but not internal chloride sources. Internal chloride, such as the calcium chloride setting agent can prevent passivation. Experience with zinc coated bar has not produced evidence of sufficient superiority over bare steel to justify its widespread application.

Epoxy Resin Coated Steel

Epoxy resin coated steel bar was pioneered in the United States, where it proved to be a solution to the corrosion of reinforcement in bridge decks subject to road wash containing de-icing salts. Originally it was manufactured on a small scale and the steel rod was simply cleaned by shot blasting, the epoxy resin powder was applied by spray and then cured but as production expanded, a regular phosphate conversion coating was introduced as a base for the epoxy resin. Greater care is needed to store and protect the coated bar on building sites, because the coating is vulnerable to damage by rough handling and degrades by prolonged exposure to extremes of weather.

Stray current corrosion is an issue of some concern for concrete structures in close proximity to electric railroads and streetcars. The effect is due to current leaking from the power lines into the reinforcement bar, anodically polarizing the metal and stimulating enhanced corrosion at the point of entry. If it is expected, it can be averted by using epoxy resin coated bar in which the steel is electrically insulated.

Stainless Steels

The high cost of stainless steels deters their general use as concrete reinforcement. AISI 304 and AISI 316 steel bars are more expensive than plain carbon steel bars by factors of 8 and 10, respectively. They are used only where the performance is justified and the cost is commensurate with the value and permanence of the project. Examples are prestige buildings such as the new Guildhall in London, England and the decks of strategic bridges.

As produced, stainless steel bar, *black bar*, is coated with oxide from the high temperature it experiences when it is hot-rolled. It can corrode in concrete if used in this condition and it must first be descaled by pickling in a nitric acid/hydrofluoric acid mixture.

Cathodic Protection

Cathodic protection is rarely a viable option because it is expensive to install and run and requires attention throughout the life of the structure. Nevertheless there are occasional difficult situations where it is prudent to make provision for it; examples are piers immersed in seawater or parts of a structure below ground with a high chloride content. The provision entails ensuring that the relevant bars are in electrical contact and are connected to terminals for application of an impressed cathodic current.

13.2.1.5 Stress-Corrosion Cracking of Pre-Stressed Reinforcement

Pre-stressed reinforcement applies compression that offsets subsequent tensile loading in the concrete. There are two methods of pre-stressing:

1. The reinforcement is stretched elastically, the concrete is cast in a mold around it and allowed to harden. The stress is imposed through the bond between the steel and concrete. The hardened product is cut into sections. A typical application is for lintels over doors and windows in brick built residential properties.

2. Steel tubes are cast into concrete laid *in situ* and when it has hardened, the reinforcement is threaded through the tubes and stretched. The ends are capped to hold the stress and cement grouting is pumped into the tubes for protection.

Stress-corrosion cracking of the reinforcement under the pre-stress is sometimes encountered. The cause is rarely the hydroxide content of the cement, as might be expected, but usually an adventitious alternative specific agent that should not be present. Cases are known where the agent was contamination by nitrate ions (NO_3^-) from biological sources at agricultural building sites and of thiocyanate ions (SCN^-) introduced into the cement by the use of certain kinds of plasticizer. Other cases have occurred when water has been sealed into cavities around reinforcement through incorrect grouting.

Stronger steel is needed for pre-stressing than that used for normal reinforcement. Pearlitic steel with 0.8% carbon, as used in the United Kingdom, is not so vulnerable to stress-corrosion cracking as quenched and tempered martensitic steels often used elsewhere.

13.2.2   Steel Frames

Steel frames are protected from premature corrosion by painting and successful protection is mainly a matter of good geometric design and good practice. Conditions at a building site are not conducive to refinements such as chemical pre-cleaning, conversion coatings, and automatic paint application. A rugged approach is inevitable and the quality of the results depends on the skill, conscientiousness and supervision of those who carry out the work.

13.2.2.1   Design

The corrosion protection starts with good design. Whether painted or not, the less time the metal spends in contact with water, the less is the chance of corrosion. Water can accumulate from rain and snow, and from internal sources by condensation. To avoid trapping it, angled sections must be orientated to drain freely, box sections are end-capped or fitted with drainage holes. Crevices must be eliminated to avoid oxygen depleted water traps for the reasons given in Section 3.2.3.2. This entails ensuring full penetration of butt welds, double sided welding for lap welds and the application of sealant between the interfaces of mechanical joints. Traps in which dust and debris can accumulate and absorb condensate must also be eliminated.

13.2.2.2   Protection

As purchased, bare rolled steel sections may carry patches of strongly adherent millscale from hot-rolling. It must be removed and the easiest way is to leave the steel in a stockyard open to the weather before assembly. If any patch of millscale remains, it can absorb water through the paint, forming an electrolyte that stimulates corrosion of the steel underneath the paint. Pre-treatment for painting consists in shot-blasting the assembled steel and applying priming paint to the fresh surface. Shot-blasting should be delayed to within an hour or two before painting, to avoid formation of rust; it is obvious that the paint is likely to be more durable if applied during a spell of fine weather than if preceded by rain or frost.

Painting is expensive, especially because it is labor intensive, and no more is applied than is needed. The treatment varies with the position within the structure. In contact with the external leaf of a building, where conditions are most aggressive, it may be necessary to use galvanized steel sections overlaid with a thick paint coating, but concealed steel sections in the interior of a warmed air-conditioned building can sometimes be left uncoated. The thickness of paint is adjusted to suit intermediate situations. Paints for steelwork are described in an international standard, ISO 12944. There is a current movement towards using *high-build* paints that can give coatings 400 μm thick.

13.2.3   Traditional Structures

Traditional buildings with load-bearing walls of bricks or cement block masonry bonded by mortar are less dependent on metals but there are some critical applications.

13.2.3.1   Wall Ties

Exterior masonry is built with cavity walls, i.e., two skins of brick or block-work with a space between. The skins are held together at intervals with metal wall ties inserted in the mortar between bricks or blocks. Mortars are rich in lime or cement providing an environment similar to that in concrete. The atmosphere in the cavity is frequently moist and the ties are designed to resist corrosion from water condensing on them. A typical tie is made from flat steel bar of 20 mm × 3 mm cross-section, protected by a thick coating of zinc, 970 mg m^{-2}. It is splayed at the ends to anchor it in the mortar and has a double twist within the cavity providing a vertical edge from which the condensate drips away so that it does not collect on the flat surface. An alternative tie design is a thick wire loop with a twist directed downwards to provide the drip facility.

13.2.3.2   Rainwater Goods

Traditionally, roof gutters, and down pipes were cast from iron with a high phosphorous content to confer the fluidity needed to flow into thin sections. They are heavy, brittle, and require regular repainting. Some remain on older buildings but they have been mainly superseded by the lighter plastic or aluminum alternatives. Aluminum is protected by chromate/organic coatings. A common cause of corrosion failure in metal gutters is neglect to repaint the inside and clear away accumulated debris that retains water and locally screens the metal from oxygen, setting up differential aeration.

13.3 Cladding

Framed buildings can be enclosed in masonry but they are more often clad with panels that are hung on the frame externally. Two kinds of panel depending on metals predominate, reinforced concrete, and aluminum alloy sheet. Plain carbon steels supplied with colored polymer coatings applied during manufacture are less expensive alternatives, which respond to environments as do polymer coated steels generally. Glass is a competitive material.

13.3.1 Reinforced Concrete Panels

The same considerations apply in principle to reinforced concrete used as panels as when used as frame sections. To reduce weight, thinner sections are needed that have less concrete cover but even so bare steel rod is usually adequate, provided that sufficient attention is paid to the quality of the concrete, the care with which it is cast and the application of suitable external coatings. With best practice good lives are obtained but there are examples of careless work where the steel corrodes, promoting premature concrete failure.

13.3.2 Aluminum Alloy Panels

Aluminum alloy panels are formed from rolled sheet and protected from corrosion by anodizing, a surface treatment that is exploited to produce a wide range of attractive finishes. The alloys used are AA 1050 and AA 5005 listed in Table 9.2, both of which develop their strength from the cold rolling applied during manufacture. The towers of the World Trade Center in New York City are an example of the impressive results that can be produced. Besides their primary application in new buildings, aluminum alloy panels are also used to refurbish depreciating exteriors of older buildings.

The sheet is cleaned in alkaline solutions, chemically brightened and anodized, usually in sulfuric acid, applying the principles described in Section 6.1.2.2 and 6.4.2. The architectural use of aluminum alloys is a highly critical application, requiring material that yields anodized finishes free from surface blemishes with prescribed reflectivity and color. To meet the standards required, the surface finishing procedures and the structure of the metal must both be carefully controlled.

Deficiencies introduced in the surface finishing operations are usually not difficult to recognize and correct. They can usually be traced to loss of control of solution compositions, temperature or electrical parameters, or to lack of care in cleaning and rinsing. Provided that the metal is suitable, a first class anodizer can produce anodic films with consistent thickness, hardness and transparency.

Deficiencies in the metal cannot be rectified at the metal finishing stage because they are caused by faulty manufacture of the metal product and often can be traced back to features of the direct chill (DC) cast ingot from which the sheet was rolled. Porosity due to excessive hydrogen contents dissolved in the metal and aluminum oxide particles or films allowed to remain in the liquid metal from which the ingots are cast persist through rolling and form elongated blemishes on the anodized finished sheet. A more subtle deficiency is associated with the metallurgical structure of cast ingots. It is well known that the surface zone of a regular semi-continuously cast (DC) aluminum alloy ingot has a non-equilibrium metallurgical structure that is different to the structure of the rest of the ingot. This surface zone is undulating due to the solidification mechanism and when the ingot is prepared for rolling by machining, away the irregular cast surface, i.e., *scalping*, areas of both type of structure outcrop at the surface. The two kinds of surface structure persist to the rolled sheet, where they respond differently to brightening and anodizing, yielding objectionable streaks. Casting and other techniques have been developed to ensure that the surface zone is either thick enough to accept the surface machining without exposing the underlying different structure or so thin that it is all removed. Both procedures yield a uniform appearance but which is used depends on the appearance preferred by the architect. All of this means that aluminum alloys for architectural use are special products that must be purchased from reputable aluminum producers that appreciate the problems.

Aluminum panels are often used in their attractive natural silvery metallic appearance but some are colored to suit the requirements of architects. The color can be imparted either by dyeing the anodic film before sealing it as described in Section 6.4.2.3. or by using a self-coloring anodizing process that can produce shades ranging from yellow through bronze to black, without the need for dying. There are several proprietary processes, mostly based on anodizing in organic acids, controlled to produce the color required.

Where appearance is unimportant, aluminum cladding is protected by less expensive chromate/organic coating systems, as for aluminum roofs considered in the Section 13.4.1 following next.

13.4 Metal Roofs, Siding, and Flashing

13.4.1   Self-Supporting Roofs and Siding

Two materials are commonly used for self-supporting roofs, galvanized steel sheet and aluminum alloy sheet, profiled by corrugating them to confer longitudinal stiffness. These roofs are suitable for buildings with design lives of the order of 20 years, such as supermarkets, light industrial premises etc. The basic surface protection, galvanizing for steel and application of conversion, and baked paint coatings on aluminum, is applied to the sheet by the metal manufacturer when flat and it must withstand the subsequent deformation in profiling.

The galvanized steel sheet is typically coated with 275 g m^{-3} of electrodeposited zinc and then further coated with a 200 μm thick film of polyvinyl difluoride on the outside and a 25 μm thick film of lacquer on the inside.

The alternative material, aluminum alloy sheet is produced from a strain-hardening alloy, such as AA 3004 in medium hard temper. The alloy selected must be free from copper to avoid exfoliation corrosion. It is protected by a chromate or chromate-phosphate conversion coating as described in Section 6.4.3.1 and supplemented by a baked paint coating. Similar material is used for siding, i.e., cladding on the exterior of low-rise domestic property. If the cut ends are left untreated, as they often may be, corrosion working in from the ends gradually undermines the protective coatings and they peel back progressively.

13.4.2 Fully Supported Roofs and Flashings

Pure lead and copper sheet are traditional roofing materials used for buildings with a long life. The sheet is not rigid enough for unsupported spans and is supported on timber or other suitable substrate. A related use of supported lead sheet is for flashings to seal valleys in pitched tiled roofs and for joints between roofs and chimneys or vents; it is well suited to this function because it is soft and easily shaped to conform with awkward profiles.

Lead roofs exposed to the outside atmosphere develop films composed of lead carbonate, $PbCO_3$, and lead sulfate, $PbSO_4$, that are insoluble and electrically insulating so that protection can be established even in atmospheres polluted with sulfurous gases. In contrast, the film formed on the underside from condensing water vapor is predominantly the unprotective oxide, PbO, so that most failures of lead roofs are from the *inside*. Because lead is so soft, it can also suffer erosion corrosion from constant flow or dripping of water laden with grit. Copper roofs, similarly exposed, exhibit the familiar green patina of basic copper carbonate and sulfate, $CuCO_3 \cdot Cu(OH)_2$, $CuSO_4 \cdot Cu(OH)_2$, that is both protective and aesthetically pleasing.

The lives of all roofs that depend on the establishment of a natural protective coating on originally bare metals are determined *inter alia* by the initial and early conditions of exposure. Aggressive species such as chloride ions contaminating the carbonate, sulfate or oxide layers during their evolution reduce their protective powers.

13.5 Plumbing and Central Heating Installations

Supply waters vary considerably, depending on sources, contact with substrates, biological activity and artificial treatment. They may be hard or soft as described in Section 2.2.9 with pH values usually in the range 6 to 8; they contain various concentrations of dissolved oxygen and carbon dioxide and other soluble species. The corrosion resistance of metals used in plumbing and central heating systems depends critically on all of these aspects of composition and different metals are selected to suit different localities.

13.5.1　Pipes

Galvanized steel, copper and austenitic stainless steels are all used for pipes. The choice between them is based mainly on experience of what works and what does not in particular localities.

Galvanized Steel

Zinc coatings are unreliable in soft acidic waters and galvanized steel is best suited to hard waters that it resists well due to precipitation of a tenacious calcareous scale supplementing the natural passivity of zinc; more failures of galvanized steel pipe in hard waters are due to furring, i.e., reduction of internal diameter by accumulated scale, than by corrosion. In cold water, zinc sacrificially protects steel exposed at gaps but this does not apply to hot water because there is a polarity reversal at 70°C and at higher temperatures, the zinc coating can stimulate attack on exposed steel.

Copper

Copper is a current standard material for tube used in plumbing and central heating circuits, usually with 1 mm wall thickness. It is tolerant of most water supplies but there are certain recognized causes of corrosion failure, type 1 pitting, type 2 pitting and dissolution in certain waters that can slowly dissolve copper.

Type 1 pitting occurs in cold water and is associated with a very thin carbon film on the inside wall formed due to lack of care in manufacturing the tube, as described in Section 4.1.3.4. The film acts as a cathodic collector stimulating the dissolution of copper exposed at gaps. The effect is well known and responsibility for it lies squarely with the manufacturer, who accepts liability, typically by guaranteeing the product for 25 years.

Type 2 pitting occurs in hot water and is associated with particular locations, where the water contains traces of manganese. A deposit of manganese dioxide accumulates during several years, forming a cathodic surface that stimulates corrosion of copper exposed at gaps.

Soft acidic waters with low oxygen contents can dissolve copper, i.e., they are *cuprosolvent*. If the effect is small the copper is not impaired but any base metals over which the water subsequently flows can suffer indirectly stimulated galvanic attack by the mechanism described in Section 4.1.3.5. This can cause failure of downstream galvanized steel in the system and of aluminum cooking utensils that are filled from it. This is a good example of where care must be taken not only in laying out a system so that water does not flow from more noble to less noble metals but also in advising clients who use it.

Austenitic Stainless Steels

Where waters are so cuprosolvent that they can damage copper pipes, austenitic stainless steel pipes are used instead. The cost differential is not prohibitive but it is more difficult to make joints in stainless steel.

13.5.2 Tanks

Many older installations used galvanized steel for both cold and hot water tanks. Causes of premature failure of cold water tanks could often be attributed to differential aeration, either at the water line or at the sites of debris that had fallen in. It is now usual to install reinforced plastic tanks. Cylinders formed from copper sheet are now standard for hot water tanks. They are of course compatible with the copper tubing used in modern systems.

13.5.3 Joints

One of the advantages of copper tubes is the ease with which joints can be made, either by fittings containing rings of solder or by compression fittings. Soldering is the most reliable and least expensive method and is preferred where the heat does no damage; current trends are towards lead-free solders. The copper must be fluxed with a material that enables the solder to wet the metal. There are various fluxes but since their function is to dissolve the copper oxide that covers and protects the metal, they must be rinsed away; corrosion can sometimes be observed in the track of flux that trickled from a joint in a vertical pipe.

13.5.4 Central-Heating Circuits

A water circuit in a central heating system is almost inevitably a mixed metal system because of differences in the functions of the components and the most economic means of manufacturing them. The boiler in which water is heated, usually by gas or oil flames, is an iron casting; radiators are constructed by welding pressed steel panels that are painted on the outside but are uncoated inside; brass castings serve for pump and valve bodies; the circuit is connected by copper tubing for ease of installation.

The mixed metal system survives because the circuit is closed. Oxygen in the charge of water is depleted by initial corrosion but is not then replenished. If the water is hard, a thin calcareous scale also affords protection. The system can usually run uninhibited but if necessary inhibitors can be added to the water. Since the system has more than one metal, a mixture of inhibitors is required such as sodium nitrite and mercaptobenzothiazole to protect the iron and copper, respectively.

Most failures occur through inadvertent and probably unsuspected aeration during service due to poor maintenance. The most common fault is an improperly balanced circulating pump that continuously expels some of the water through the overflow; another fault is neglect in sealing slight leaks that drain the water charge. Either of these faults opens the closed system to a constant supply of fresh aerated water to replenish that which is lost.

13.6 Corrosion of Metals in Timber

Building entails extensive use of metals in contact with and in close proximity to woods. Woods can promote corrosion in two different ways:

1. Providing an aggressive environment for metals in contact with it, especially fasteners e.g., nails, screws, and brackets.
2. Emitting corrosive vapors.

13.6.1 Contact Corrosion

Woods are botanical materials that vary in properties both between and to a lesser extent within species. One of their chief characteristics is the ability to absorb and desorb water with corresponding dimensional changes. They are neutral or acidic media with pH values generally in the range 3.5 to 7.0. Among other solutes they can contain acetic, formic and oxalic acids and carbon dioxide solutions derived from bacterial transformation of starch and sugars. Although woods vary in chemical characteristics even within the same species, there is a recognized hierarchy in their ability to promote corrosion. Generally, harder woods are more acidic and more corrosive than softer woods; some qualitative examples are given in Table 13.1. Electrochemical processes causing corrosion of contacting metals proceed in the aqueous phase in the wood and the more water that is present the more damage ensues. Woods are at their most corrosive when they are damp, when they are new and when the atmosphere is humid. It is advisable to maintain the moisture content of timber below that in equilibrium with 60 to 70% relative humidity. New oak and sweet chestnut are among the more aggressive woods and ramin, walnut, and African mahogany are among the least. Iron, steel, lead, cadmium, and zinc are the most susceptible metals and stainless steels, copper and its alloys, aluminum and its alloys and tin are less vulnerable.

TABLE 13.1
Qualitative Comparison of Environments in Some
Common Woods

Material	Representative pH	Corrosive Influence
Oak	3.6	Strong
Sweet chestnut	3.5	Strong
Red cedar	3.5	Strong
Douglas Fir	3.8	Significant
Teak	5.0	Significant
Spruce	4.2	Mild
Walnut	4.7	Mild
Ramin	5.3	Mild
African Mahogany	5.6	Mild

Treatments given to woods in contact with metals can exacerbate their aggressive nature. Some preservatives with which they are impregnated to protect against biological attack are water-borne and increase the electrolytic conductivity. Alternative formulations based on oxides or organic solvents are less harmful. Fire retardant preparations based on halogens used to impregnate wood can also be aggressive to metal fixings.

When steel nails, screws, or bolts, corrode in wood, there are two concurrent damaging processes that weaken the fixture. Not only does steel lose cross-section but the voluminous corrosion products, iron hydroxides, and iron salts, disrupt and soften the wood, an effect sometimes called *nail sickness*. For this reason, unprotected steel should not be in contact with wood exposed outside. Nails used to secure battens and clay tiles to wooden roof trusses should at least be galvanized but it is better to use stainless steel or brass.

13.6.2   Corrosion by Vapors from Wood

Some woods emit acidic vapors that can corrode metals in their vicinity. There are several situations in building where problems can be anticipated and appropriate precautions taken. Red cedar is a popular material for use as shingles, i.e., wooden tiles, on roofs or walls, but its emissions are particularly aggressive to metals in the immediate vicinity. New oak is an attractive wood for interior fittings such as panelling, shelving and window surrounds but its vapors can damage associated metal fittings and the metal parts of adjacent equipment and furnishings.

13.7 Application of Stainless Steels in Leisure Pool Buildings

Stainless steels are applied extensively in swimming pool buildings, both for structural members and for accessories like balustrades and ladders. The austenitic stainless steels, AISI 304 and AISI 316 have a good service record in traditional unheated swimming pools providing facilities for exercise and sport. Public swimming pools are now evolving into more comprehensive leisure centers based around the water. More people use them and spend longer times in the water imposing the following changes in the environment that have increased its hostility towards materials of construction:

1. The water is heated to temperatures in the range 26 to 30°C.
2. The water is turbulent in features such as water slides and fountains.
3. Higher concentrations of chlorine-based disinfectants are used.

The first two factors stimulate evaporation and hence condensation on cooler surfaces, particularly when the pool is closed.

Greater use of disinfectants increases the aggression of condensates by reaction with organic species in body fluids discharged into the water. Chlorine and some materials containing chlorine interact with urea and other substances to produce chlorinated nitrogenous substances of the generic type, *chloramines*, based on the simplest member chloramine, NH_2Cl; in more complex chloramines the hydrogen atoms are replaced by organic radicles containing carbon and hydrogen atoms. They are formed by overall reactions represented tentatively by:

$$CO \cdot (NH_2)_2 \ (urea) + 2Cl_2 + H_2O = 2NH_2Cl \ (chloramine) + CO_2 + 2HCl$$

$$(13.2)$$

The chemistry of these interactions is complicated and the nature of the particular products formed is sensitive to the pH of the water. Chloramines are very volatile and unstable; their presence is manifest by a pungent odor characteristic of swimming pools. Two aspects of the problems they cause, safety-critical damage and area degradation of the building have stimulated reassessment of the selection and use of the steels.

13.7.1 Corrosion Damage

Safety-Critical Damage by Stress-Corrosion Cracking

A particular concern is stress-corrosion cracking. Attention was drawn to the problem as recently as 1985, by the collapse of a suspended concrete ceiling in Switzerland through failure of the stainless steel supporting structure. The volatile chloramines can carry chlorine species to condensates in parts of the building remote from the pool, where they decompose into more stable species that can be concentrated by repeated evaporation, e.g.:

$$2NH_2Cl + H_2O = 3HCl + HClO + N_2 \qquad (13.3)$$

Typical structures at risk are roof supports, wire suspensions and bolt heads. The stress may be applied by external loads or imparted internally by fabrication or pulling up and tightening bolts. The danger is the insidious progress of incubation preceding crack initiation.

Area Damage

Area damage is due to depassivation of the steel by the chloride condensate. On open panels, the effect is unsightly rust staining from dissolved iron. Undetected pitting on hidden surfaces can develop into perforation of sheet in ventilation ducts and other services. Corrosion is confined mainly to areas where evaporation can concentrate condensates or fine spray; metal that is fully immersed or frequently washed is less vulnerable.

13.7.2 Control

As with other structures, corrosion control begins with good geometric design to eliminate not only traps for liquid water but also traps for condensate remote from the pool with special attention to load-bearing structures and devices. Where possible and appropriate, the materials should be stress-relieved after shaping.

Steels can be selected to suit different situations. The less expensive austenitic steels, AISI 304 and AISI 316 still have a useful role in non-critical applications in direct contact with the pool. More specialized steels are needed for critical structures and some other areas sensitive to condensation. Steels with higher molybdenum contents are less vulnerable to stress-corrosion cracking. These include AISI 317, an austenitic steel with 3 to 4% molybdenum, and duplex steels with 3% molybdenum, listed in Table 8.3. Duplex steels have an advantage in the more resistant ferrite they contain but AISI 317 may prove to have the best pitting resistance.

Condition monitoring of the structure is now strongly recommended, especially for buildings that were erected before the full extent of the problems were fully appreciated. The first concern is safety and although stress-corrosion cracking cannot be anticipated during its incubation period, the onset of cracking can be detected before it becomes catastrophic, provided that inspection is targeted, detailed and at short intervals. Other damage can be reduced by inspection for condensation on open and hidden surfaces and cleaning them regularly to remove aggressive substances.

Further Reading

Glaser, F. P. (ed.), *The Chemistry and Chemistry Related Properties of Cement*, British Ceramic Society, London, 1984.

Portland Cement Paste and Concrete, Macmillan, London, 1979.

Page, C. L., Treadaway, K. W. J. and Barnforth, P. B. (eds.), *Corrosion of Reinforcement in Concrete*, Elsevier Applied Science, London, 1996.

Berke, N. S., Chaker, V. and Whiting, D. (eds.), *Corrosion of Steel in Concrete*, ASTM, Philadelphia, PA, 1990.

Wernick, S., Pinner, R. and Sheasby, P. G., *The Surface Treatment of Aluminum and its Alloys*, ASM International, Metals Park, OH, 1990.

Standards for Anodized Architectural Aluminum, Aluminum Associaton, Washington, D.C., 1978

Short, E. P. and Bryant, A. J., A review of some defects appearing on anodized aluminum, *Trans. Inst. Metals Finish.*, 53, 169, 1975.

Emley, E. F., Continuous casting of aluminum, *International Met. Reviews*, 21, 75, 1976.

Franks, F., *Water*, The Royal Society for Chemistry, London, 1984.

Butler, J. N., *Carbon Dioxide Equilibria and Their Applications*, Addison-Wesley, Reading, MA, 1982.

Oldfield, J. W. and Todd, B., Room temperature stress corrosion cracking of stainless steels in indoor swimming pool atmospheres, *Br. Corros. J.*, 26, 173, 1991.

Index

A